"Engle and Armstrong devotees won't be disappointed with the stories about their heroes in this tome."

—*American Space*

"*The X-15 Rocket Plane* is an engaging account of America's push into space before pilots became astronauts, and America began a new era of exploration beyond the Earth to the Moon."

—Anthony Young, *Space Review*

"The importance of the x-15 on the early exploration of space and the subsequent development of the shuttle program cannot be overstated."

—Book Bit, WTBF-AM/FM

"This book gives a fascinating and superbly detailed look into x-15 technology and the dedicated people who first took a winged craft beyond the atmosphere. Along the way it teaches valuable and pertinent lessons for those of us in the private space sector now working to build on that phenomenal legacy."

—Col. Rick Searfoss, U.S. Air Force (Ret.), former space shuttle commander and pilot

"In this gripping book, Michelle Evans brings to life the x-15 and the aerospace pioneers who made it a success. For those already aware of the program, this will bring back fond memories and renew an appreciation for the remarkable people who conceived, operated, and supported this incredible craft. For those who aren't, prepare for an incredible journey of discovery."

—Richard P. Hallion, former historian of the U.S. Air Force (1991–2002), the Air Force Flight Test Center (1982–86), and NASA Dryden Flight Research Center (1976–82)

"Long before the space shuttle, the United States was flying astronauts with the courage of lions into space on board wings of steel. This is the story of the astonishing X-15, America's first space plane, which broke records nearly every time it flew. It is a magnificent tale, well told in this meticulously researched book. Everyone with an interest in aviation, space, or high-flying adventure should read it."

—Homer Hickam, author of *Rocket Boys*

The X-15 Rocket Plane

Outward Odyssey
A People's History of Spaceflight

Series editor
Colin Burgess

Michelle Evans

Foreword by Joe H. Engle

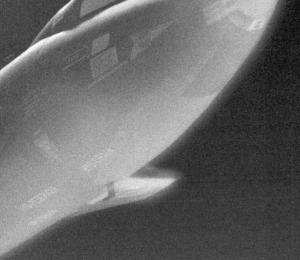

The X-15 Rocket Plane

Flying the First Wings into Space

UNIVERSITY OF NEBRASKA PRESS • LINCOLN

The University of Nebraska Press
is part of a land-grant institution
with campuses and programs on the
past, present, and future homelands
of the Pawnee, Ponca, Otoe-
Missouria, Omaha, Dakota, Lakota,
Kaw, Cheyenne, and Arapaho
Peoples, as well as those of the
relocated Ho-Chunk, Sac and Fox,
and Iowa Peoples.

Library of Congress
Cataloging-in-Publication Data
Evans, Michelle, 1955–
The x-15 rocket plane: flying the
first wings into space / Michelle
Evans; foreword by Joe H. Engle.
pages cm. — (Outward odyssey:
a people's history of spaceflight)
Includes bibliographical references
and index.
ISBN 978-0-8032-2840-5 (cloth: alk.
paper)
ISBN 978-1-4962-2984-7 (paper-
back)
1. X-15 (Rocket aircraft) 2. Research
aircraft—United States. 3. Aerodynamics, Hypersonic—Research—
United States—History. I. Title.
TL789.8.U6X514 2013
629.133′38—dc23   2012047260

Set in Adobe Garamond
and Futura by Laura Wellington.

This work is dedicated to all the men and women who made the x-15 possible, in flight, on the ground, and behind the scenes, and to Milt Thompson for giving me that first interview.

And a very special dedication to Cherie, my partner in life, my wife, my muse, without whose love and support this book would have never found light. And to Fluffy and Max, who stood vigil at my computer for so many years.

[Man] is a tough creature who has traveled here by a very long road. His nature has been shaped and his virtues ingrained by many millions of years of struggle, fear, and pain, and his spirit has, from the earliest dawn of history, shown itself upon occasion capable of mounting to the sublime, far above material conditions or mortal terrors.

Winston Churchill (1874–1965)

Contents

Illustrations

Additional photographs pertaining to specific incidents portrayed
in this book may be found on the author's website at
http://www.mach25media.com/x15index.html.

Foreword

As test pilots within the Flight Test community, my colleagues and I tend to associate significant advances in our world of flight in terms of hardware, specifically the airplanes. We give those airplanes names and personalities, and we speak of them respectfully (or sometimes not so respectfully) as living things. We often overlook the fact that it was people who conceived the ideas and goals, designed the often beautiful yet functional mold lines, the often critical but not visible systems, and gave these airplanes the ability to fly—with a little help from the laws of physics.

In this truly unique book, Michelle Evans has focused on those people, giving us a wonderful insight into who they were and what it was that drove them to dedicate their careers, and sometimes their very lives, to expanding our knowledge of flight. As I read the manuscript, I was familiar with many of the stories and incidents, but the accompanying details that Michelle's thorough research unearthed has made them so much more meaningful, fascinating—and fun! (I can't wait to retell them.) I've often joked that my x-15 stories get better each time I tell them. Now, at least for those told here, it'll be true.

Thankfully, the book is not limited to the pilots, who often receive the bulk of the credit and attention. We are introduced to engineers, technicians, mechanics, managers, administrative support folks, and people who, at times, were carrying out tasks far beyond (and sometimes beneath) what they were originally hired to do. But they did whatever it took to make the flight program safe and successful. The people you will read about in this book are a wonderful reminder of why this nation enjoys such a proud aviation heritage, and why we have been able to take the next steps in our continual quest for more speed and to go ever higher—even into space. They also represent why Edwards Air Force Base is universally recognized as the world's greatest flight test facility. Sure, the ideal weather and an abundant array of dry lakebeds providing emergency landing areas doesn't hurt, but it's the people of Edwards and their spirit, unmatched anywhere in the world, that made it happen.

The x-15 was the greatest airplane I have ever had the privilege to fly. It was an honest, beautifully handling airplane and, most of the time, a real joy to fly. It was also the most rewarding airplane I have flown, while requiring the pilot's continuous, undivided attention throughout the entire flight profile. At subsonic, and even low supersonic speeds, the x-15 handled like a really good fighter. Because of its incredibly effective vertical stabilizer, and the fact it used differential elevons instead of ailerons mounted out on the wings for roll control, it exhibited little of the yaw-roll coupling experienced on most aircraft. The lower ventral part of this very effective directional control surface, which had to be accommodated and even removed during the high angle of attack entries on altitude flights, is discussed in this book.

The information and techniques developed in the x-15 program literally laid the groundwork for the Space Shuttle. Hardware, flight control systems, operational flight techniques, physiological data, crew operations with a full pressure suit, data monitoring of both aircraft and biological parameters, simulation, and real-time mission control were all direct beneficiaries of the x-15 program. But perhaps the most significant contribution was the airplane's demonstration of the ability to routinely manage energy of an unpowered, low lift-to-drag winged vehicle through reentry from space to a precise touchdown on a runway or lakebed.

During the design development phase of the Space Shuttle, many key engineers and managers at NASA felt it was necessary to have deployable air-breathing engines to provide go-around capability, or at least a shallower and more benign final approach to touchdown after reentering from space. Because the lift-to-drag final approach angles and the approach pattern and touchdown speeds of the x-15 and Space Shuttle were nearly identical from about Mach 5 to touchdown, those concerns were answered and put to rest. The significant resources, development time, and weight impact of air-breathing engines on the useable payload of the Space Shuttle were thus avoided.

It would be incorrect to suggest that if we did not have the information and experience of the x-15, we would never have been able to fly the Space

Shuttle. However, the x-15 was a critical step toward that goal, allowing us to get there safer, quicker, more efficiently, and with more confidence.

It will be a very long time—if ever—before we once again have a research aircraft so capable. This magnificent airplane taught us how to fly at hypersonic speeds and to routinely fly out of the atmosphere, conduct experiments, then reenter to make a precision landing.

The x-15 was a cutting edge airplane with a powerful rocket engine. This combination gave us the ability not only to go incredibly fast but to attain high enough altitudes to fly our first wings into space.

Maj. Gen. Joe H. Engle, U.S. Air Force (ret.)

x-15 pilot (sixteen missions)

Space Shuttle commander: ALT-2, ALT-4, STS-2, STS-51I

Acknowledgments

So many people helped make this book a reality. As with any work of a historical nature, it will always be true that an author cannot work in a vacuum. There are interviews, research, travel, and more interviews. Each person you meet along the way helps get you to that finish line. However, the biggest hurdle in any book is to get off the blocks. With that in mind, I acknowledge Dr. Richard Hallion, who answered my first phone call on the day I got the idea to pursue this project. His response was to come directly to Edwards Air Force Base so he could point me in the right direction. I grabbed my gear and headed up to meet with him the next day. Dr. Hallion went on to make introductions with several key people, who all eventually sat down to endure my questions.

Milt Thompson was the first in a long line of interviews. He and Jack Kolf visited with me several times. I will be forever saddened that neither of these fine gentlemen survived to see this work published.

Much later, I found a good friend in Dave Stoddard, formerly of the NASA rocket engine shop. We met after a long hiatus from my writing. He opened more doors than anyone else. For that, and so much more, thank you, Dave.

Others I must note include Sheri McKay-Lowe, daughter of x-15 pilot Jack McKay. One of eight children, Sheri spent a lot of time with me in her home, and on the telephone, sharing memories of her parents and her life with seven siblings.

Francis French was the one who first put me in touch with my editor, Colin Burgess. I am forever grateful to Francis for getting the ball rolling that night at "Killer Pizza from Mars" and to Colin for believing in me, even when he could have easily given up and gone elsewhere. Colin, you put up with a lot from me throughout the gestation process, reassuring me I was the only person who could tell this story. You are amazing.

Thommy Eriksson first contacted me many years ago, telling me of his love of the x-15 and of his special skill in creating photo-realistic computer images. The marriage of those two skills is seen throughout this book with the amazing images he conjured while skillfully manipulating pixels. My gratitude in what he created is as high as the altitudes reached by the x-15.

Special thanks go to all those people who shared their time and stories of the x-15: Brent Adams, Freida Adams, George Adams, Bill Albrecht, Johnny Armstrong, Neil Armstrong, T. D. Barnes, Florence Barnett, Larry Barnett, Roger Barniki, Paul Bikle, Phil Brandt, Dean Bryan, Stan Butchart, Vince Capasso, Scott Crossfield, Sally Crossfield Farley, Bill Dana, Meryl DeGeer, Joe Engle, Frank Fedor, Fitz Fulton, Billy Furr, Charles Gerdel, Byron Gibbs, Don Hallberg, Bob Hoey, David Knight, Pete Knight, Jack Kolf, Eldon Kordes, Terry Larson, Wade Martin, Charlie McKay, John McKay, Mac McKay, Mark McKay, Sheri McKay-Lowe, John McTigue, Phil Moore, Edward Nice, John Painter, Forrest Petersen, Bob Revert, Ralph Richardson, Daniel Riegert, Jim Robertson, Bob Rushworth, Harry Shapiro, Glynn Smith, Dave Stoddard, Harrison Storms, Bill Szuwalski, Milt Thompson, Daryl Townsend, Jim Townsend, Donald Veatch, Grace Walker-Wiesmann, Gene Waltman, Lonnie Dean Webb, Bob White, Ray White, Walt Williams, and Jim Wilson.

Thanks to so many others who helped along the way with this journey: Pat Baker, David Ball of the Civil Air Patrol, Phil Broad, Jim Busby, Greg Cooper, Christine Daniels, Al Esquivel, Bryce Evans and Celee Evans, Dr. Christian Gelzer from the Dryden History Office, Dennis Gilliam, Cathie Godwin, Richard and Robert Godwin from Apogee Books, Bob Holland from Reaction Motors, Eric Jones from the *Apollo Lunar Surface Journal*, Bob Kline, Tony Landis, Joseph and Carolena Lapierre, Pam Leestma, David Livingston from *The Space Show*, Pete Merlin, Jon Page Risque, Stephanie Smith from the Edwards AFB Flight Test Center, Kaya and Mary Tuncer from Space Camp Turkey and Global Friendship Through Space Education, Larry Turoski, Karen Vance, Thomas Walker, Paul Wood, and Jim Young of the Edwards AFB History Office.

Special thanks to Robert Lanktree and Genie Parrish for reading through early versions of the manuscript to offer input and corrections. After their perusals, and those of Cherie and Colin, any leftover mistakes are certainly on my own head.

A separate acknowledgment must assuredly go out to my favorite kiwi, Ray Montgomery. He stood by me, or maybe more accurately, pushed me forward, at a time when so many others had abandoned me.

And finally, from the University of Nebraska Press: Rob Taylor for bringing me into the fold, Katie Neubauer for all her help in the early stages, and Courtney Ochsner and Sara Springsteen for their aid as the project neared completion.

Introduction

The x-15 was the first winged rocket ship to take astronauts into space and back again. It was designed in the mid-1950s, at a time when, to the public, rocket ships meant gleaming silver stilettos with swept-back fins, filled with astronauts in bubble-headed spacesuits, doing battle against aliens bent on the overthrow and subjugation — or annihilation — of Earth.

Although the x-15 was sleek from a distant perspective, a closer look revealed construction much heavier than might be expected. Protuberances such as bug-eyed cameras and antennae bulged from the heat-resistant hypersonic skin, while surfaces at the rear were corrugated for strength rather than aesthetics. Seeing the intricate details of the craft reminded one more of an industrial boiler rather than of the sculpted visage people were used to seeing in the science fiction of that period.

Yet the x-15 was still a beauty in its own right, not created to please an audience, but instead was the vanguard of a far-reaching research program that dealt with the real idea of being able to fly a fully reusable spacecraft out of the atmosphere and land it safely back on terra firma under a pilot's control. And even though the experimental data garnered from more than nine years of flight testing often lent itself to technical journals and scientific publications, the program also inspired people about the real excitement and promise of air and space exploration.

A generation earlier, a silver, single-engine, high-wing monoplane, with only a periscope for forward viewing, swung in over the ocean and landed at a small field. Local military men, with arms linked, tried to hold back the swarm of onlookers who attempted to rush the field. Inside the *Spirit of St. Louis*, Charles Lindbergh shut down the engine and organized his materials and thoughts, then climbed out of his airplane to the waiting jubilation of the people. This was not the evening of 21 May 1927 at the Le Bourget airport outside Paris, but instead in St. Thomas, U.S. Virgin Islands, more than eight months later.

In early 1928 Lindbergh was finishing a tour that extended throughout Latin America and the northern parts of South America, culminating with stops in the Caribbean. He was riding the crest of fame and admiration for his feat of the first solo crossing of the Atlantic. As with so many places around the world, the entire island of St. Thomas had been enthusiastic followers of Lindbergh's exploits. Although not on his original itinerary, the territorial governor of the Virgin Islands, Waldo Evans, sent a special invitation for Lindbergh to visit St. Thomas before he returned to the United States at the end of this tour. He was hoping not only to have the famous aviator in his territory, but that Lindbergh's presence might help foster aviation throughout the islands and the resultant tourism that would entail. Governor Evans's request had originally been misplaced by the State Department; however, the governor was finally able to make contact while Lindbergh was in Panama. With just a few weeks until he would arrive, Lindbergh accepted the invitation, landing on St. Thomas on 31 January.

The following day was full of festivities, including horse races, tours, and official government receptions. The U.S. Navy was responsible for most of the arrangements, which included having their official photographer record the events. He was Milton Barron McKay, or simply "Cap Mac" to close friends and family. His twin sons were Jim and John, although John preferred the nickname Jack. The boys were just past their fifth birthday when Lindbergh landed.

Eighty years later, Sheri McKay-Lowe, daughter of Jack, said that after the great aviator's visit, both boys immediately fell in love with the idea of flying into the clouds. They started making small popsicle-stick airplanes. As they grew, so did their ambitions to emulate what they had seen in Lindbergh. They built proper model airplanes, and both hoped for later careers in aviation. Jim eventually became an aeronautical engineer, working for the National Aeronautics and Space Administration (NASA) at the Flight Research Center. Jack was able to enter test flying, becoming the fifth pilot of the X-15 research aircraft.

My passion for exploration began at about the same age as the McKay twins. When I was a young child in kindergarten and on into grade school, my father worked for a company called Sangamo Electric, setting up expensive, multitrack data tape recorders at government locales such as the Naval Weapons Center at China Lake and the Flight Research Center at Edwards Air Force Base in the Mojave Desert of California. His instruments acquired telemetry from military tests of missiles and weapon systems; most of what he did he could not share with me. One morning before going to school, my father asked if I would like to tag along on one of his day trips to Edwards. I jumped at the chance, although for a five-year-old kid, it was a scary proposition to head into the desert with fighters, bombers, and everything else under the sun thundering through the clear, bright sky. I waited outside my classroom at the appointed time for my father to pick me up to begin my adventure—one that continues to this day. The trepidation of that first trip quickly turned to awe as I saw the wonders in store. It was the first of many such excursions over the next several years.

This was the era of spaceflight, when we could go everywhere and do everything. The moon was within our grasp. The rockets being launched at Cape Canaveral were wonderful, but I saw them only on television; whereas, on these trips to the other NASA at Edwards, I could see the real stuff up close. I also had a young child's delight in getting to meet the men who flew the test flights, and lots of other people on the ground who truly made it all happen.

Since my father also had work to do, he often found a friend who could walk me around on behind-the-scene tours. We'd wander through the hangars or out on the flightline to see what was on the ramp being prepared for a future flight or what may have just come back from a sortie. I vividly recall watching with fascination as the X-15 was slung under the wing of a B-52 and made ready for flight. There was so much going on: the jumble of equipment, personnel going about their jobs, everyone often doing things where I had no idea of their purpose. It was paradise for a curious kid with lots of questions and a yearning for excitement.

On one such trip, my father's designated friend decided to take me to see the X-15 simulator, known as the "Iron Bird." He knew it was currently in use and thought I might like to see its operation. It was a long contraption, with pipes and wires and sheet metal, looking like something a child such as myself might cobble together out of an Erector Set. We stood and watched for a bit, then the pilot finished his "mission" and exited the cockpit. Instead of walking directly away to some debriefing or to his office, he saw me there and decided to come over, say hello, and shake my hand. I have no idea who my guide was that day, but I will always remember that moment when he introduced me to the research pilot who had just exited the simulator. His name was Neil Armstrong.

When I was first getting the idea to write the book you now hold in your hands, I contacted Neil for an interview, writing a letter asking to set up the appointment. Neil called me one morning to let me know he was willing to get together. After a couple months of planning, I drove into a tiny Ohio town on a rainy Thursday afternoon in early May, looking for his office. Somehow the address was off by a digit, and I ended up blocks away from where I should have been. The place just didn't look right. I knocked on the door and for a long time no one answered. Had he forgotten our appointment? Finally, the door creaked open and a very old man was standing there. When I said I was there to see Neil, he thought for a long moment, scratched his head, then said, "Oh, you must be looking for that astronaut fella." I told him I was, and he explained the mistaken address number, directing me to the right location down the street.

Soon afterward I was knocking at the correct door, and his secretary, Vivian, immediately answered and invited me in out of the rain. My first glimpse of Neil in his office that afternoon—the first time I had seen him in person since that day with the x-15 simulator—was him standing in the middle of his office with a wastebasket in hand, trying to place it to catch the rain coming in through leaks in a bad roof. The can was just one of several I saw scattered about the room. Introducing myself, we shook hands, then he set the can on the floor under a drip. Looking around the room at the mess, he suggested we find a new location for the interview, so we walked across the street to a neighborhood ice cream parlor. We found a corner table near the front window and sat down to talk for a couple hours. I recall that meeting fondly and can now listen to that tape of his mild but assured voice answering my queries. In the background are the various sounds of the parlor: chairs scratching on the tile floor, the cash register ringing up, the low murmur of other patrons, the tinkling of the tiny bell as the door is opened and closed.

This is just one of my many vivid memories from all the interviews conducted for this book, part of my personal journey with regard to the x-15.

Earlier on that same research trip to speak with Neil, I had an appointment in the suburbs of Virginia to see the only navy pilot in the program, Forrest Petersen. Early Saturday morning I was looking for his street near Georgetown Pike and Dolly Madison Boulevard when I recognized the entrance to the Central Intelligence Agency on my left. Just a few blocks later I turned onto the small dead-end street where Admiral Petersen lived and pulled to a stop. When he invited me in, I immediately saw it would be an informal interview as he had apparently just awakened and was wearing a light blue robe covered in cat hair. We conducted the interview at his kitchen table, as his wife puttered around making coffee.

Each interview was special: Joe Engle in a dimly lit sitting room with a porthole at the Queen Mary cruise ship in Long Beach; Scott Crossfield on a bench in the echoing hallways of the Rayburn Senate Office Building in Washington DC; another Armstrong, this time Johnny, opening his desk drawer at Edwards, pulling out a piece of the X-15A-2 scramjet he found in the desert after Pete Knight's high-speed flight in 1967; Jack Kolf and Milt Thompson at their desks at Dryden on Christmas Day with brownies Cherie and I brought to share as we conducted the research; Robert Rushworth with his doormat reading "Aliens Welcome," when we went to see him on my birthday; Pete Knight fielding phone calls as he was deciding whether to run for political office for the first time; Walt Williams in a dark house, hooked up to his oxygen tank; seventeen technicians and engineers from the program in a back room at a restaurant, speaking together of their X-15 exploits, "to keep each other honest," they quipped; Billy Furr sharing a pizza in an out-of-the-way place up in the Sierras; John McTigue proudly displaying a leftover rear landing skid in his living room; Jim Robertson talking of working on the external tanks, then pulling out numerous display cases of creepy and exotic bugs he had collected from his trips all over the world; Jim Townsend, becoming animated with stories, just weeks before he passed away; Grace Walker and Freida Adams, at opposite ends of the country, but both with mist in their eyes as they talked of their X-15 pilot husbands now long gone.

Although I was never able to become a pilot myself, at a very young age the X-15 and that first meeting with Neil Armstrong had inspired me onto my career path in aerospace, as the *Spirit of St. Louis* and Charles Lindbergh had for the McKay twins a generation earlier.

The x-15 Rocket Plane

1. The Whole Nine Yards

There is no such thing as an accident. It was
either designed wrong, built wrong, or used wrong.
Generally, it's used wrong.

A. Scott Crossfield

Scott Crossfield started the x-15's rocket engine and moved the throttle forward to 50 percent thrust. The bright exhaust lit up the surrounding area as a long string of Mach diamonds formed and stabilized. Scott then throttled forward to 100 percent. Firmly secured to the test stand, the aircraft strained to pull away with more than 50,000 pounds of force. At a level of 140 decibels at the cockpit, the noise generated by the LR-99 rocket was crippling if unprotected. Inside the nearby control room there was a loud muffled roar, something felt through to the bones.

The purpose of the test was to check the ability of the rocket to restart. Crossfield shut down, then primed the engine for a second ignition. As he brought it back up to 50 percent, the safety system automatically stopped the sequence. Scott called over the radio, "Malfunction. Throttle off." His hand moved toward the instrument panel to make another attempt to relight the engine. He called, "Reset." The moment Scott hit the switch, the x-15 exploded.

Eleven weeks previous to the accident, on 28 March 1960, the first XLR99-RM-1 powerplant for the x-15 was delivered to Edwards Air Force Base (AFB) in the California desert. It was shipped in from the Reaction Motors plant in Denville, New Jersey. By late May the engine was mounted into aircraft no. 3, and the entire assembly was trucked down to the Propulsion System Test Stand (PSTS). It was backed into the fixture, with the exhaust nozzle of the LR-99 butted up against the rectangular yellow and silver structure at the rear of the stand. A four-foot-diameter hole in the fixture allowed the exhaust gasses to pass through during the test. The aircraft was firmly secured in position for the first engine firing, which was successfully accomplished on 2 June at the PSTS. On Wednesday, 8 June, another run was scheduled to test the restart capabilities of the engine.

People from North American Aviation (NAA), NASA, and the military started arriving at the facility. Vehicles seemed to be parked haphazardly. Hoses, toolboxes, barrels, and ladders cluttered the area. Power carts and a fuel truck added to what looked like a chaotic scene. For those people whose job it was to make the X-15 ready, everything appeared in perfect order.

Crossfield approached the aircraft wearing business attire rather than the expected silver spacesuit. In this instance, he was not planning to fly but to remain firmly attached to the ground, so there was no need to put on the cumbersome and hot multilayered garment. He climbed up an aluminum ladder into the cockpit, while engineers and technicians continued their work.

Preparations proceeded well into the waning light of the early evening hours. Surrounding the aircraft, two banks of three light boxes atop tall poles provided bright, even illumination. Crossfield put on an oxygen mask, since he did not have his spacesuit supply to breathe. Finally, the canopy was closed and secured over Scott. Even bolted to the test stand pad, with access panels removed for instrumentation wiring and miscellaneous plumbing, the X-15 looked ready to leap into space. It was easy to imagine the speeds this rocket plane would achieve in just a few short years.

One by one, and in groups, everyone cleared the area, entering the protection of the blockhouse about fifty feet away. The rectangular command structure had steeply angled exterior walls, giving the casual impression of a truncated pyramid. Inside were racks of electronic instruments and television monitors, and even a periscope for more direct observation. Most important, the blockhouse provided protection for everyone in the event of a malfunction of the highly volatile rocket motor. It was fueled by 18,000 pounds of liquid oxygen and anhydrous ammonia. The concrete structure was a recent addition to the rocket test facility, having been installed barely a month previously. Strapped into the vehicle, Scott had no such protection.

Crossfield lit the rocket, and the first run was completed successfully.

He reset for the second, which triggered the powerful explosion. As the aft end of the aircraft disintegrated behind him, Scott, inside the fuselage, was hurtled twenty feet forward, with a force estimated at fifty gs — fifty times that of gravity. He later said of the noise and fire, "It was the biggest bang I had ever heard. It was like being in the Sun."

Sally Crossfield Farley is the second daughter, the fifth of six children, of Albert Scott Crossfield and Alice Virginia Knoph. She talked about her father, who was one of the best-known rocket pilots during the 1950s and 1960s and who was the first to fly the x-15. "My dad missed most of my life growing up. When I go back into my memory bank, I don't have as many memories as I would have liked. . . . It never felt like he lived with us until I was sixteen, and by then I resented him."

Like many in Crossfield's profession as a test pilot, he was a driven man. Scott paid the price for that single-mindedness. Sally continued, "Years passed and understanding grew, and we healed our relationship. People like him, who pursue the improbable, the once considered impossible, often have to focus so intently on the goal, that it is inevitable something gets lost in the process. It is that very intensity and sole pursuit of the goal that also saved his life on occasion. I understand that now, but regret not having him during those days. I am very proud of him, but that pride had a price."

Pride is what Scott Crossfield had in all he accomplished as a test pilot and even more so as an engineer. He firmly believed he was responsible in great measure for the success of the x-15 program because of the fact he was part of it from the very beginning. Scott was right in having that pride. At that time, he put the success of the program ahead of his personal life, ahead of his family, ahead of all else. Scott explained why he had such a focus on his job: "One thing I have never been able to get people to appreciate, I was a test pilot only in that it was an inherent part of building better airplanes. I wasn't a test pilot to be a test pilot; I was a test pilot because I'm an engineer, a designer. I think that single purpose helped the x-15 be as successful as it was. I was much more involved in the program than I was in just the flying. I designed a lot of that airplane, contributed a lot to that airplane."

There are many people who might disagree with how responsible Crossfield was in this regard. It took hundreds of people at North American Aviation to complete the design and construction of the x-15, and hundreds more in the U.S. Air Force, NASA, and the U.S. Navy to bring the potential of the rocket plane to fruition. Without Scott Crossfield, however, there is little doubt it would have been a different aircraft.

Sally spoke of what drove her father; it was what he called "The Crossfield Way." She explained, "[My father] believed in rules but wasn't necessarily confined by them. He would see a problem, situation, or possibility, and work out a solution — often a very creative one. He could think outside the box, yet didn't always go for the difficult answer. Sometimes a very easy one is sitting right in front of you and it gets missed." But that wasn't all. Sally continued, "He didn't let anything stop him once he set his mind on something. He was always a positive thinker. He could build or repair absolutely anything. . . . If he needed a bolt, he would go down to his lathe and make it. He was such a unique individual. He wasn't by any means perfect . . . [but had] strength of character [and] personal integrity."

Crossfield's mother, Lucia Dwyer, was born in Mexico. His father, Albert Scott Crossfield Sr., originally came from Browns Valley, Minnesota. His mother and father wound a convoluted path toward each other. By Scott's account, as a child Lucia helped the family fight off Mexican bandits with a bullwhip before she and the rest of the ten children were sent north to El Paso. Lucia eventually moved west to attend college at the University of California, Berkeley. Across the Pacific, Scott's paternal grandfather, Amasa, served as chief of the Customs Department in the Philippines under the governor of the islands, William Howard Taft. Amasa's three children, including Scott's father, spent many formative years in the island territory. They returned to California, where Scott Sr. graduated high school and moved on to college at Berkeley, where he met Lucia. Scott's parents married in 1916, and on 2 October 1921, the future test pilot was born. He explained his complicated family heritage in his book, *Always Another Dawn*: "One chemical result of this union was me, Albert Scott Crossfield, Junior, one-quarter Mexican, with a good sprinkling of English, Irish, Boston Brown, and the good Lord only knows what else."

Scott's father was a chemist, who instilled a love of science. "Outwardly, my father was the coolest man I have ever known," he wrote. "He took great pains to disguise his courage." Working for the Union Oil Company in Wilmington, California, his father was, at one time, responsible for firing many people during the Great Depression of the early 1930s. He found rampant discrimination from above, being told to fire anyone without an Anglo-Saxon name. With the Mexican heritage in their family, he couldn't abide this and left the company himself. Soon after, he formed a dairy, but the family money was gutted by price wars, and the business ultimately failed.

The family moved to a farm in the Boistfort Valley, about ten miles west of Chehalis, Washington. Scott's father turned the scientific methods he had been taught at Berkeley into a meticulous plan for a successful poultry and dairy farm. The hard work and long hours to make a go of the new family business proved beneficial to Scott. He became determined to design and build radio-controlled model airplanes, then to finish his flying lessons and finally solo in a Curtiss Robin at the Chehalis airport. Scott had melded his two loves together into what became his career in engineering and flying.

Once he graduated from Boistfort Consolidated School in June 1939, Crossfield intended to immediately leave the farm to head for the University of Washington. Before he graduated, his younger sister, Mary Anne, died of polio. Her death devastated Scott and the family, so he forestalled his plan for a year. Once into the engineering program, he still did not sail right through but instead took several turns along the way over the next eighteen months. Crossfield described his route: "I had entered the University, graduated from a civilian aviation school, officially soloed, and obtained my private pilot's license, withdrawn from the University, worked for Boeing Aircraft Company, quit to join the air force briefly, worked for Boeing again, quit again to join the navy."

Scott's first attempt at joining the air force following the attack on Pearl Harbor was short-lived due to the service not being able to find a class in which to put him. Restless to get into the fray of World War II, he joined the navy instead. That plan also went awry when he was assigned flight instructor duties, which kept him stateside. He didn't get close to actual combat until he was moved to Hawaii as part of the preparations to invade Japan, which were canceled after the atomic bombings and the Japanese surrender in August 1945.

While Crossfield was working at Boeing, he met Alice Knoph and they became engaged. Sally explained some of the qualities her father saw in Alice, "My mom [is] a beautiful blond, blue-eyed Norwegian, even to this day, though the blond's gone silver. She is strong, determined, and has the best sense of humor of anyone I've ever known." When Scott and Alice first met, she was engaged to another man, but this never deterred him. Sally went on, "She would put up with nothing from my dad or any of us. I've never felt more loved by anyone, and I know my dad loved her more than life itself."

Following four years in the navy flying Grumman's F6F Hellcat and Chance Vought's F4U Corsair, Crossfield returned to the University of Washington, receiving his bachelor's degree in aeronautical engineering in 1949. He went directly into the master of science program, completing his thesis the following year. It was titled "A Semi-Empirical Method of Obtaining Static and Dynamic Aerodynamic Parameters of Swept-Back Wings Analyzed on the Basis of Plan Form." As Scott explained, "A thesis had to respond to the requirement to be a new and innovative contribution to the art. I devoted all my effort to coming up with a very simplified way to determine the characteristics of wings. . . . It was only twenty pages long, including all the figures and contents. . . . When I took it to the University library, the librarian hefted it and said, 'That's a thesis?' I just thought the quality was better than the quantity."

By June 1950 Scott decided to resign his commission in the naval reserve and applied for a job with the National Advisory Committee for Aeronautics (NACA) at the High Speed Flight Research Station at Edwards AFB. His timing was fortuitous: within three weeks after he left the navy, his unit was called up for duty in Korea. Upon arrival at the NACA facility, Crossfield took the slot being vacated by pilot John Griffith, who was leaving to take a job at Chance Vought. Scott said, "This was really the place to be. . . . I was awestruck. When Walt Williams showed me the airplanes they had, there was no question in my mind where I belonged."

Scott's reputation at NACA grew quickly, and he became the first pilot to rival Col. Charles E. "Chuck" Yeager for notoriety. He flew a steady stream of experimental airplanes, some with the coveted "X" designation, such as the XS-1, the tailless X-4 Bantam, and X-5, which first used variable sweep wings in flight. He entered the cockpits of many other aircraft, including the XF-92 Dart, which was used to test theories concerning delta-shaped wings and tail surfaces.

With the variety of aircraft in the inventory, many of his first flights had anomalies: an xs-1 went into a spin, the xf-92 ran off a runway onto a dirt road before it could be brought to a stop, and an f-107 flipped around and nearly upended when the brakes caught fire during high-speed taxi. The one he is most remembered for occurred on 8 September 1954. Scott brought an f-100a Super Sabre in for a dead-stick landing, found he had no brakes, and crashed the front end through a hangar wall. Except for his ego, he was uninjured, and the airplane was eventually repaired and returned to flight status. In an nbc News interview, Crossfield explained how Yeager took this mishap as an opportunity to jab at Scott's piloting abilities, which highlighted their growing rivalry, lasting until after Scott's death in 2006. He said of Yeager, "Every time he had a chance in public, Yeager would always say, 'The sonic wall was his, the hangar wall was Crossfield's.'"

The one experimental aircraft Scott flew more in his career than any other was the sleek, white Douglas d-558, Phase 2. It was built for the U.S. Navy in their effort to compete against the U.S. Air Force for government flight research funds. Also known as the Skyrocket, the severely swept wings and tail lived up to the name's expectations.

In 1947 Charles Yeager broke the previously impenetrable sound barrier of Mach 1. Six years later, as the fifty-year anniversary of the Wright brothers' first powered airplane flight neared in late 1953, the next Mach number had yet to be pierced. The race between Yeager and Crossfield heated up to see if either could get there first.

Crossfield made a bid to accomplish the flight using the d-558-2 but was hampered primarily because the aircraft was designed for a maximum speed of only Mach 1.5. Modifications to the lr8-rm-6 rocket nozzles added nearly a half-Mach number but was just short of the requisite speed. Scott felt he knew the plane intimately enough so he could coax it through Mach 2, with the navy's support. In a nasa interview in 2003, Scott said, "Everybody on base knew we were going to make the try, but very few people thought we were going to make it. . . . Frankly, we had our own doubts that we just were asking the airplane to do more than it was ever designed to do." At the same time, Yeager and the air force prepared the x-1a for his bid, knowing this aircraft was made for those high speeds.

On the morning of 20 November 1953, Scott made his final attempt to break the record. He said of the event, "We cold-soaked that thing. Wiped off every fly speck. There wasn't any excess drag on her at all. We started loading [the fuel], oh, six or eight hours earlier than we [usually] would, and let that liquid oxygen soak it until it was so damn cold you couldn't touch [it]. Your hand would freeze on it, anywhere. It was a good, cold morning, miserable morning. I had the flu, [and the] weather was bad." The combination of the cold-soaked aircraft, along with the frigid and dry winter air in the desert, was exactly what Scott and the engineering team calculated was needed to increase the D-558-2 performance past the threshold.

The navy had modified a B-29 Superfortress into the mothership used to carry the Skyrocket aloft for launch. The new designation was the P2B-1S and had been nicknamed "Fertile Myrtle" because of its ungainly look with the D-558-2 attached inside the bomber's belly.

Once at altitude, Crossfield dropped away and fired the four chambers of the LR-8 rocket engine. He quickly climbed and accelerated. When he reached approximately 72,000 feet, Scott dipped the nose and started a shallow 10,000-foot descent, which increased his speed to 1,291 miles an hour. This equated to Mach 2.005, a tiny fraction of a percentage past where he needed to be, but enough to nab the record. Scott said, "[I] wanted to be first, particularly if I could needle Yeager about it." He later explained, "NASA claims to be pure in their technical programs, but they don't mind making a record now and then. I started that with the Mach 2 flight."

Yeager didn't like having someone beat him to a record. Three weeks after Crossfield's flight, Yeager took the X-1A easily past Mach 2. The previous best speed on this aircraft was Mach 1.9, and Yeager pushed it by more than a half-Mach number, maxing out at Mach 2.4, nearly 275 miles an hour faster than the D-558-2 had attained. But he'd pushed too hard and lost control due to inertial coupling, the phenomenon where several aircraft axes coupled together and caused a dangerous tumble. Yeager spiraled downward at a rate of 1,000 feet a second. The thicker air at low altitudes slowed his speed to subsonic, where he was able to recover from an inverted spin at 25,000 feet. Yeager found the lakebed and landed as quickly as possible. Scott credited Yeager's skill as a pilot in saving the aircraft, and his life. "It was probably fortunate that Yeager was the pilot on that flight," he said.

Just five years after he arrived at NACA, in addition to all his other experimental flights in a weird menagerie of aircraft, Crossfield had racked up eighty-seven rocket-powered flights, ten of those in the Bell XS-1 and the rest in the Douglas D-558-2. This was a record number of such flights for any pilot. Coupled with his fourteen still to come in the x-15, Scott has a record that will be hard for any modern day pilot to exceed.

In the late 1930s a German engineer, Eugen Sänger, who was unknown to aeronautical scientists in the United States, started thinking about the concept of extremely high velocity human flight. Long before any aircraft had gone supersonic, Sänger, along with Irene Bredt, designed a vehicle that could travel at hypersonic speeds (above Mach 5). Using the upper atmosphere to literally skip over intercontinental distances, their *Silbervogal*, or Silverbird, was envisioned by the Third Reich as a weapon that could bomb targets as far away as America. The technology at that time was not yet up to the task, and the project was canceled in 1942.

Following World War II, the work of Sänger and Bredt came to the attention of John V. Becker and Hartley A. Soulé at NACA's Aeronautical Laboratory in Langley, Virginia. Becker wrote in 1968, "Until the Sänger and Bredt paper became available to us . . . we had thought of hypersonic flight only as a domain for missiles. The concept of manned rocket aircraft flying efficiently at hypersonic speeds for very long ranges was new and highly stimulating. The remarkably detailed analyses of many aspects of their new concept . . . gave real substance to the idea. . . . These studies provided the background from which the x-15 proposal emerged."

The primary barrier against building such a hypersonic vehicle was heat, with expected skin temperatures in the range of 1,200 degrees Fahrenheit. This problem was solved with the introduction of a nickel-chromium alloy called Inconel, initially manufactured in the 1940s as part of the emerging development of jet engine technologies. Creating small parts, such as turbine blades inside the engines, was much different than expanding this concept to encompass the outer skin of an entire aircraft. The largest technological leap was the ability to machine the alloy and to shape it into large pieces. Weight was also a major factor, in that the Inconel-X, which was eventually settled on for use on the x-15, weighs approximately three times more than standard aluminum. Inconel-X was seen as a key piece of technology; instead of trying to protect the aircraft from high temperatures it was turned into what was called a hot structure, where these temperatures were embraced and made part of the overall research effort.

1. Scott Crossfield checks out the cockpit on the original engineering mockup of the x-15. Courtesy of North American Aviation.

Seeing the devastating potential of missile technology through the use of the v-2 against England during the war provided an impetus toward engineering more powerful rockets in the United States. As Becker wrote, "In 1954, nearly everyone believed intuitively in the continuing rapid increase in flight speeds of aeronautical vehicles. The powerful new propulsion systems needed for aircraft flight beyond Mach 3 were identifiable in the large rocket engines being developed in the long-range missile programs. There was virtually unanimous support for hypersonic technology development. . . . The x-15 proposal was born at what appears, in retrospect, as the most propitious of all possible times for its promotion and approval."

While these discussions were happening in the early 1950s, Scott Crossfield recalled a fateful trip, saying, "Coming home with Walt Williams from a fishing trip one night, [Gerald M.] Truczynski and [Joseph R.] Vensel were asleep in the back end, and Walt and I were just talking. We heard over the radio there was a successful firing of the Viking engine at [the Rocketdyne facility in] Santa Susanna, [California.]" Scott pondered the idea of mating such an engine to a manned aircraft. "I picked a piece of scrap paper from the glove compartment and came up with what was awfully close to the x-15 specification of how much energy you'd have available, how high, [and] how fast. We were estimating that an airplane . . . would go to Mach 6 or 7." The idea was sound, and the Viking engine was used in some x-15 baseline studies.

It didn't take long to get moving with the idea, since the groundwork had now been laid in metallurgy and rocketry. Scott said, "By 1954 it was very active, and the contract was let in 1955, which is pretty fast for a hare-brained idea."

The x-15 was right in the middle of a map set up by NACA on the road they envisioned into space. The three sections were called simply Round One, Round Two, and Round Three. The first was supersonic flight, such as with the xs-1 and its follow-ons, the second became the x-15's assault on the hypersonic regime, and the last was to take astronauts into orbit. All three expanded on winged technology, but by the time the third round was ready with the development of the x-20 Dyna-Soar, wings were upstaged by the brute-force methods of ballistic rockets.

Since coming to NACA, Scott had seen how the research airplanes often got hijacked by military priorities and how they would often come in over budget and under performance. During the development of the x-2, he had gone so far as to consider a transfer to Bell Aircraft to overcome these challenges. Scott recalled, "I felt, if I could go to Bell and provide the single purpose they needed to put back into this program, I could get that airplane out of there and get rid of a lot of the make-work. . . . That's why I left NACA to do exactly that on the x-15. I think the results were pretty evident. It came in on-time and on-money."

Scott went on to explain the merits of the competing designs from the various prospective contractors. "The Bell airplane was, to my mind, the best proposal, but maybe not the best airplane . . . [but] there was ample evidence they had been in the rocket airplane business before. . . . The Republic airplane followed traditional Republic design as being heavy and complicated. . . . North American showed some engineering ingenuity in their approach to it. . . . They actually proposed an airplane that had much more capability than the basic specification. The Douglas airplane . . . was

2. Three-view layout of the standard x-15 rocket plane as it appeared in December 1960. © Thommy Eriksson.

a magnesium airplane and it would do exactly, and only exactly, what the design asked it to do."

Harrison Storms at North American Aviation remembered how his company won the competition. "We picked up on one feature NACA wanted. They were not too particular about how fast it went, but they wanted to make sure that it was plenty strong and stable. Those were the things we concentrated on, the stability and the strength."

By December 1955 Scott left NACA to start his new job at North American, shepherding the x-15 through construction and into the flight program.

Paul Bikle took over as head of the x-15 after Walt Williams departed to work on the Mercury program. He explained who was who when it came to the x-15. "My understanding was, when Scott went to North American, he felt he would have far more impact on getting a good airplane from within the company than from without. I think that's probably right due to his normal ego, and then, due to his close association with it for a good part of his life. . . . Harrison Storms, he was really the top gun on the x-15 from

3. x-15 no. 2 proudly displayed on 27 February 1959, the day of its rollout from the North American Aviation plant in Los Angeles. Courtesy of North American Aviation.

the contractor standpoint. Decision-wise, of course, Charlie Feltz was the chief engineer. I think he was the brains behind almost everything."

Inside the North American Aviation building, at the corner of Imperial Highway and Aviation Boulevard, near the southeast corner of Los Angeles International Airport, the x-15 took shape. NAA had been reluctant at one time to even accept the contract, but soon transitioned to the proud parent of the rocket plane. A neon sign was mounted on metal trestlework on the roof of the building, proudly proclaiming "Home of x-15" for all to see.

Across the country, at the Reaction Motors plant, work was slow on the LR-99. Many critiqued the company and the design of the world's first rocket engine able to be throttled. The entire science of building rocket engines was relatively new, and no one had ever done anything remotely like the LR-99 previously. The closest idea was the LR-11 engine (and the navy equivalent, the LR-8), which had four chambers that could be independently ignited. This did change the thrust but was far from the same as doing this within a single large chamber.

Fires and explosions plagued the engine program, and there was talk of bringing in Rocketdyne as a second contractor to build an alternative engine in case Reaction Motors failed. Crossfield understood the problems and the reality of the situation, saying, "I don't really think there was all that much difficulty. It was a new engine, it was a totally new concept, and it was hard to do. You have trouble with an airplane and you made dents in it. If you had trouble with the rocket, you blow up the whole stand."

The rocket was delayed because Reaction Motors overestimated their ability to do it from the beginning. According to Scott, "You can't speculate on what's going to cause trouble until you go and do it. Because of that, I don't think Rocketdyne could have really done the job any better. . . . They didn't know how to make a throttleable engine any more than Reaction Motors, and RMI had actually built manned rocket airplane engines longer than anybody in the world." By 1953 more than 250 manned rocket flights had used the RMI engines on the X-1 series and the D-558-2. Crossfield reiterated, "Naturally, you're going to go to the people with that track record, and that was RMI, which was bought out by Thiokol [in 1958]."

With the production version of the engine falling behind, the question then became one of what to do with the X-15. The construction of the rocket plane was proceeding on schedule at NAA, but there was little chance of the LR-99 being ready at the same time. Should the X-15 just sit as a hangar queen? With so much invested in the program, that was never considered a viable option. The idea was floated to use not one, but two, upgraded LR-11 engines, the XLR11-RM-13. Each chamber of this version of the LR-11 produced 2,000 pounds of thrust, making for 16,000 pounds total. Hardly the 57,000 pounds expected at altitude for what was called the "big engine." However, this compromise provided the opportunity to get X-15 testing underway more than a year before it would otherwise have happened.

One possibility of which Crossfield was paranoid was fire in the cockpit. He was instrumental in having it designed to be filled with inert nitrogen when sealed. The pilot obtained oxygen only from his spacesuit supply and, even then, just in the face mask area. The rest of the suit was filled with nitrogen and was segregated by a rubber bladder that fit snugly around the face, with a second around the pilot's neck. Harry Shapiro, an engineer at NAA, recalled Crossfield's unyielding nature concerning flame:

"Scott got very disturbed down there when the guys were building the airplane. There were guys smoking, and he said, 'I don't want to see anybody smoke in here.' He put 'No Smoking' signs up and had them paint white lines around the airplane."

Harry also talked of a prank that made it look like someone was ripping the finely honed cockpit apart. One of the guys in the shop played the joke when Crossfield came by, possibly to get back at him for taking away their cigarettes while on duty. "One of the mechanics was sitting in the cockpit and conspiratorially said, 'I'm going to shake up Scott a little bit.' He had this extra wire with him, and he cut it up into little pieces. Then he started throwing it out of the cockpit to fake out Scott, shouting, 'We don't need this wire! The heck with it!' Scott was livid."

After years of plans and designs, fabrication and construction, ready or not, the x-15 was finally unveiled to the public on 15 October 1958. Just two weeks earlier, NACA had been officially converted to a new civilian space agency, the National Aeronautics and Space Administration, better known as simply NASA.

The assembled crowd consisted of military officers, civilian contractors, and invited guests, along with several of the pilots who had been selected for the program. Scott Crossfield was joined by Joe Walker from NASA, Bob White and Bob Rushworth from the U.S. Air Force, and Forrest Petersen from the U.S. Navy. The highest-ranking dignitary was Vice President Richard M. Nixon. Harrison Storms explained that with these political concerns, "You had to set that date maybe six months ahead of time. As far as the airplane schedule goes, it couldn't care less who's coming and what's set up. If it ain't done, it ain't done!"

A large stage was constructed in front of the giant door to the building where the aircraft would appear. A barrier of fabric and flat panels was behind the stage to shield the view. The VIP speakers sat in an arc of two rows behind the podium. Once the ceremony began, one after another came forward to extol on the virtues and promise of the x-15. Just over a year previously, the Soviet Union shocked the world, and frightened many in the United States, with the launch of the first Sputnik satellite. The news of Sputnik had shaken people to the core when it came to America's supposed infallibility. Nixon was there to instill new confidence in beating the Russians. The following is an excerpt of what he said to the assembled crowd:

It is a tremendous and exhilarating experience to see, as I have for the first time, this vehicle destined to carry the first man into space. . . . We are here today to mark a major step in man's greatest adventure in exploration beyond our world. . . .

No one can question the accomplishment of Russian scientists in placing a vehicle in orbit. . . . The achievement could not be taken lightly. . . . Americans can proudly say today that we have moved into first place in the race to outer space. More than that, we are on the threshold of even more exciting adventures into space, of which the x-15 is but one manifestation.

But the x-15 is perhaps the most exciting because of the fact that it is designed to carry man into space for the first time. . . . [It] is an integral part of an orderly and reasoned space program. . . .

These [pilots] are men to be envied, for theirs will be the ultimate goal toward which men's dreams have stretched for centuries since the first medicine man gazed into the heavens and sought from them their secrets. It is useless for me to cite their bravery. We can only envy them for the opportunity which is theirs.

Pilots Bob White and Bob Rushworth talked about reactions at the rollout. White said Nixon's private thoughts on his first view of the x-15 were somewhat different from his public speech. "It's one of these things where you hope the microphone isn't open. When it came rolling out, Nixon made a side comment to someone up on the stage, saying, 'Gee, that's a funny looking thing isn't it?'" Considering the type of language Nixon was often known for in private, White smiled and said, "That may not be a direct quote." Bob Rushworth admitted he was initially shocked at the black color. "When I first saw the x-15 the day before it rolled out, they were still hard at work. It was just like any other airplane, aluminum-colored, silver, beautiful. Then they opened the door and rolled it out and it was black! I thought, 'What the hell was that?'" The black paint was added as a way to control what is called "emissivity," or the control of the heat rate of the structure during flight. Once the x-15 was soaked in high heat during flight, the Inconel itself changed color to a deep blue-black, negating the need for paint.

One major component required to flight test the research aircraft was a mothership that would take it aloft for launch. The x-15 was unable to take off from the ground under its own power, much the same as earlier experimental rocket planes. With its bigger size and weight, the x-15 needed a large carrier, which originally was planned to be the Consolidated Vultee B-36 Peacemaker.

The carry point in early designs had been inside a modified and enlarged bomb bay, but questions emerged about the viability of the B-36 over the span of the program, especially since the bomber was ready to be phased out of military service. Next in line was the Boeing B-52 Stratofortress. The difficulty with this idea was the X-15 could not be carried beneath the bomber but must be mounted to a pylon on the wing. This meant the pilot would have to ride to altitude already secured inside the cockpit, unable to transfer to the mothership if a problem arose. The wing-pylon concept had first been tried as early as 1947 in the Soviet Union with their answer to the XS-1, the T-tailed Samolet 346. It was mounted under the right wing of the Tupolev Tu-4 bomber, itself a reversed-engineered copy of the American B-29.

For the X-15, launching from a higher altitude and faster speed were major considerations in selecting the B-52 over the B-36. Proving the choice a good one, the Peacemakers were retired in early 1959, four months before the first X-15 flight. Walt Williams related, "Crossfield was not very enthusiastic about changing from the B-36 to the B-52, and neither was Boeing. In fact, North American did the modification [of the bomber], not Boeing, because they really didn't want to do it. [Boeing] knew it would end up taking early test aircraft away from them—which it did."

Two B-52s were eventually selected for use with the X-15: NB-52A no. 52-0003 and NB-52B no. 52-0008. Due to the three zeros in each tail number, the aircraft were often informally called "Balls 3" and "Balls 8." Officially, no. 003 received the name "The High and Mighty One," while early in the program no. 008 used the nickname "The Challenger." Nose art was added to both aircraft reflecting those names, but only no. 003 retained its name through retirement.

Modifications were needed allowing the X-15 to be attached and carried. The most notable changes were that a pylon was added between the right side of the B-52's fuselage and the first jet engine pod on the right wing, along with a trapezoidal section that was cut out of the rear of the wing to allow clearance of the X-15's vertical tail. This cut precluded the B-52 from using its flaps for takeoff and landing, necessitating higher velocities for both actions. Because of this, a special set of high-speed landing gear was also added.

Capt. Charles C. Bock Jr. and Capt. John E. "Jack" Allavie were the primary pilots first assigned to the B-52s. A total of twenty-four men flew in the cockpit over the course of the program, most notably Maj. Fitzhugh L. Fulton Jr. He came along soon after Bock and Allavie and ended up flying a record ninety-four X-15 mothership drop missions.

Fulton earned his pilot's wings in 1943 and saw combat in the Korean War flying the Douglas B-26 Invader. One of the most notable aspects of his early career was flying the C-54 Skymaster on 225 supply missions during the critical Berlin Airlift of 1948 and 1949. He admitted he would have been happy to have flown the X-15 but knew that heavy aircraft pilots were not going to get those sorts of invitations. Fitz said of the crews of the motherships: "We always flew with two pilots, a launch panel operator [LPO], and flight engineer. We were adamant that we did not take any more. There were a few rare missions where we might have taken a photographer or an advisor along, but we really discouraged that. . . . Everybody should have an ejection seat, and there were only four in the airplane."

Three X-15s were eventually going to fly, so a system had to be devised to properly signify and record each flight. The numbering system settled on consisted of three elements. The first designated the aircraft number, the second specified the number of the flight (or used "A" for aborted missions or "C" to signify a captive flight), with the final number recording how many times that particular aircraft was carried aloft by a B-52. An example is flight 3-29-48. This meant X-15 no. 3 completed its twenty-ninth flight, after being carried forty-eight times by the mothership on attempted missions.

With the numbering system complete, the B-52s modified, the first of three X-15s completed and delivered to Edwards AFB (trucked by convoy on the back of an air force flatbed, wrapped in plain brown paper), and an interim rocket engine selected for installation (with a much bigger one yet to come), the pieces were all in place for the start of the flight test program envisioned to expand the frontiers of aerospace knowledge, while taking pilots to hypersonic speeds and spaceflight altitudes.

Crossfield nurtured a vision of becoming the sole pilot on the X-15 program. This was true even though North American assigned a backup pilot who also hoped to fly the rocket plane, Alvin S. White. Nearly three years

Scott's senior, White was never given the chance. As Scott laughingly said, "Al was laying awake nights figuring out how he could break my legs so he could get into the x-15." Storms agreed, "Well, he just never got by Crossfield, that's all. . . . I know it was one of the big disappointments of Al's life, but there were only so many flights, and once you got going, there was no reason to change. . . . The situation never arose where Al could be worked into the cockpit."

Eventually, White was given a new assignment as the chief pilot on the massive xb-70 Valkyrie bomber program, another aircraft in Storms's arsenal of futuristic designs. The opinion differed on how he got the job, with Crossfield explaining, "I gave that to Al because . . . it gave him something to do. I wasn't all that enamored with bombers." Storms said, "I gave him the b-70 because I wanted him to fly it." Considering Storms's position at NAA as Crossfield's boss, his statement carries more weight.

Arriving at Edwards AFB two days after the rollout ceremonies in October 1958, x-15 no. 1, tail no. 66670, remained earthbound. By the spring of 1959 the contractor was beginning to feel the pressure of canceled and aborted flights. A scheduled captive flight on 10 March was the first time the aircraft left the runway under the bomber's wing. During three attempts to launch in April and May, the rocket plane had been unable to drop from the b-52 and glide back for a landing, let alone fire the rocket engines for a press into the sky. The primary culprit was the twin auxiliary power units, critical for supplying electricity to the x-15 systems during flight. The APUs remained problematic throughout much of the program.

The media was clamoring for evidence that Americans were going to upstage the Soviets by firing the x-15 into space. Reports, fueled by the publicity department at North American, kept telling people the x-15 would quickly reach 100 miles altitude, even though the design specifications called for only half that. They also ignored the concept of an incremental build-up to high altitude and speed, seemingly expecting a man in space within the first few flights.

On the flightline at Edwards, the expectations were more realistic. Most everyone had been involved with rocket planes previously in some capacity and knew they could be temperamental beasts. Repeated delays as the weeks, then months, passed were causing what Harrison Storms called "cancelitis." Combating this as much as possible, Crossfield spent large amounts of time in the x-15 cockpit after it was mated to the B-52 in order to provide inspiration to everyone working diligently to get off the ground.

When Scott was asked about his preparations for flight, he grinned and explained his two rules: "Stay sober the night before, [and] get at least four hours sleep." Then Scott got serious, saying, "The test flight has to be designed and implemented just as carefully as you design a wing spar, because you leave no known chance, not even a calculated risk. . . . You have a lot of time to think about the design and building, and when you use it, it gets more dynamic, so there's more room to make mistakes. . . . You deviate from the plan with great reluctance. But that's why you have a pilot there, too, so he can use judgment."

Before dawn on Monday, 8 June 1959, the ramp lights bathed the B-52 and x-15 in harsh contrast, creating an oasis in the dark of the desert. Personnel crawled into the bomber to check its systems, while near the pylon and x-15, others on ladders, jack stands, and portable stairs gave the once-over to the rocket plane. Ground fog appeared as liquid oxygen was pumped into the tanks, lending the scene a surreal appearance.

With the sun breaking the horizon, Crossfield arrived in the suiting van. His silver spacesuit, made for a seated position inside the x-15, forced him to walk toward the aircraft slightly hunched over. Scott climbed the stairs, and technicians helped him into the cockpit, connecting him to the internal supplies, then closing and sealing the canopy.

All the paraphernalia was pulled away, the area cleared so the bomber could start its engines. Radio communication between everyone verified readiness to taxi. By the time they reached the threshold, the mission was still a go. At 8:00 a.m. B-52 no. 003 rumbled down the runway, spewing clouds of black jet engine exhaust as it took off. Air force x-15 pilot Maj. Bob White in an F-104, along with two F-100s with Capt. James Wood and Capt. James Roberts, took off to join as chase planes.

The entourage stayed relatively close to the base, climbing counterclockwise to an altitude of 37,500 feet before preparing for the drop. The plan called for a simple glide flight by Crossfield back to Rogers Dry Lake. This was only a test to verify the low-speed handling characteristics, with no rocket fuel on board, just water ballast.

As they approached the launch point about fifteen miles from Edwards, the B-52 was heading northeast, passing over the southwestern edge of Rosamond Dry Lake. Crossfield ran through the last few items on his check-list. Over the radio Scott notified Q. C. Harvey, the x-15 test director and supervisor for North American Aviation, along with the B-52 crew and chase planes, that all was ready. He flipped the ready to launch toggle on his left-hand control panel, then said, "I'm ready when you are, buddy."

At this point, control of the launch of flight 1-1-5 was in the hands of B-52 pilot Charlie Bock. He quickly went through his checklist and set the appropriate switches on the launch panel to the left of the pilot's yoke. He said, "Master arming on. System arm is on. Fifteen seconds . . . ten seconds . . . Starting countdown . . . three, two, one . . . release." At 8:30:40 a.m. the x-15 was dropped from its shackles, its wings providing lift for the very first time.

The nose of the aircraft dipped slightly, and the right wing dropped in what would become a signature of sorts due to the position on the pylon and the B-52 slipstream. From the lead chase plane, Bob White called to Crossfield, "Clean break." Scott replied, "Looks pretty good here." The heavy Inconel-skinned airplane, with small wings and no power, dropped quickly.

Crossfield noted all the flight parameters and aircraft response, then quickly had to get ready for the landing. He asked White in the chase plane, "Where am I sitting now?" to which Bob replied, "Right off runway 4." The seconds quickly passed as Scott rolled into a right turn, then back toward the left to line up on the lakebed. Crossfield could tell he was coming in a bit faster than anticipated. "Going to land a little long on the lake," he said.

About four minutes after release, the x-15 had dropped 20,000 feet. Scott called, "Wish I had guts enough to do a barrel roll here. Feel like I'm back in the saddle again, buddy."

The x-15 was designed with two nearly equally sized vertical tails, one above the fuselage and one below called a ventral fin. At high angles of attack during future flights out of the atmosphere, the upper tail could be blocked from the airstream, necessitating the large lower tail. Because of its size, the lower half of the ventral had to be jettisoned, to clear the main landing skids. A small parachute was supposed to drop it safely for reuse, but that often turned out to not be the case.

As touchdown approached, Bob White, in his F-104, stayed close to Crossfield to help him through the landing. White had to deploy his own landing gear and drop his power to stay with the rapidly descending x-15. He reminded Scott, "Don't forget your ventral." Crossfield replied, "Okay, wait till I clear the edge of the lake here." Once clear, Scott called, "Coming off now," and the ventral popped away.

Soon after the jettison, Scott remarked, "She handles nice right along here." He spoke too soon. As the speed dropped, his inputs started feeding back into the plane's control system, developing a PIO, or pilot induced oscillation. With little time to correct the problem, it could have turned deadly. The nose started a deep pitching up and down. Crossfield's breathing intensified and was heard on the radio as his adrenalin surged to work the problem. He called, "Gear down." White stayed with him, telling Scott his altitude above the lakebed, calmly offering encouragement. "About thirty feet. . . . Just hold it steady and set it right there." As the rear of the x-15 dipped to the bottom of a cycle, Crossfield expertly put the skids onto the dirt, the nose gear slamming down a second later. After 4 minutes, 56.6 seconds, the first x-15 mission was successfully completed.

The press was out in force that day at Edwards. Scott's daughter, Sally Crossfield Farley, recalled, "I saw him on the news with a bunch of microphones in his face, but then he came home and he was the same dad. He made it a point to keep his professional life separate from his personal [one. It] was just his job." Of course, most other fathers didn't fly rocket planes for a living. "In the x-15 days, he started flying his [Beechcraft] Bonanza to and from [Edwards] because we had moved from Lancaster to Los Angeles, and it was the only way he could get any home time. The powers that be didn't like him flying his personal plane because they thought it was too dangerous. How's that for ironic?"

During the debriefings following the flight, Crossfield thought a design oversight was responsible for the wild pitching as he tried to land. Storms knew better and told as much to Scott. "Oh, that was ridiculous. I didn't have to prove it to him. I just told him. I was his boss. . . . You've got to remember that Scotty was a little excited that day. A few minutes before, he was sitting on a time bomb under a B-52, then he got dropped with no power, and now he's ready to land. His heart rate was very high, so he's not what you'd call a relaxed, normal pilot at that point." Other than an adjustment made in the pitch damper system, and Storms's talk with Crossfield, it was time to add fuel and apply rocket power.

4. Scott Crossfield drops away from the B-52 mothership. Courtesy of North American Aviation.

Crossfield was proud of his first flight achievement and later talked of a special award presented to him following the flight. "A soaring society gave me a beautiful streamlined brick trophy for the record of having the shortest time from 38,000 feet to the ground."

Six weeks later, mission 2-C-1 was taken aloft. As a scheduled captive flight to check all systems with a full fuel load, this was the first time aircraft no. 2 was under the B-52's wing in flight. It had arrived at Edwards from North American on 10 April, and the process began toward the first powered mission. Two LR-11 rocket engines were mounted, one on top of the other, in the X-15 engine bay.

It wasn't until 4 September when the first attempted launch was made but was stopped when a vent caused fluctuating pressure in the liquid oxygen tank. By 17 September the problems had been resolved. As with all flights Crossfield made for the contractor, the B-52 stayed in the local area of Edwards AFB and the safety of either Rogers or Rosamond Dry Lakes. Once in position, Scott finished his checklist and again turned over launch control to Bock in the B-52. The drop was smooth, and Crossfield quickly flipped the toggles, turning on each of the eight chambers of the two engines. Unchained at last, the x-15 leapt forward as flight 2-1-3 commenced.

Scott felt the controls respond. They were perfect as he accelerated, outpacing his backup pilot Al White in an F-104 chase, as well as Joe Walker and Bob White following in the other chase aircraft. Keeping his speed to a maximum of Mach 2.11, or 1,393 miles an hour, Scott had to hold the x-15 back from its full potential. After burning the engines for 224 seconds, Crossfield shut off the switches, the LR-11s going silent. Beside the intermittent radio chatter, the only sounds were the whine of the auxiliary power units, servo motors, and the whistle of the supersonic air over the Inconel-X skin.

The x-15 flew as Crossfield imagined it would after that late night ride home from his fishing trip with Walt Williams, Joe Vensel, and Gerry Truczynski. He was exhilarated by the performance, letting his passion show by doing an unplanned barrel roll, as he said he wished he could have done on the first glide flight. This time, at a higher altitude, there was more latitude as the plane flipped a full 360 degrees. The maneuver caused him to drop even quicker than before, sliding past the chase planes. This same maneuver would have repercussions for air force pilot Joe Engle four years in the future, on his first flight. Crossfield had full confidence in the ability of the x-15. He worked for the contractor, not NASA nor the military, so he felt no qualms in what he did. Soon after the roll, he touched down, sending counterrotating vortices of dust up from the skids at the point of contact.

The landing gear on the x-15 was unique. A standard undercarriage with wheels at the center of gravity was weight prohibitive, so they were changed to skids made from Inconel-X. The position was moved to the rear of the aircraft, which meant that at the moment of touchdown, the nose wheel slammed down immediately. Scott explained how the gear worked: "They were gravity-fall and no actuation. Fell down, kerchunk! It was a very simple way to do it. On landing, lakebed dust would collect the heat from the skid rather than the skid itself heating up. [It was like] putting something to a grind wheel. You put your hand on the track [the skid made] and it was hot, but the skid itself was not."

Two flights were aborted in October, followed by a successful launch on mission 2-2-6 on 17 October, one month after the first rocket flight. Trying to get the third powered flight resulted in two more aborts, the last, 2-A-8, on Halloween morning. This was the first time a flight was canceled due to bad weather after the B-52 was airborne.

Flight 2-3-9, or simply 2-3 for short, quickly changed from routine to dangerous. Moments after being released from the B-52, Crossfield ignited all chambers of the LR-11s. The four nozzles for each engine were arranged in a diamond shape, with numbers one through four assigned starting with the left chamber and moving clockwise. The number two chamber on the lower engine exploded, destroying several inches at the rear of the chamber, along with the nozzle. Immediately a fire started, putting the aircraft and pilot in peril.

Chase pilot Bob White called to warn Crossfield: "Got a little blowout in back, and fire, Scott." He acknowledged, then asked for more information: "Got a fire warning. How much fire have I got?" White replied, "Doesn't look like too much. A little back by the rear of the engine."

Scott shut down the engines and started to jettison propellant. It appeared the fire was out, but the condition inside the fuselage was unknown. Crossfield remained calm as he went through his emergency checklist, making the decision to land on Rosamond Dry Lake, west of Edwards. He kept up a commentary as he approached for landing. "Felt the explosion incidentally, whatever it was. . . . I have good control. . . . Lost the roll damper on launch. . . . I still have a fire warning light. . . . Fire warning light just went out."

With the nose-down attitude, jettisoning the propellant was difficult. White called, "You're still jettisoning, but it's showing signs of giving out." Scott unnecessarily apologized for the shortened flight: "I'm sorry I'm going to miss those couple of data points coming in here." Then: "I'll be heavy this time, so I may make just a little faster descent than we're used to." He stopped the jettison to prepare for landing, then dropped the ventral. As he closed on the lakebed, the chase verified the gear release and altitude, then Scott touched down—hard.

Paul Bikle, head of the X-15 program, delineated what happened in the next moments as the skids hit the lakebed. In his report for NASA headquarters he wrote, "At initial main gear contact [talking of the rear skids], the . . . shock struts compressed about sixty percent, then extended to about twenty-three percent. This was followed by nose gear contact and bottoming of the nose gear oleo [the strut made to absorb the shock, where the tires were attached]." The excessive landing loads exposed a weakness in the structure and the fuselage buckled just forward of the liquid oxygen tank. The aircraft literally broke its back. Bikle finished, saying, "The joint opened and sheared about seventy percent of the bolts. The fuselage contacted the ground and dragged for the remainder of the runout, which covered a total distance of about 1,500 feet."

Bob White in the chase plane called to test controller Q. C. Harvey after Scott came to a stop: "Q.C., he is in the middle of the dry lake at Rosamond." Harvey asked, "Is he okay, Bob?" White replied, "Yes, looks okay." Flying overhead, Bob could see the x-15 bent in the middle, but otherwise intact. Scott was opening the canopy, so it appeared he was fine. The aircraft was another matter, yet to be determined by the investigation that followed.

Scott later said, "There was a design oversight in the nose gear, and that's what caused it to break in two." Harrison Storms disagreed, saying, "Actually, the whole thing was caused by the engine fire. Let's put the blame where it really belongs. . . . The fuselage broke further aft—it was supposed to break right at the cockpit."

Considering the fire and fuselage break, the damage was easily repaired back at the North American plant. x-15 no. 2, tail no. 66671, was back in the air with Crossfield at the controls three months later, on 11 February 1960. Flight 2-4 also marked the highest altitude Scott attained in the program—88,116 feet. After all his push for the x-15, sacrificing his career at NACA to work directly with NAA, he never broke through 100,000 feet in an airplane designed to fly 50 miles high.

The program was maturing quickly. After one additional flight by Scott in aircraft no. 1, that vehicle was turned over to NASA and the military so they could begin their own flight tests to find the true capabilities of the x-15. Crossfield flew the four remaining LR-11 flights of the contractor program in aircraft no. 2 before moving on to test the LR-99 big engine.

Exactly one year after the first glide flight, Crossfield was surrounded momentarily by the fireball of the horrendous explosion at the test stand on 8 June 1960. The force pushed the front end of aircraft no. 3's fuselage far enough forward that it was relatively safe from the rest of the conflagration. Scott understood he was in a vehicle designed to withstand very high temperatures, although those were supposed to be generated during high-Mach reentry from space, not from sitting on the ground.

Wearing only his civilian clothing, Scott mentioned how he got wet during the extraction from the airplane, later making a comment to a reporter, "The only casualty was the crease in my trousers." Scott immediately regretted his choice of words, thinking someone would then run the headline: "Space Ship Explodes; Pilot Wets Pants." He said in an interview in 1981 that an East Coast newspaper did run with that headline.

Thirty years later, at a NASA X-15 symposium, Harrison Storms pointed out what likely saved Crossfield's life: "I was very glad that we had a nitrogen cockpit pressurization and cooling system. Had we employed oxygen, we would no doubt have also lost one pilot, one crew member, and a cockpit." In his NAA report, Crossfield said what was on everyone's mind, "The first reaction we had was that the engine had blown up, but like many first impressions, this was wrong." With all the delays and problems during development, a natural reaction was to think the LR-99 was at fault for what happened. If that was true, the entire X-15 program was in jeopardy.

Without the LR-99 to take over from the LR-11s, pushing the X-15 to its full limits, there wasn't much point to the program. Mach 6 and 50 miles altitude were out of the question. The answer had to be found quickly. Scott said, "Just before the blow-up a cloud of vapor appeared ahead of the engine, so the search was concentrated on this area."

Very quickly, the LR-99 was cleared of any culpability. The problem instead was found in a small valve that was supposed to provide pressure relief to the ammonia tank. That relief system had been hooked up to a line that ran about a hundred feet away, venting excess ammonia vapors into a water-filled trench. This caused a back pressure in the line due to a differential temperature between the tank in the X-15 exposed to the sun and the cooler water in the trench. NASA engineer John McTigue explained, "It built up enough pressure that it was over the limits of the ammonia tank, and that tank ruptured. When it did, it blew it back against the peroxide tank [and] ruptured the peroxide tank. The peroxide and ammonia then became an explosion that made it look [at first] like the engine had blown." The entire pressurizing and relief systems were analyzed, redesigned, tested, and retested before everyone was convinced the problem was solved.

The rear of the no. 3 airframe had been destroyed by the accident. Harrison Storms said the decision to rebuild "didn't take very long. We needed three airplanes." John McTigue detailed how much had to be replaced. "Everything behind the wings [was new]. We salvaged some equipment out of the back end. . . . The big problem, which we lived with for the rest of that airplane's life, was that every single wire at the end of that wing going aft had to be spliced. . . . Later if you had a problem, you knew where it was [because of those splices]." Storms recalled that the rebuild amounted to approximately 60 to 75 percent of the aircraft.

In the meantime, the LR-99 program needed to move forward. Testing was then transferred to aircraft no. 2, since no. 3 was out of commission for close to a year. Four months after the test stand accident, on 13 October 1960, the attempt at a first flight with the big engine was ready. Problems with the auxiliary power units and leaking liquid oxygen aborted this flight, as well as the second attempt on 4 November. Finally, on 15 November the flight was able to proceed.

With Captain Allavie at the controls of the B-52, and Capt. Charles Kuyk in the right seat, an hour was spent as they circled and climbed high enough to be able to drop Crossfield for flight 2-10. At 9:59 a.m. the X-15 dropped from the pylon, and Scott pushed the LR-99 throttle forward. After years of development and testing, and a ground accident that could have stopped the program cold, the complete X-15 was finally taking flight.

Crossfield, in X-15 no. 2, powered directly east from the launch point, about halfway between the towns of Palmdale and Rosamond. The plan was to achieve Mach 2.75 at an altitude of 78,000 feet. Soon after passing through Mach 1, he deployed the speed brakes to help add drag and keep the velocity below his contracted maximum of Mach 3. Passing south of Rogers Dry Lake, he performed a left bank, turning north, then proceeded with a 20-degree pull-up toward his peak altitude. After running the engine for 137 seconds, he shut down the rocket. Scott was still relatively low, at approximately 60,000 feet, but was continuing to climb, peaking at 81,200 feet. He then made another left turn of 180 degrees to line up for landing.

Contrary to many reports that stated he never exceeded the contractual speed limit, Scott confided in our personal interview: "I did go over Mach 3 by the cockpit instruments. I don't know what the internal documentation was." Officially, his speed was recorded at Mach 2.97, or 1,960 miles an hour. Would anyone have quibbled giving Crossfield another official

Mach number? When I asked Scott about this, he said, "Civilian professionals aren't allowed to make records. My contract very clearly called out the limit of Mach 2 and 100,000 feet, which would break no records. But you couldn't keep it down to Mach 2, so they opened that up to Mach 3." If his flight was recorded as achieving this speed, it would have not made any difference. Lead NASA pilot, Joe Walker, had already surpassed that number three times in the X-15, while the air force's Robert White had done so once. Scott's daughter, Sally, also confirmed that her father told her he had surpassed Mach 3.

Crossfield made his decision years previously to move to North American Aviation and thus gave up any chance of seeking the true speeds and altitudes of which the X-15 was capable. He had hoped to change their minds, but it was clear that would never happen. In a news interview Scott said, "I got just about all any man could get out of a single program. . . . I got all eight yards. Didn't get to nine yards, that's all."

At this point, the only thing left for Crossfield was to meet North American's obligation to demonstrate the LR-99's restart capabilities in flight. This was the same type of test that went wrong when the ammonia tank ruptured and exploded on the test stand on 8 June. If something happened in the air, Scott would not have firefighters and other emergency personnel standing by to help him from the cockpit.

One week after the first LR-99 flight, mission 2-11 went perfectly. The engine restart task was accomplished, and specific throttle settings of 50, 70, and 100 percent were all run with no problems. To make sure the speed was kept within limits, the flight plan called for Mach 2.5. Two weeks later, on 6 December, Scott repeated the test at a slightly higher speed, Mach 2.85. At 1:29 p.m. he launched away from the B-52 one last time. He throttled up, then purposely shut it down to restart. The engine did not catch, so he calmly went through the sequence again. This time he got the requisite ignition. He then did one more successful restart. Slightly more than eight minutes after the flight began, Crossfield came to a stop on the lakebed, his time on the X-15 and the contract flights for North American Aviation completed.

Harrison Storms spoke about Crossfield's desires with regard to the X-15: "Scott was primarily looking for a way to haul the X-15 from the cradle to the grave." Storms said that once the NAA portion was finished, "[Scott] then really wanted to leave the company and finish the program with NASA, except they didn't want him. Bikle said he could come back, but not on the x-15." Scott apparently asked his boss to put in a good word for him at NASA, but Storms decided against doing so. "We weren't going to advise them on personnel."

Several other pilots weighed in on the controversy. One of the last x-15 pilots, Pete Knight, said, "Nobody got along with Crossfield too well, because he thought the x-15 was his airplane. He didn't understand why he couldn't do the envelope expansion, and figured he was the most qualified to do that. [He felt] it was ridiculous to leave the airplane and the program to NASA and the air force. . . . There was always a professional friction between Crossfield and the government pilots." Scott said he understood he would not be able to continue to fly after the basic requirements had been met. Pete explained, "He knew that, but he was always lobbying to get it changed."

Bill Dana flew the x-15 at the same time as Knight and had this to say: "Crossfield was an extremely abrasive individual, and I don't think that Walker or White liked him. I was never very fond of Crossfield, either. . . . He [was] certainly a colorful individual. He really accomplished a lot, probably not as much as he'd have told you he did. . . . Crossfield was educated, [but] he let his education overtake his natural flying ability. . . . That got him into some trouble with the x-15."

Paul Bikle added this about Crossfield: "He got to where he thought the program couldn't go on without him. . . . I guess he visualized, in his own mind, something along the lines that he was going to not only be one of the pilots throughout the program, but, because of his vastly greater experience, he was the key voice in deciding what the operation was going to be. That didn't happen."

There was also the fact that NASA had a job to do, and couldn't really get started until the contractor was finished. "I didn't select pilots to start with," Bikle continued, "but I was probably as much responsible for selecting Scotty out of the program as anyone. . . . We wanted our own program and our own pilots, and we wanted to get the contractor out of there as quickly as we could. . . . While they were there, it made it look like we couldn't handle it."

Many people around Scott felt very positively toward him. B-52 pilot Fitz

Fulton shared his thoughts, saying, "My contact with Crossfield was he was always a friendly, happy pilot. Maybe if you worked for him it might be different." Sheri McKay-Lowe, daughter of x-15 pilot Jack McKay, fondly recalled, "I always liked his personality. To me, he really stood out as a remarkable person."

In the end, Crossfield stayed at North American, although his attitude also gave him difficulties in that regard. Harrison Storms said, "By that time, he had alienated himself from the other [NAA] pilots. . . . See, he was either working on the x-15 or working on the engineering, but he wasn't getting the time with other airplanes. There was a little bit of feeling [of] 'Hey, you've got your airplane, now why are you coming to look at ours.' So I took him with me over to Apollo."

The end of the contractor portion of the x-15 program meant the team that had been assembled, and had worked so well together for several years, was to be broken apart. Harry Shapiro from North American Aviation recalled, "The x-15 team split in two. One became the Apollo group and the remaining people stayed as the x-15 group. It was about a fifty-fifty break. The people who stayed on had the responsibility of three x-15 airplanes, two B-52 carriers, and two engine test stands. . . . It was a lot of work, but nobody was ever overworked."

After Crossfield had poured his heart and soul into the rocket plane for so many years, he recalled that in the end, "I transferred over to Downey, I believe, in January or February of 1961. It was a couple of months after the x-15 [flight testing] was completed. I went on to the Saturn/Apollo program and didn't pay any attention to the x-15 after that." Scott explained what transpired next: "Harrison Storms . . . was made president of the Missile Division in Downey, which was about to become defunct and close its doors. He was a very ambitious guy, and he took about twenty of us over there that everybody called the 'Storm Troopers.'"

With the Apollo lunar program getting under way, there was a push to be a part of this national effort. Scott said that Storms was a driven man. "He was going after the second-stage Saturn booster, which was the first of the whole lunar program proposals to go out. Everybody said, 'No way this stinking little division was going get it.' Well, we got it! Actually, that was probably the biggest technological challenge on the whole Apollo program—the second stage booster. Then we went after the Apollo, and everybody said, 'No way you're going to get two major contracts back-to-back.' Well, we got the contract for the Apollo Command and Service Modules, and the Earth Escape System. So Apollo was really North American's." Scott was magnanimous enough to add, "Douglas cut a little tin for the third stage and Boeing cut some tin for the first stage."

It was a tough but rewarding time, as Scott recalled: "When you worked for a man like [Storms], the rest of the world doesn't exist. You worked twenty hours a day, seven days a week, and we put together the Apollo program. I was responsible for all the quality assurance, the reliability engineering, and systems test. . . . We set up [at the Kennedy Space Center] and from there on it deteriorated because of the difficulties with NASA. It turned to politics, so I got fired."

Crossfield left North American Aviation on New Year's Eve 1966. From there, he moved to Eastern Airlines, becoming their vice president of Technical Development. He said, "Primarily, I was responsible for everything forward of the cockpit bulkhead. . . . Eastern probably [flew] the best cockpit in the air. A lot of airlines specified Eastern cockpits." Scott also went to Boeing in Seattle to fly acceptance tests for aircraft coming off the assembly line prior to being transferred to the airline. His stay at Eastern lasted until 1972.

In 1974 and 1975 Crossfield had a short stint at Hawker-Siddley Aviation, before feeling the need to get more directly involved with the roots of aerospace technology in America by becoming a technical consultant for the Science and Technology Committee of the U.S. House of Representatives. In this capacity, "Our intent [was] to weigh what resources [were] available and hope we got the most of importance out of our research buck. All of the space program comes out of that committee," Scott said.

His specialty on Capitol Hill was, naturally, in aeronautics, and Scott was a primary mover behind the x-30 Aerospaceplane project. The idea initially appeared to take flight by catching the public's imagination with a vehicle that could go to any point on Earth in an hour or less. It was nicknamed the "Orient Express."

Crossfield explained the problem, which continues to occur: "There came out of that period of the 1960s a terrible disdain for technology and what it could do. It's what I call the 'Volkswagen Mafia.' They came up with absurd reasons to cancel everything. I call them that because, no matter what you wanted to do, there'd be a bunch of Volkswagen drivers who would come up, and a group of broad-beam women would get out with signs and start marching around." Scott had truly thought the X-30 would make it through, but unfortunately, like so many other programs that pushed aerospace technology after the X-15, such as the American supersonic transport, the Aerospaceplane eventually fell to the budget ax. Scott was critical of the airline industry in particular: "It's a crime this country is still flying these aluminum clouds around. They have the same technology as the B-17, just better engines." In the end, a smile crossed his face as he laughingly said, "We have the worst system in the world—except for all the others."

By 1993 Crossfield decided to stop trying to change the system and retired. The NASA administrator at that time, Daniel Goldin, presented him with the Distinguished Public Service Medal for his half century of service to the aviation community and flight research. Scott was quoted as saying, "Our biggest risk for the future is to *not* risk for the future."

One question that pressed was, with his love of flight and the early possibility of going into space in the X-15, why would Scott not apply to the NASA astronaut office? He explained, "[Dr.] Randy Lovelace and General [Donald] Flickinger were on the selection board. They took me to supper one night and asked me not to put in for astronaut. I asked them, 'Why not?' and they said, 'Well, we're friends of yours. We don't want to have to turn you down.' I asked, 'Why would you have to turn me down?' and they said, 'You're too independent.'"

During our interview I asked Scott how much of his work on the X-15 he had kept over the years. "I don't have any of the personal files that I kept." He went on to explain: "A navy A-7 [Corsair II] hit an apartment house out near Alameda and went right down to the basement; burned the whole thing down. My ex-secretary from North American lived in that apartment house, and she had all my papers. I'd asked her to organize them and put them all into some kind of useful form. They were just the way we'd packed them up in boxes when we left Los Angeles. She'd gone out to dinner and this airplane burned the place down. All of those papers are gone, every note I ever took on the X-15, every bit of correspondence is gone."

The incident occurred at 8:13 p.m. on 7 February 1973. Two Corsairs were on a training flight out of Sacramento, heading to Lemoore Naval Air Station. The flight leader was Lt. John B. Pianetta, with his wingman, Lt. Robert L. Ward. While flying over the San Francisco Bay, Ward's plane suddenly veered off and disappeared. Pianetta turned back and descended, just in time to see an explosion on the ground. The A-7E came down almost vertically, hitting the Tahoe Apartments in Alameda, killing Ward and ten others on the ground.

A flash fire occurred in the cockpit, centered on the pilot's face mask and oxygen hose. It was demonstrated that the most likely cause of the fire and subsequent accident was that Ward lit up a cigarette and was smoking, a bizarre and lethal choice in the closed confines of a fighter jet cockpit with pure oxygen being pumped in. The thirty-six-unit apartment building, including all of Crossfield's X-15 records, were consumed in the flames. The fact his secretary decided to go out for dinner that Wednesday evening saved her life.

Even after all this, Scott Crossfield was not done with aviation. Soon after the turn of the twenty-first century, plans started to form concerning the centennial of the first powered flight by the Wright brothers. A re-creation of the 1903 *Wright Flyer* was in the works, and Scott was asked to participate as director of Flight Operations. This translated into Crossfield being the trainer for the pilots who would attempt to take it into the air from the same location at Kill Devil Hills, North Carolina, exactly one hundred years later.

The Experimental Aircraft Association partnered with a group called The Wright Experience, headquartered in Warrenton, Virginia. Ken Hyde was the president, in charge of the reproduction. The press event to unveil the *Flyer* on 18 March 2003 was held in Washington DC, at Reagan National Airport. Hyde said, "It's pretty easy to build a *Wright Flyer* replica that looks like the first plane, but it's very difficult to build one that is an exact reproduction. Building this *Flyer* was the ultimate reverse engineering job, with a major catch — we had to ignore what we had learned over the past 100 years and embrace the Wright brothers' way of thinking."

While Orville and Wilbur could choose exactly the correct moment to bring their invention into the wind and start the engines, the clock ticked down on the centennial, and the weather proved uncooperative. The Wright airplane was created to fly in winds of 15 to 20 miles an hour, but on 17 December 2003 it was cold and rainy, with no wind. A friend with me at the event that day said, "I would have been drier had I jumped in a lake with all my clothes on." President George W. Bush arrived and made his statements of history to the crowd, then departed with a ghostly pass of Air Force One through the low, black clouds.

About two hours after the appointed time, the rain subsided long enough to tow the *Flyer* into position in front of a crowd estimated at 35,000. With the engine clattering and propellers spinning into a blur, Kevin Kochersberger, the pilot trained by Crossfield for the task, throttled up the engine and slid down the rail. The *Flyer* appeared to jump from the track momentarily, but lift from the nonexistent wind never materialized. It could not be sustained in the air. Kevin fought the controls, while the *Flyer* skidded to the right, landing inelegantly in the mud. Sally Crossfield Farley said of her father's disappointment, "[My father had] the best time working on that project. He admired the Wrights tremendously. He and my mom hand wove the rope that towed the *Flyer* in the Warrenton training." Ken Hyde said of Crossfield, and the entire team, "I have nothing but pride about the people involved in this project."

Scott Crossfield's career spanned a large part of that first hundred years of human flight, so it was appropriate he was a part of this historic event. He helped design the world's fastest rocket plane, then, toward the end of his life, trained the pilots to fly the world's slowest.

When Scott was working for the Science and Technology Committee, he invited me to sit with him at the Rayburn House Office Building. We were just across Independence Avenue from the Capitol and a couple blocks east of the Smithsonian National Air and Space Museum, which houses x-15 no. 1 in its Milestones of Flight gallery. He said, "I've done what I wanted. The reason I'm working here is, aviation's been very good to me. I feel I owe it a debt."

Years later Sally said, "My special memories growing up were of our Los Angeles home, mostly. Playing in the pool with my brothers and Dad, watching them race the go carts that my dad made. . . . Family dinners with all eight of us, each one sneaking the carrots to the next plate until they all ended up on my dad's." Switching gears, Sally continued, "The x-15 had always been a member of my family, like the seventh kid. . . . [My father] always said he got more out of that program than any man could ask for, and he marched on afterwards. . . . I think he was heartbroken when he lost the x-15. He never said that, but I saw it in his eyes. Yet again the 'Crossfield Way' wouldn't let him dwell on it or wallow in it. He moved on and looked for new challenges and opportunities."

Scott Crossfield's legacy remains. Johnny Armstrong, an x-15 flight planner who came to the program after Scott left, talked of how he was inspired while a sophomore in high school. "One of the reasons that motivated me to come to Edwards was to work on rocketships like the D-558. When [Crossfield] went to Mach 2, that really got my attention." Johnny said, "Scott started in a very different time period. He was there for all the unknowns in the development [of the x-15]. He had a totally different picture from the rest of us. By the time we walked in, it was already a piece of hardware."

2. A Record High

I felt just as sharp as the young fighter pilot I had
once been, and overhead, a cloudless sky of
vivid blue awaited me.

Joseph A. Walker

"Joseph [Albert] Walker, scientist, war hero, holder of world records for speed and altitude for controlled flights, and one of those fighting for our place in outer space—this is your life!" The music swelled with those words from Ralph Edwards, in front of an NBC nationwide audience at home, watching their fuzzy, black-and-white television screens. It was 10:30 p.m. on the evening of Sunday, 4 June 1961, although the filming had taken place in early April. America was about to meet the man whom fellow X-15 pilot, Pete Knight, later called, "Just an old farm boy that was a good pilot." Walker's boss, Walt Williams, described Joe as "a hillbilly from Pennsylvania with a PhD in physics."

The show that night, as with all other episodes of *This Is Your Life*, was saccharin sweet. Although the program had lasted thirteen years on radio and television, many had come to dislike the ambush style of confronting people with family, friends, and colleagues, as Ralph Edwards took the subject down memory lane, bestowing accolades and prizes for their discomfort.

It was obvious throughout the show that Walker was hesitant and would rather have been anywhere other than as the center of attention. However, NASA had provided its full cooperation in the deception against Walker, to receive free public attention for the X-15 and its role in America's space program. When Joe first realized he was the subject of the show, he actually flinched away from Ralph Edwards, saying, "You startled me."

The main hangar at the Flight Research Center served as the makeshift studio, with lights, microphones, and rows of chairs for the audience, primarily made up of people from the three agencies working on the x-15: the air force, navy, and NASA. The backdrop for the proceedings was a full-scale mockup of the rocket plane, recently used as a stand-in for the real aircraft in a movie about the research program, slated for release in November of that year. The reality of filming in such a location showed how difficult it could be when the roar of jet engines threatened to drown out the host and speakers.

The extra publicity and the television show about Walker was due to his flight on 30 March, when he reached 169,600 feet. This was the third time above 100,000 feet for the program but was much higher than the two flights accomplished by Bob White in May and August of the previous year. Part of the reason for Walker's higher altitude was the first NASA use of the big LR-99 engine. This increasing height, as the x-15 headed for space, caught the attention of a lot of people in the public and also in government. For Joe it was simply a joy to be in a position to observe the sights from such heights. With his usual "ah, shucks" attitude, he said, "As a whole, the flight was a rousing success."

Following the mission Walker was asked to provide a special report to Congressman Roman C. Pucinski from northwest Chicago, who had a great interest in aerospace. Walker described the view to Pucinski, saying, "The most impressive observation initially was the aspect of the sky overhead. The color, I would describe as being a very deep violet blue, not indicative of a black shading, but an extremely dark bluish cast." Joe went on to tell the congressman:

No difficulty is experienced in observing and identifying geographical features on the surface of the Earth, particularly in areas with which one is familiar. An outstanding aspect of this is the appreciation of relative heights or elevations. . . . Mountains still stand out as mountains. . . . Areas which are heavily forested or under agricultural development, could be separated from those areas where nothing was growing. . . . Looking down vertically or near vertically, features are very distinct. As one's gaze swings further and further out toward the horizon, of course, features become more blurred.

It was a disappointment to me that there seemed to be an almost continuous string of low stratus along the coast, all the way from left to right, so that my efforts to identify prominent coastal features, in order to arrive at an estimate of how far I could see to the side, were frustrated. However, judging a little from the apparent angle towards San Diego, it was obvious that I was looking well down the coast of Baja, California, and I could see equal distance up northerly along the coast. The curvature of the Earth was very apparent.

Getting to the point of seeing Earth from such a perspective was a journey that took Walker more than forty years to accomplish, starting with his birth to Thomas and Pauline, at their farm in Washington, Pennsylvania, on 20 February 1921. The town is about thirty miles southwest of Pittsburgh and is the seat of Washington County, named for the first U.S. president.

This honor was put to the test in 1791, a decade after the town's founding, under the administration of that same president. When a special excise tax was imposed on the distilling of whiskey, the small producers felt especially put upon by the American government because they usually had to pay more tax for each gallon produced than larger companies. This led to the first armed rebellion against the idea of a central authority in a country not yet twenty years old. Tax collectors were forcibly evicted from their homes, and some residences were burned, as happened in Pennsylvania. George Washington sent militia west to put down the insurrection in 1794, but by the time they arrived, most of the violent actions had dissipated.

By the time Joe Walker was a young adult, this chapter in the annals of his hometown was long gone and taught only in school history class. He was probably fascinated by the events that shaped the area where he grew up, but science was the bigger draw. His father recalled that Joe had a practical side, too: "Joe always was good in books, but he was mighty handy with machinery and tractors and tools and everything like that." This combination was perfect for someone who eventually found his niche as an experimental test pilot.

Joe finished Trinity High School, then won a scholarship and enrolled as a physics major at Washington & Jefferson College. The small liberal arts and science school is located—appropriately for the legend of George Washington and his penchant for taking an ax to an unsuspecting tree—at the east end of Cherry Avenue, near the heart of town. Walker excelled at the college.

During his senior year, he made a decision to begin flying lessons. Joe entered the flight school of R. E. "Buck" Springer, under the Civilian Pilot Training Program, instituted to interest men in the vocation as World

War II was getting started. Joe's first time aloft was in a venerable Piper J-3 Cub, like nearly all pilots who entered the government program at that time. Buck said later of his student, "Joe [had] the seriousness and tenacity I knew would make a good pilot." This was proven when Walker scored "the highest marks on record" for his aviation cadet mental acuity exams during the application process to join the Army Air Corps. He went on active military duty following college graduation with his bachelor of science degree in 1942.

Walker was deployed to the North African theater during the war and was assigned photo and weather reconnaissance missions flying the twin-boomed, twin-tailed Lockheed P-38 Lightning. Joe was part of the 15th Air Force, originally headquartered in Tunisia, on the central African coast of the Mediterranean, under the command of Gen. James Harold "Jimmy" Doolittle. Within two months of its formation, the 15th was transferred to an air base in the Province of Foggia on the east coast of Italy, where it spent the rest of the war.

On the forward fuselage of the green, camouflaged plane, Joe affectionately painted the name "Chuggin' Charlie" in large, billowy, white letters inside a puffy smoke cloud emanating from a cartoon caricature of a steam locomotive. Besides seven Air Medals, in July 1944 he was awarded the Distinguished Flying Cross from Gen. Nathan Farragut Twining, who had taken command from Doolittle. To receive such recognition, Walker showed "heroism or extraordinary achievement while participating in an aerial flight."

Walker returned stateside in March 1945, taking advantage of a special program with the National Advisory Committee for Aeronautics. If NACA felt someone was valuable to the agency, they were able to secure release from the military. In one way, it could be said Joe was drafted by NACA. With Joe's physics and aviation background, he was the perfect candidate. Joe Vensel grabbed Walker and brought him to the Aircraft Engine Research Laboratory in Cleveland, Ohio, which changed its name to the more familiar Lewis Flight Propulsion Laboratory in 1948. Although hired as a scientist, not a pilot, that status changed within four or five days of his arrival as Walker took over a P-38 de-icing program. This was a quick job transition, but according to Joe, "It was two days longer than I wanted."

Within three months, Walker had his first accident. A Republic P-47G Thunderbolt he was flying ditched into a field near a house outside Cleveland on 23 June 1945. He avoided the power and telephone lines, but took out some small trees as he skidded along. He came to rest pointing directly at the home, about fifty feet in front of him. He was unhurt but embarrassed as a crowd gathered. The accident did nothing to dampen his enthusiasm for test piloting.

While in Cleveland, Joe met Grace. She was a teacher in her second year at Lakewood, just down the Lake Erie coast. There was a Presbyterian church group that often went bowling as a mixer. Grace admitted she "went with trepidation" because she didn't really want to meet anyone. A friend talked her into the outing, and she saw Joe. "He was kind of a shy guy, really. His mother had red hair and, at that time, I had red hair. . . . We ran into them after the meeting [and] Joe called me for a date the next week."

Grace and a friend lived with a retired schoolteacher, who volunteered her home to host the date. Grace said, "We invited Joe and another guy over to dinner, and we used [the teacher's] crystal [glassware]. She sat in the living room listening to all this, and it was hilarious as a date. A real uptight meal if you've ever had one." From that point, their future together seemed a foregone conclusion, at least for everyone except Joe. Grace explained, "Joe ran around with a group of fellows. They were all married except him, so they were working hard to get him into that state. He was a little reluctant."

Everything changed one weekend when Joe decided to take his sweetheart home to meet his parents. This went well, so Grace returned the favor soon after, although she recalled, "It was kind of stiff and stilted. [But then] we were sitting in the living room, and all of a sudden the radio exploded. . . . It went 'Pow!' That made everything okay, [because] they had to figure out what happened—and Joe did. My father was a similar kind. He was a Tennessee mountain boy and liked Joe right away."

From their first meeting, their relationship developed quickly. "[We] met, I think, in February, got engaged in July, and were married the next April [1949] at Easter vacation time. I was still teaching, and you couldn't teach if you were married in Lakewood District, so that was the end of my teaching career."

Joe and Grace had their first child, Thomas, on 5 August 1950. About that time, Joe Vensel left Lewis for the High Speed Flight Research Station at Edwards AFB. He asked Walker to join him, so it was time to convince Grace this was a good move. She said, "Joe was very intense about what he was doing. . . . I wasn't going to understand jets, and all I knew was a lot of them augered in." Thomas was just six months old, but Grace agreed to go if that was what her husband wanted. In February 1951 Joe, Grace, and Thomas arrived in California.

As with other wives uprooted from less arid locales, coming to the desert towns around Edwards was a shock. Grace remembered, "When we first came out here, I'd never been west of the Mississippi. . . . It was a big adventure for us both. We wound up in Mojave in some beat-up old apartments from World War II. . . . It was a grand time. It was almost like pioneering." She talked about precautions against the incessant sand: "The windows were taped shut on the houses with duct tape. We took that off because it was too warm for us in February as compared to Ohio, then we got piles of sand on the floor. . . . I was homesick for greenery. We'd go up to Tehachapi [to] their little tiny park and sit. It's a pretty little town when the wind isn't blowing."

Four more children joined the family over the following decade, James, Robert, Joseph, and Elizabeth, although they lost Robert when he was less than twenty-seven months old.

Walker was soon flying all the aircraft coming under the auspices of NACA. By the summer of 1954, a bizarre-looking, needle-nosed, white aircraft with twin jet engines, a minuscule tail, and thin silver wings similar to those later used on the x-15 came to the agency for evaluation. It was called the Douglas x-3 Stiletto and was supposed to study flight while cruising at Mach 2, but it barely ever reached above the speed of sound because of its underpowered engines.

NACA decided to use the aircraft for transonic and low-supersonic roll research. Joe took on the challenge. On 27 October 1954, his first roll at Mach 0.92 experienced an unexpected pitch-up and yaw. After a momentary loss of control, Joe righted the x-3, put it into a shallow dive to pass Mach 1, and went into a second test. Violent up-and-down pitching movements at seven gs came dangerously close to overstressing the airframe. Any higher levels and the Stiletto probably would have broken apart. The phenomenon experienced by Walker was, at that time, called roll coupling, and had been the same problem that almost killed Chuck Yeager in December 1953 in the x-1A. Today it is more commonly referred to as inertial coupling.

It was a year before flight testing resumed on the x-3, and it never again approached the types of conditions that had nearly taken Walker's life. However, the problems Joe encountered that day led to the understanding of inertial coupling. Engineering data like this was used extensively on the design of the x-15.

A second incident for Walker involved the same x-1a used by Yeager in his bid against Scott Crossfield to be the first to break Mach 2. Hiding inside the liquid oxygen tank of the rocket plane was a bomb. The explosive device was not created on purpose; instead, it came into being through engineering and chemical ignorance.

On the early afternoon of 8 August 1955, with about a minute left before drop from the b-29 mothership, Walker was sealed into the x-1a cockpit, setting his switches for launch. Flying at 30,000 feet, both planes were suddenly jolted by an explosion inside the rocket plane. The oxygen tank blew open, expelling its contents into the surrounding area, including inside the bomb bay. Maj. Arthur W. "Kit" Murray was flying close by in an F-86 chase plane and was momentarily enveloped by the cloud. Debris flew by, cracking his windshield and a wingtip light. Walker felt the explosion and saw most of his electrical systems go dark, including radio contact with the outside world. Abruptly, the landing gear of the x-1a extended, as the gear doors were blown off by the overpressure.

Stan Butchart, who was the mothership's pilot, said, "There was kind of a 'gentlemen's agreement' we had between the pilots of either airplane, that if we ever had a real problem in either the b-29 or the [rocket plane], we would separate them and let them fight their battle by themselves." In this case, Stan held back on immediately jettisoning Walker. "Something told me to wait a minute, hold on, and it gave us time to settle down."

Stan, along with copilot Jack McKay, started an immediate descent in the b-29. At the same time, two crew members, Charles "Duke" Littleton and Jack Moise, went into overdrive to help Joe out of the stricken x-1a's cockpit. Stan recalled, "The two of them crawled back in [the b-29's bomb bay] and got Joe out, then brought him up front. When Joe came up, he was without oxygen from the time they'd unhooked him back there until he got up to me. He came crawling up along the floor, looking up at me like a little puppy for help. I immediately unhooked my oxygen and plugged it into Joe's [receptacle] to get him going on his feet again, until they had time to find another hose in the area."

The x-1A was a valuable investment, and, at first, Stan was hoping to save that aircraft, along with his own. "That was a tremendous explosion. I wanted to land with the thing [to save the x-1A], but [Joe Vensel in ground control] talked me out of it. . . . It had sheared most of the bolts holding it to the [mothership], and it moved that thing forward, so it wasn't hanging in there very tough. And it blew the gear down. I thought, 'Well, we'll land and it will just squash the gear up.' But [Vensel] said, 'No, get rid of it.'" Stan was disappointed but resigned himself, radioing for instructions. "Give me a spot to drop it." With Walker safely aboard, the release lever was pulled, and the x-1A dropped away toward the desert. Murray, and fellow chase pilot Neil Armstrong in a p-51, said the plane entered a tail-first spin, then exploded on impact.

The cause was eventually found to be an organic compound called tricresyl phosphate, used to make leather gaskets in the aircraft. In the presence of liquid oxygen, the compound was unstable, and when an impact occurred between the gasket and the surrounding metal, such as in turbulent air, an explosion was all but inevitable. Four research aircraft (x-1 no. 3, x-1D, x-2 no. 1, and now the x-1A) and two irreplaceable lives (Jean Ziegler and Frank Wolko) had been lost due to this gasket problem over the years. The x-1A incident finally provided enough clues to show investigators what was happening and to stop it from occurring again.

For their actions during the emergency, Littleton and Moise were awarded the highest NACA honor, the Distinguished Service Medal, while Walker and Butchart received the Exceptional Service Medal.

One rocket plane possibly saved from destruction by the explosive gasket was the x-1E. The last of the x-1 series, this started life as a twin to the aircraft that Yeager first pushed past Mach 1. Since that aircraft was already retired to the Smithsonian, an idea was floated to modify the second airframe into the x-1E to test a thin, new airfoil wing design at Mach 2, as had been the hope for the failed x-3. Also, a high-speed turbopump for the LR-11 rocket engine was substituted for the nitrogen pressure-fed system in the older model. In outward appearance the biggest difference was replacing the conformal cockpit windows and side entrance hatch of the x-1 with an upward-opening canopy, which afforded much better pilot visibility and safety. Walker was assigned the project, eventually completing all but five flights in the modified aircraft. He had to leave the program following his twenty-first X-1E flight in September 1958 to take up full-time duties on the X-15, which rolled out 15 October.

Walker waited patiently through the first eight X-15 flights and twelve aborted attempts by Scott Crossfield before he finally had a chance to take to the air himself.

Several years earlier, Crossfield and Walker had a rivalry when they both worked at NACA, and both felt they were the chief test pilot. Stan Butchart explained, "Part of the problem was that Vensel, our boss, had never really designated who was in charge of the office, who was going to be the chief pilot. Joe [Walker] felt he was, because he had more seniority [and] because he had been at Lewis Lab. Scott felt he was because he arrived at Edwards first. Joe [Vensel] never settled the dispute, and it didn't make for a happy home." Stan put it bluntly when he said, "They didn't get along. Joe Walker, when he and I were in B-47 school, he turned to me a couple of times and said, 'I don't know how in the hell you can be friends with Scott.' They just fought tooth and nail. Oh man, Scott had a very odd way about him, and he crapped on so many people, but he always treated me just beautifully."

On the flip side, Stan said Walker could be just as ornery sometimes. "Joe had an awful temper, really. Oh, man. I bowled with him on a NASA team . . . and if he'd get a split he'd just be red in the face. You didn't dare speak to him either, you just kind of looked away. Next time Joe'd go up and he'd get a strike, then he'd come back and have a big grin."

As soon as Joe was inside the cockpit, everything was different. As the prime X-15 pilot for NASA, it fell to Joe to be the first to make a flight for the government, doing so successfully on flight 1-3-8 on Friday, 25 March 1960. The B-52, with Jack Allavie and Fitz Fulton at the controls, was scheduled to roll at 8:00 a.m. Various problems kept delaying the takeoff until 2:42 p.m. Electrical difficulties with the connection between the B-52 and X-15 created further problems once they were airborne. The combination had to fly around a racetrack pattern an additional time before everything was ready for Joe's launch, which occurred at 3:43 p.m., late in the afternoon for lighting conditions that time of year.

For someone who had experienced many drops from the B-29 in earlier tests of rocket planes, Walker said, "There is no question but that the pylon arrangement and B-52 for [the] mothership is of several orders of magnitude improvement over the bomb bay B-29 system used for previous research aircraft."

When Joe dropped away, he stated that the X-15, "snapped abruptly to the right approximately thirty degrees, and was righted immediately by the pilot, who proceeded to start up the rocket engines. The lower engine lit off okay. . . . However, the upper engine . . . went out in overspeed. Upon reset and attempted restart, it again went out in overspeed. The third attempt was successful, and the flight continued on eight chambers."

As he reached Mach 2, Walker shut off three of the LR-11 chambers to maintain his speed at 48,000 feet. He swung around on the planned course, coming in at the south end of the Rogers lakebed on runway 35. Joe released the landing gear. In his post-flight report, he said, "Initial skid contact is barely noticeable. However, the crash when the nose gear hits the ground, leaves no doubt that you are on the ground." He concluded his first report with a stilted test pilot's accolade to the X-15: "I think we have a real good airplane here, and we should be able to prosecute the research program without undue drawback."

With no nose wheel steering, and simple skids in the rear of the aircraft, it was always assumed there was no way to control the rollout with the vertical tail rudder once touchdown occurred. Twenty-five days after his first mission, Walker made his second flight and discovered that, contrary to that belief, he could affect his direction. He wrote in his pilot's report: "During the glide-out after touchdown, it was observed that a small amount of left rudder caused the rear end of the aircraft to start skidding to the right slightly. . . . It is apparent then that, while pointed into the breeze on the ground, the rudder control action of this aircraft is the same as in flight."

Robert White served the same pilot capacity as Walker, but represented the U.S. Air Force. As Scott Crossfield and the contractor phase of the program went on with aircraft no. 2, Walker and White alternated missions between the agencies, while the U.S. Navy and their pilot, Forrest Petersen, waited in the wings. Five flights for Joe and Bob went back and forth, each at first nearly mirroring the other's parameters in the air. A notable exception to this was when Walker made his first Mach 3 flight his third time aloft. White did not achieve the same speed milestone until his fifth mission, although Bob set his own records by exceeding 100,000 feet on two of those flights.

Joe's third flight, 1-7, was the first on the x-15 to move out of the local Edwards AFB area. It was a cautious flight, with the B-52 gaining most of its altitude while circling above Rogers Dry Lake. Once at 38,000 feet the B-52/x-15, along with three chase planes with three future x-15 pilots—Bob Rushworth, Pete Knight, and Jack McKay—all headed northeast toward the town of Baker, on the road toward Las Vegas, Nevada. They darted from the protection of one dry lake to another—Cuddeback and Three Sisters—until reaching their destination, near Silver Dry Lake. Walker stated, "The whole operation proceeded very smoothly and expeditiously." At launch, his exuberance was evident as he radioed, "Hot ziggity!"

Although the flight was scheduled for Mach 2.8, things got away from Walker as he accelerated. He later wrote in his report, "A recheck on the Mach meter showed an indicated number of 3.2. With considerable shock it was realized that the intended program of reducing to half power at the same time as pulling out level had been delayed too long."

When Joe discovered how effortlessly the x-15 handled at that speed, so much so that he had not even noticed how fast he was flying, he reached to shut down three chambers on the LR-11. The three stopped, but so did the other five, shutting down all rocket power. Immediately, Joe asked, "Are we in the Edwards area?" Neil Armstrong, in the control room, replied, "Negative. You're in Three Sisters area."

Walker had no difficulty controlling the no. 1 aircraft that morning and had plenty of momentum to bring it safely back home to Rogers. As he touched down, Joe had not only accomplished the first mission launching away from Edwards but had also become the first man to successfully penetrate past 2,000 miles an hour and Mach 3. The last pilot who had done so, Capt. Milburn G. Apt, died in the x-2 on 27 September 1956. Apt achieved a nearly identical speed to Walker's but lost control when he experienced roll coupling as he attempted to turn back to Edwards after the speed run. It was Apt's first and last flight as a test pilot.

Essential for expanded x-15 operations was the flight corridor known as the High Range. With earlier rocket plane research, such as the x-1 series and D-558, Phase 2, the flights were able to stay within the local Edwards AFB area, zooming through the airspace around the Rogers and Rosamond dry lakes at up to Mach 3 and more than 120,000 feet. The x-15 was scheduled to double both those marks—and more. Because of this, flight planners needed a much larger canvas on which to work. The range extended northeast from Edwards, 450 miles to the Bonneville Salt Flats area of Wendover, Utah, located near the northwest corner of the state, on the border with Nevada.

With such an area to monitor on long-distance flights, Edwards could not do it alone. Two additional manned radar tracking and telemetry stations were established. Both were in Nevada, along the line between Edwards and Wendover. First was Beatty, then Ely, at 150 miles and 350 miles, respectively. These stations sat atop mountain ridges to provide clear horizons for receiving information from the x-15 as it passed along the route. Besides picking up signals from the rocket plane, the stations also sent navigational assistance to the x-15, b-52, and chase planes.

Getting to the radar sites was an adventure in and of itself. Beatty is about twenty-five miles east of the heart of Death Valley, across the California state line into Nevada, in the Bullfrog Hills of the Amargosa Desert. The entrance to the early 1900s ghost town of Rhyolite is passed as you go up and over a final hill into Beatty. Zipping through the tiny town and heading north, the turnoff for the radar site was about fifteen miles up the highway on the west side. From there, it was a three-quarter-mile trek up a straight, steep dirt road to a sharp bend to the south along the ridgeline for another half mile, before getting to the site, located 4,920 feet up in the hills. To the west, the spot overlooked the southern end of the Sarcobatus Flat.

A rectangular, flat-roofed building, topped with a large, swiveling radar dish, housed the control room. Surrounding that were several smaller, outlying buildings, radio and microwave antennae, a gas storage tank, and a wide, flat parking area for visiting vehicles. Inside, it was packed with electronic equipment, plotting boards, telecommunications, and a large radar screen, all housed in government-gray racks, consoles, and desks. Squeaky swivel chairs provided limited seating. On the ceiling were acoustic tiles and round lights with covers that looked like leftover models from a cheap 1950s UFO invasion film. When fully manned for a mission, with ten or more people, it got crowded. Construction was accomplished by the Electronic Engineering Company of Los Angeles.

Thornton D. Barnes, better known to all as simply "T.D.," served at Beatty. He said there was something unique at the site: "The area was covered by chuckawallas, these huge lizards, I mean, monster things. We also had a lot of sidewinders, so we kept a couple bull snakes up there in the summer, and that kept the sidewinders away. Then, one winter [the bull snakes] didn't survive, so every time you stepped out of the building, you had to watch to make sure there weren't any vipers there."

T.D. went on to talk of the people he lived with, saying, "We drove up when we first got there, stopped at the station, gassed up, and found out where we were supposed to go. It must have been 104 degrees that day, and there was a woman there who had a stocking cap on. I was getting ready to leave, and she leaned down and said, 'Barnes, hotter than Hell's hubs, isn't it?' She knew my name! I'd never seen her before! Panamint Annie was her name. Her folks came over in a covered wagon, and she was real sickly."

There were many other characters like her in the town. "We had Pissy Pants Richard. He would get thrown out of the casino because he'd be there playing blackjack and peeing his pants. Ma Vincent, she had a third set of teeth. All the tourists who came through, she'd show [them] her teeth. It was such a town. It was so different."

T.D. had dozens of stories, including one about Paul Laxalt, who ran for Nevada governor in 1966 and won. He lived in Beatty and was a good friend of T.D.'s. "He'd come up there and sit at the bar [at the casino]. . . . To get your driver's license, they would set up at the end of the bar to give the test. He'd always get mad if somebody would come up with a bottle of beer in their hand to take the test. He'd go, 'Damn-it, at least leave your beer at the table.'"

The town was certainly unique. "It was so different, you just can't imagine. We loved it. The kids were young. We got us an old military Jeep, and we'd go out in the hills. You'd find mines where there hadn't been [anyone] since the miner walked away."

The Ely station was even more remote, although at 9,263 feet there was little worry of snakes. The town, which lent the remote X-15 site its name, was approximately 250 miles north of Las Vegas. Ely is the easternmost town in Nevada along Highway 50, also known as "The Loneliest Road in America." Even being so far off the beaten path, there were plenty of gambling halls and other discreet distractions to keep the locals interested. The Hotel Nevada was a favorite of many Hollywood celebrities, including Jimmy Stewart, Gary Cooper, and Mickey Rooney.

5. The blistering cold aftermath of an ice storm on the radar dish at the High Range station in Ely, Nevada. Courtesy of the Armstrong Flight Research Center.

Heading northwest along the highway from Ely, which started life as the Pony Express trail in 1860, the High Range crew went about eleven miles to a southern turnoff. From there, they drove up into the mountains, often snowbound even into June. It could be a grueling trip, six miles of dirt, with switchbacks for the last third of the route. Once at the site, the main building looked identical to that at Beatty, although the area seemed more at odds with the elements. Almost all photos showed frigid conditions, with personnel in parkas shoveling snow. The after-effects of a severe ice storm revealed the frozen tableau where high winds had blown parallel to the ground. As I sat with T.D. on a warm, sunny day in southern Nevada, he explained, "Now, the Ely station, those poor people, we'd have weather like this, and they'd have snowplows."

Also essential for High Range operations, and used to define the limits of its terrain, were a series of dry lakes on either side of the centerline. Their names were varied, usually derived from pioneering days in the west. Heading up the corridor from Edwards, the closest was Cuddeback, then Silver, Hidden Hills, Delamar, Mud, Railroad Valley, and Smith Ranch. Others, such as Three Sisters, Ballarat, Racetrack, and Grapevine, were surveyed for possible use, but launches never occurred from them. All the lakebeds served as safe havens in the event of an emergency after launch or along the way back to the home base.

Directing operations throughout the High Range was the NASA control room at Edwards. The room was located on the top floor of the main NASA building at the High Speed Flight Station and held very similar equipment to that found at Beatty and Ely, but without as much constraint in the square footage. Amenities at the center were also more plentiful than could be found at the remote sites in Nevada.

The heart of the control room was the man in charge of communications. Technically he was "NASA I." With very few exceptions, he was always one of the x-15 pilots, so there was never a problem getting across any point necessary. He served as the test conductor and coordinator for all aircraft and ground stations used during any mission. Although the B-52 and chase planes could talk directly with the x-15, it was the controller's job to be the primary spokesman on the radio loop, making sure everyone was in sync. Others in the room monitored incoming telemetry and the flight path chart, and, if necessary, passed along technical suggestions to NASA I, to then transmit to the x-15 pilot or others uprange.

Like so many things pioneered by the x-15, from the centrifuge simulator to the pilot suiting van to the control room, these served as successful models that were later taken by other NASA facilities at Cape Canaveral in Florida and the Manned Spacecraft Center in Houston for use in the orbital programs of Mercury, Gemini, Apollo, and beyond. The famous Mission Control Center in Houston, and specifically the person who serves as "CapCom," or Capsule Communicator, was a direct outgrowth of the lessons learned with NASA 1 during the early days of the x-15.

The idea of the High Range was incorporated on a larger scale required for spaceflight operations. In fact, many of the people who set up the system for the x-15 were given the task of doing the same for the worldwide tracking and communications grid of ground stations and ships at sea to stay in touch with American astronauts, until finally replaced by orbiting satellites.

Less than a year after x-15 rollout, on 27 September 1959, the High Speed Flight Station was upgraded in status within NASA to the Flight Research Center, largely because of the prominent role the x-15 rocket plane established. Joe Walker was one of the best known of the pilots on the program, and his upcoming flights captured the imagination of the public, almost as much as the exploits of astronauts on their orbital missions from Cape Canaveral.

Joe's next flight after the one highlighted by Ralph Edwards on television accomplished another first for the research program by launching from Mud Dry Lake, outside Tonopah, Nevada. Not only was this nearly double the distance of previous remote lake launches, it was also the first time an x-plane started its mission outside California.

Walker was anything but satisfied with his performance on flight 2-16, as he rocketed past Mach 4 and nearly touched Mach 5, heading straight south on his way back to Edwards. After the 25 May 1961 mission, he said, "You will find that this was not the best flight . . . either from the adherence to the program or the steadiness of the aircraft while it is on its way." This is a rare admission from a test pilot that he was not as on top of things as he should be at all times. There was nothing spectacularly wrong, just a lot of nagging issues where Joe felt he was not keeping to the profile.

The date of this flight was significant in that it came less than three weeks after Alan Shepard became the first American in space, as his *Freedom 7* Mercury capsule was lobbed by a Redstone rocket on a ballistic arc out over the Atlantic Ocean. On 25 May, the day of Walker's flight, Shepard's actions had prompted President John F. Kennedy to deliver one of his most famous speeches to a special joint session of Congress, "that this nation should commit itself to achieving the goal, before this decade is out, of landing a man on the moon and returning him safely to the earth." At that moment, the strides being accomplished by the x-15 toward winged spaceflight were officially put on the back burner for more than a decade. Nonetheless, the research being accomplished in the California desert was important and eventually led to the thirty-year career of the Space Shuttle, the longest of any manned spacecraft designed, developed, and constructed in America.

x-15 testing kept moving apace, with several new pilots coming into the program and getting check flights. Because of this influx, Walker's cockpit time diminished significantly, allowing him to make only two flights in the next eleven months. His schedule picked up in April 1962, with two flights in eleven days. The second of these, 1-27, was set up for the purpose of outdistancing America from the Soviet Union, which had taken the world's altitude record for their own. The United States wanted it back, so Walker and the x-15 took the job. The Federation Aeronautique Internationale (FAI), which certifies all aviation records, was contacted to monitor the flight, scheduled for 27 April.

Cloud cover above Mud aborted the first attempt; however, all went well the second time, on 30 April. Walker said, "I went through all checks in apple-pie shape." The countdown seemed to sneak up on Joe at 10:23 a.m. "I took notice that we were very rapidly going through zero seconds, so I grabbed the launch switch and went off. I intentionally hesitated, and then hit full throttle." Aircraft no. 1 quickly headed upward, peaking several minutes later at 246,700 feet. This altitude was more than double the Soviet record of 113,891 feet, set the previous year by Georgi Mosolov in a rocket-boosted MiG-21.

With the new FAI-certified altitude, nothing could dampen the enthusiasm for Walker's accomplishment. On board the x-15, he marveled at the sight from such heights. "I identified Monterey Bay, the Gulf of California, and well down the lower California peninsula. I couldn't see clear to the tip, due to a cloud layer." He also experienced the familiar sensation of, at first, not understanding his perspective from that height. "Up on top I thought that [Los Angeles] was already under me, and I was mentally bracing myself for an unfortunate flight call, but once I got the nose pitched down, I noticed we had a ways we could go yet." Finally oriented, he said, "I had the field [at Rogers] in sight way up there . . . then, after the call from NASA 1,

I tilted a little to the left and looked at things, pulled in the speed brakes, and flew the pattern in a lazy 360[-degree] overhead [turn] with absolutely no strain."

Most flights used a fairly routine pattern of dropping from the B-52, then heading directly toward home, with major turns restricted for the landing pattern, as Joe spoke of at the end of flight 1-27. An exception to this was his next mission, 1-29, on 7 June. Above the state border, south of Beatty and west of Hidden Hills, Joe launched and directed the nose of the aircraft well east of the ground track that would take him to Edwards. The idea was to truly demonstrate the advantage and maneuverability of a rocket plane versus a simple ballistic capsule during return from orbital flight.

Soon after LR-99 burnout, Walker placed the X-15 into a steep, sixty-degree right bank above the old silver mining area turned tourist destination of Calico Ghost Town. He continued banking to the right above Daggett and passed south of Barstow, before heading west to Edwards. The sustained bank definitely had an effect on the airframe. Joe noted, "I kept hearing from NASA 1 that [dynamic pressure] was decreasing, but the airplane was creaking and popping, and it continued to do this. . . . I looked out, and the nose was where it should be relative to the horizon. It appeared to me that it was lower than it had to be to maintain level flight. I didn't want to push the nose down, thinking that I would wind up in a screaming dive."

Regardless of strain on the aircraft, the pilot, engineers, and flight planners all got what they wanted: proof of the viability of wings in space. "I'm convinced this flight plan was exactly what we advertised," Joe said, "illustrating one of the benefits of having aerodynamic surfaces for a reentry vehicle."

Three weeks later, Walker achieved his highest speed in the X-15, as he topped out at Mach 5.92, or 4,104 miles an hour. The plan had been to hit Mach 6, but he didn't quite make the mark. The reason for this may have been tied to a discrepancy with the altitude given Joe by the radar system. He said, "I must have been a couple of thousand feet low on altitude compared with the radar." T. D. Barnes knew that Joe was correct.

The radar systems at Edwards, Beatty, and Ely were there to not only provide information on the altitude and position of the X-15 but to make sure these data were correct for safety. The last thing a pilot in a hypersonic aircraft wanted was to not know his exact position. T.D. spoke of how the system was supposed to work, saying, "When we would have a mission, the planes would be telling us what altitude they were [with the on-board beacon]. The Beatty transmission system always showed the same altitude. When the plane said it was at 45,000 feet, that's what our little chart pins would be showing." Then the problem showed up that could have cost the life of a pilot. "Anytime you switched your data to Edwards or Ely, the altitude would jump 2,000 feet. . . . I had the NASA guy beside us one flight, [and he said], 'We've got us a problem here.' I told him, 'It's a problem in the system, ever since they put the radar in.'"

It continued to bother T.D., but it seemed no one cared enough to find out what was going on. "One day, Norm West called up and said we were going to be down for two or three weeks because the weather was bad. I got on the horn to my counterpart at NASA, John Hayden, I think it was. I said, 'John, this would be a good time for us to figure out Ely's radar problem.' Our station manager was kind of weak, and he whirled around, turned white, and said, 'You're on your own on this one!' I explained what was happening, and John said, 'That's an inherent problem—and it's Beatty's problem.'" Between the site manager's attitude and being told that his radar was the one that was incorrect, T.D. went on the defensive. "I said, 'John, first of all there's no such thing as an inherent problem in the radar, [and] Beatty agrees with the plane every mission.' John, you could tell, boy he was mad. He said, 'Well, just forget about it. It's always been there and always will be.'"

The radio conversation was being carried live on the x-15 network, and Joe Walker was listening in from the Edwards pilot lounge. "Now, Walker was hearing this . . . and he immediately got on the net intercom and said, 'Effective now, there will be no more flights until that problem's fixed.' So, they approved calibration flights to Beatty, because they were still suspecting us. [Once overhead,] the pilot was saying, 'I'm at 30,000 feet,' and I've got NASA inspectors looking over my shoulder, and I'm showing 30,000 feet [on our radar]." The same was not true at Ely and Edwards. T.D. was vindicated. "They shut the range down and worked on it for about two weeks, tearing those radars all the way down at both stations. When the radars were made . . . they had a problem in them. Somehow or other, the one at Beatty had gotten fixed. . . . They had not changed the other two radars."

Out of the three stations, Beatty was the only one reading correct altitude data, yet T.D. had been initially treated as if he was the problem. His wife, Doris, said, "You almost got fired!" T.D. admitted, "Initially, oh boy, I was the black sheep. If they hadn't found a problem, I'd probably be looking for another job. But it really was going to kill someone, because you couldn't tell them how high they were. . . . If I wasn't agreeing with the planes, I would have accepted that, but I monitored the planes very closely for what they were showing for altitude." Once the problem was identified, and T.D. was shown to be correct, things changed. "I ended up taking over the station, and the company got a huge bonus that quarter for what I'd done."

The technology of these radar sites was old when the High Range was initially built for the x-15, but they made do. T.D. said, "We had the old Mark II radar the air force had [thrown] away. We got it out of their trash pile. You'd pull the chassis out and the wire would just crumble. You'd be in the middle of a mission and the screen would black out, so you'd hit the chassis and, poof, it'd work again. . . . It makes you wonder how we succeeded. It was serious stuff, because we had a guy at the edge of outer space that was trying to get home. . . . Every one of those flights was an adventure of its own."

Once all the radar problems were fixed, it was back to business as usual. The x-15 mission to go higher and faster was still on track as Walker and White traded off the envelope expansion flights. White ended up with the speed flights, while Walker took the ultimate altitude.

NASA operations engineer John McTigue remembered, "I thought Walker was a really good person. But he was, in many cases, obstinate. He had to have it his way or no way." When it came time for his first flight in the no. 3 aircraft, he told John there was a specific way he wanted the cockpit instruments laid out on the main control panel. John thought this was the perfect time to play a prank on the irascible pilot. McTigue said, "I had [a technician] take the panel to the shop and had new holes made where Joe wanted to have stuff done, then repaired the other ones. Before we put it back, I had it painted pink!" John called Walker down to the hangar to check out the new panel layout, then hung back to watch his reaction. "We had to walk around a certain way before you could see it, and finally Joe saw it, and his face got red—he knew he'd been had."

McTigue obligingly put the panel back to its normal color, and Walker went ahead with preparations for flight. He had three in a row in no. 3, building up altitude and performing other experiments, such as checking out the stability augmentation system and testing different reentry techniques with and without the lower ventral fin in place. By 17 January 1963 Walker was ready to make a bid to surpass the 50-mile limit for recognition as an astronaut in a winged vehicle.

As of the previous month, Walker was left as the sole original pilot of the four at the program's outset, now replaced by a new set of pilots. North American Aviation's Scott Crossfield, the U.S. Navy's Forrest Petersen, and the U.S. Air Force's Robert White had now all gone on to other assignments. Walker knew his time was also limited, so he was determined to make the best of it.

With flight 3-14, Joe was actually not yet supposed to hit 50 miles, but that became the case, nonetheless. It was his eighteenth x-15 mission.

At 10:59 a.m., from 45,000 feet, in the vicinity of Delamar Dry Lake, Walker launched away from B-52 no. 008. In his eloquent and joking manner, Joe later stated, "I apparently faked everybody out by getting rotated and on the way expeditiously." The LR-99 thrust was smooth, and his climb was quicker than anticipated. A post-flight comment from NASA engineer Elmor J. Adkins explained the situation: "The engine was unavoidably operated for about five seconds longer than planned, so that the maximum altitude was about 20,000 feet higher than expected." Walker said of his misstep: "I did my level best to shut off the engine at the right time, but a large gap was inserted between the attempt and the accomplishment. I managed to hook the throttle once with [a] finger, and got enough room to get . . . it shut down—finally." This gave him enough energy to peak the x-15 at 271,700 feet, 7,700 feet higher than needed to make this Walker's first astronaut qualification flight. It was not officially acknowledged as such for more than forty-two years.

The ballistic control system rockets in the nose and wingtips kept the spacecraft oriented while it was too high for the wing and tail surfaces to find enough air to function. Joe left the system in automatic mode after he was satisfied it was doing the job properly. This gave him an opportunity to do something many other x-15 pilots said they never had time for—to look out the window and enjoy the view. Joe said later there was one thing that startled him at first, until he figured out the source. "Some kind of searchlight was shining in the right-hand side of the cockpit instead of the left, and I discovered it was the moon." He went on to describe it in test pilot detail: "Half moon, by the way, the line was straight on the disc rather than curved convex, or concave." Over the radio, as he flew through the top of his arc, he told everyone on the network, "The moon sure looks pretty."

Walker also took this time to apologize for being late on shutting down the engine, saying, "I couldn't get my hands on that throttle for beans!" The NASA 1 controller couldn't resist: "What do you want to eat beans up there for?"

Even securely strapped into the tiny cockpit, Joe was able to enjoy the sensation of being weightless. He said, "There was a right smart amount of miscellaneous material floating inside the cockpit, and it lasted for such an extended length of time that I decided it was probably the longest zero g I'd ever had."

After nearly ten minutes on the mission clock, Walker safely extended the landing gear, touched down, and came to a stop on the Rogers lakebed. In that short time, he traveled more than 246 miles, at a maximum speed of Mach 5.47. Joe was ready to turn it around and go again immediately, but five more research flights, and various problems, got in his way before going higher than he had that winter morning.

One of the problems encountered as the altitude increased on X-15 flights concerned the auxiliary power units. There were two APUs on each X-15, and without them, the rocket plane had no power to run its systems. John McTigue, who was in charge of the no. 3 aircraft, said, "All of a sudden, we're going up in altitude, and we started losing APUs. Turned out the APU wasn't really designed for the differential pressure at altitude. As you went to almost zero pressure on the outside, the unit started leaking. And when it started leaking, you lost your oil, and therefore, all your gears. Everything inside it just turned blue, and the APU would stop. My airplane was the first one to experience things like that." John explained the fix was fairly simple. "[We] finally devised a method of putting the damned APUs inside a box and pressurizing the box. A very simple solution when you think about it." Joe's twentieth flight, 3-16, to 209,400 feet, was primarily to check these modifications, which worked as advertised.

Walker was now entering his last phase on the program with two final flights, specifically aimed at finding the theoretical height limit of the air- and spacecraft. Before leaving the program, Bob White successfully took the x-15 above 300,000 feet. This was accomplished only four times throughout the nearly decade-long history of the program. Joe Walker was now poised to fly two of those missions, back-to-back.

Mission 3-21 launched from Smith Ranch, the first to do so from the High Range lake farthest from Edwards. Aboard the x-15 were several experiments, including a unique one where a balloon was to be ejected from the back of the craft, then towed along when in zero g before reentry. The balloon experiment was to test the deployment mechanism planned for use on two later flights of the Mercury program. The mylar balloon panels were of different, bright colors to test the ability of an astronaut to distinguish those colors in space. Walker was to verify it for the first time, although in the supposedly more benign environment of ballistic flight versus being on orbit. Other experiments included infrared and ultraviolet sensors, along with a horizon scanner. None of these experiments operated successfully.

At 10:20 a.m. on 19 July 1963, Walker dropped from the pylon, lit the LR-99, and headed south. He nosed up at a steep angle, aiming for 315,000 feet, close to the mark set by White the previous summer. By the time he was passing through 80,000 feet the ballistic control rockets were already operating to keep him on the flight path. At eighty-four seconds, and approximately 170,000 feet, he reached to shut down the engine. He said later, "[In] the middle of the word 'shutdown,' the engine quit." Joe continued upward, with more than half his altitude yet to go.

For this flight the orientation of the x-15 was altered as it progressed upward. Normally, the aircraft pointed into the relative direction of flight, except when the nose was brought up to reenter. In this case, once out of the atmosphere and not under rocket power, the x-15's nose was brought down twenty degrees below the horizon. Walker said of the maneuver: "There is no sensation, other than realizing that this is crazy to have the nose stuck down, and know the airplane is still going up. But there is no physical sensation that enters into the picture to confuse you at all. As a matter of fact, it is kind of interesting to be able to have a longer look out at the world."

Arcing over the top, Walker was at 347,800 feet, more than 6 miles higher than anyone else who had not orbited the earth. To get some idea of his altitude, consider that 6 miles by itself is an average height above the ground for a passenger airliner.

From Joe's perspective after the flight: "The overhead aspect is just like one of these dark velvet photographer's cloths as far as the sky's appearance, but there weren't any specks of light shining through that I could pick out." He remarked live on the radio: "By gosh, it's dark up here today. . . . Quite a different view. What a beautiful day this is!" It can probably be said that, of the twelve x-15 pilots, Joe enjoyed the ride more than anyone else. He said, "You've got a long time to where you find yourself having done your looking around, and figuring, well, if I'd had better sense, I would have looked more."

As he fell back toward the ground, his speed increased, hitting a maximum of Mach 5.5, or 3,710 miles an hour. By 240,000 feet he could tell reentry was having an effect. Zero gravity was gone, and the gs steadily built as he brought the nose upward to approximately twenty-five degrees. At three gs he brought the nose down slightly to twenty-two degrees, until the maximum of five gs pressed against his body and the structure of the x-15. Joe continued his descent, and at 145,000 feet his control inputs caused a momentary pitch oscillation. He quickly figured out his mistake and let the aircraft regain its own equilibrium. Walker entered the final leg of the journey, approaching Edwards. Touchdown and rollout were perfect. Support vehicles raced across the lakebed and pulled up next to no. 3. Inside, Joe was checking his systems and gathering his thoughts.

Usually, the first man on the scene was Roger Barniki, head of the life support group. He was out on the lakebed for all 199 flights of the x-15. He related that after landing, "I got to the airplane, was going to open the canopy, and no more than popped it, before Joe said, 'Put that down, I'm not done yet!' Joe wrote a memo [after the flight], that there's stuff he has to shut down and he doesn't need everybody in the world leaning in the cockpit—especially me. . . . This guy just went to 350,000 feet and he lands on the lakebed, then he's got everyone driving at him. The only dangerous part are all these vehicles, and us idiots, coming out there!"

From the experience gained on this mission, Walker felt little doubt there was more height left to be gained in the x-15. The LR-99 could provide enough power to shoot upward to 450,000 feet, but the strain on the airframe during reentry would have torn it apart. Joe told the engineers when he returned, "Based on today's experience 30,000 feet more seems like no strain."

Slightly over a month later, after three aborted attempts earlier in August, Walker was ready for his twenty-fifth and last x-15 flight. It was 22 August and the crew had a surprise waiting to honor Joe as he walked to the air-

6. Joe Walker brings the X-15 back into the atmosphere after flight 3-22-36. He achieved 354,200 feet, or more than 67 miles, in altitude on 22 August 1963. © Thommy Eriksson.

craft. In the late 1950s, when he flew the X-1E, the nose of the white rocket plane was embellished by a small painting of a pair of dice, each showing two dots, with the words "Little Joe" next to them. In the vernacular of gamblers, double deuces means "Good Luck." Crew chief Duke Littleton decided to again pass along that luck to Walker and the X-15. Duke changed the graphic slightly, this time saying "Little Joe, The II," along with a much larger pair of double deuce dice.

Maj. Russell P. Bement and Maj. Kenneth K. Lewis Jr. powered up B-52 no. 003 once Walker was secured into his X-15 cockpit. Chocks were removed from the giant centerline wheels of the mothership, and mission 3-22 was free to roll. By 9:09 a.m., the behemoth converted bomber was rising above the desert, once again heading three hundred miles north to Smith Ranch Dry Lake, with Capt. James Wood and Bill Dana in chase planes nearby. The mothership and its escorts turned toward the south, and at three seconds before 10:06 a.m., Walker dropped from the shackles, ignited the rocket, rotated the nose of the X-15 upward, and powered toward space.

7. Joe Walker and his sons, James and Thomas, with their friend Scott Smith at the Tropico Gold Mine. Courtesy of the Grace Walker Collection.

The flight was a near repeat of the 19 July mission, only this time Joe was aiming for 360,000 feet. For once, he fell short of his goal and topped out at 354,200 feet—67.1 miles above southwest America. Walker's unofficial record stood for more than forty-one years, broken finally by pilot Brian Binnie, rocketing the Burt Rutan–designed, privately funded SpaceShip-One to an altitude of 367,442 feet on 4 October 2004.

Following a 293-mile flight that morning, Walker slipped down onto the hard-packed lakebed at Edwards AFB at 10:17 a.m. "Little Joe" had completed his time on the X-15. He not only became the first person to ever go into space twice but had now done so three times. Only Joe Engle matched his feat of three astronaut qualification flights during the program, and that came more than two years in the future. As for Joe Walker, he had a new airplane to fly, the triple-sonic XB-70 Valkyrie bomber.

Grace Walker spoke about the events of Wednesday, 8 June 1966. "It was the cleaning lady's day to come, and I'd been to the grocery store and came home. A knock at the door, and it was the young . . . second minister at our church. He just kind of stood there. He didn't know what to say or why he was there. I invited him in, and he said, 'Well, there's been an accident at the base, but I don't know exactly what's happened.'"

A few days previous to this, on 4 and 5 June, Grace and Joe spent the weekend alone in San Francisco. She said it was "a rare, rare, rare occasion. We'd driven up, and I don't know what all we did, just mostly roamed around to have get-away time."

Two years earlier, the first of two giant XB-70s rolled out of the North American Aviation facility at Plant 42 in Palmdale, California. It was 11 May 1964, and the bomber prototype had a long and troubled history, even before the first piece of metal was cut. The vehicle was 189 feet long and 105 feet across its wings, a distance nearly equal to that of the Wright brothers' first manned powered airplane flight in 1903.

At the ceremony that day, no one cared about the past, only seeing the magnificent white, delta-winged vehicle and the technology it represented. Guests that day included Joe Walker, who wore a striped suit coat and stood on a podium having his photo taken next to actor and pilot Jimmy Stewart. A year later, to not nearly as much fanfare, the second test aircraft joined the first. This vehicle's fate was intertwined irrevocably with that of Walker's.

At 7:15 a.m. that terrible late spring morning of 8 June, XB-70 no. 2 took off for a series of tests that lasted about ninety minutes. The crew consisted of North American Aviation's pilot, Alvin S. White, who had once been the backup pilot to Scott Crossfield on the X-15, with copilot Maj. Carl S. Cross.

Once the flight tests were completed, permission had been granted for a special airborne photography session on behalf of jet-engine manufacturer General Electric. Four other aircraft joined the bomber. A T-38 Talon and F-4 Phantom took station off the left wing of the XB-70, while a NASA F-104 Starfighter with Joe Walker at the controls and an F-5 Freedom Fighter finished the V-formation on the right side. Between the five aircraft were a total of thirteen GE engines, a great public relations opportunity.

Until 9:25 a.m. everything appeared fine. Pilot and aerial director Clay Lacy, and his in-flight photographer John Zimmerman, flew nearby in a Gates Lear Jet. The formation of aircraft was beautiful against the blue sky. At 9:26 a.m. the cry of "Midair! Midair! Midair!" was shouted across the radio.

The geometry of the XB-70 and F-104 led to disaster. The bomber's high-ly swept delta wing came back to a sharp tip, which created huge, invisible vortices roiling backward. The tall T-tail of Walker's Starfighter was forty feet behind his cockpit and had become dangerously close to the bomber's drooped wingtip. From Joe's position, it appeared he was at a safe distance from the wing, but when he slid in a little too close, the vortex caught his Starfighter like a gnat, the top of the T-tail coming up from below, contacting the bomber's wingtip, breaking away, and tossing Joe and his plane up and over the top of the Valkyrie.

Walker was almost immediately upside down and sideways to the air stream. His F-104 slid back along the top of the bomber, striking the twin vertical tails. The XB-70's right tail broke apart when struck by the rear fuselage of Walker's plane. Continuing on toward the left side of the bomber, the F-104 struck the left tail, shearing it completely off the aircraft. The cockpit and forward fuselage of F-104 no. 813 separated and, at 2.4 seconds into the sequence, contacted the XB-70's left wing as it rolled off into the abyss. Walker probably died at that moment, never having a chance to eject. The rest of his fighter continued off the Valkyrie's back end, exploding into a fireball within seconds of the collision.

Aboard the stricken Valkyrie, White and Cross first heard the frantic call on the radio, without even realizing it was their plane that had been involved. White said later they assumed it must have been a collision between two of the outlying fighters. The bomber continued to fly straight, but then started to lose altitude and veer downward. The other aircraft peeled away to clear the area.

Sixteen seconds after the accident, the XB-70 started to roll onto its back. White said later, "What happened was, it yawed right and rolled right. Then it was upside down and nose down, and then right side up and nose up. It did this twice, and the second time around, a big piece of the left wing [probably weakened by Walker's impact] broke off."

The forces exerted were horrendous as White and Cross both attempted to eject, using the bomber's unique capsule system. Because of those gs, White was barely able to leave the plane, his right elbow actually caught between the capsule's clamshell halves as they closed. In excruciating pain, Al finally worked his arm inside. The doors slammed shut and the capsule shot away from the aircraft. Cross never ejected and died as XB-70 no. 2 impacted the desert, east of Edwards. As he floated down under the parachute, still en-cased inside the capsule, White said, "I heard the xb-70 hit. It made a ter-rible explosion." The airbag that was to cushion the capsule's landing failed to open. White hit the dirt at an estimated forty-three gs, injured but alive.

Walker was scheduled for his first checkout flight as chief NASA program pilot in the massive bomber the following week. It had been the first flight in the xb-70 for Major Cross.

Grace said, "I think they wanted to pin it on [Joe] because it's standard procedure if the pilot's dead, you try to get him to be the scapegoat." Many people felt that way, but the true nature and danger of a wingtip wake vor-tex was unknown at the time. For Grace, it was a harrowing experience. "They were looking for anything, I guess. And it was nobody's fault, it was a lot of sloppy things happening. I believe that definitely."

The worst, Grace said, was, as the investigation continued, "There was a lot of stuff that happened that was kind of scary. . . . My house was en-tered, but nothing was taken." But others tried to help, as Grace related: "Don Bellman, who was on the committee as [lead] investigator, had the [accident] report, and he said, 'I'm going to give it to you for the weekend, you return it and I'll take it in on Monday morning.' So I read through it, pictures and all. I took it to friends and we copied it over the weekend."

A justifiable paranoia took hold for Grace. "I got so I wasn't calling from my home phone because I knew it was being [recorded]. Any time I wanted to make a long distance call the operator would make me wait, and I knew I was being checked on, so I went down to the gas station and, lo and be-hold, I'm calling these friends that are going to do this copying, and this van pulls up by the gas station. I had seen this same van parked down the block in front of my home, and they were nibbling on whatever was said on my telephone." It was a bad time, but she took it as well as she could, saying, "I guess they must have known that I did copy [the report]. . . . I gave it to my attorney, and I never got it back."

Grace shared a personal sorrow: "I never did see him to know that he was dead. And that was a bad thing. You should see the finality because I wasn't ever sure it was real, especially with all the back-peddling that went on after this accident."

Three days after the midair collision, on Saturday morning, 11 June, a me-morial service was held for Joseph A. Walker at the Community Methodist

Church in Lancaster, California. Gill Robb Wilson, one of the founders of the Civil Air Patrol, a World War I aviator in the Lafayette Flying Corp, a poet and writer, and personal friend of Walker's, gave the eulogy.

If it seems strange, that there was such a close bond between a worn out old man and a young man at the zenith of his genius, I might explain. We were born in the same hills. We inherited the same traditions. We graduated from the same college. We were brothers in the same fraternity. And our lives successively followed the same dreams. . . . From the earliest days of flight, I have followed the lives of the airmen. And I have stood countless times, my neck bowed with the storms of sorrow, as one by one, they completed their work and let down for the last time into the great eternity. . . .

We are hardened, and then someone like Joe exposes the sham and the fraud of our defenses. And we find out at the last that we are just human beings, who feel, and love, and fear, and dare, with all the frailty of any other, and in any other profession. . . .

We shall remember you, because your dreams, unfettered by the sod, fled to the stars. . . . But most, we shall remember you in that you walked serene among your fellow man. . . . Your hopes a dream that all might share. We shall remember you, because in all creation's plan, there is no wonder here on Earth, or in the stars . . . so noble as the humble man.

Exactly three months after Walker's accident, Gill Robb Wilson succumbed, ten days short of his seventy-fourth birthday. Joe Walker had been just forty-five.

For Grace, she had a vision of Joe, saying, "I've been a movie buff from the age of five — usually Saturday cowboy matinees — so I was in love from then on with cowboys. . . . People used to think Joe was a Texan because he had such a drawl, but he was from southern Pennsylvania, and they've got their own accent down there. He was a farm boy, so he knew how to do a lot of things. The first time I saw him on a horse, I nearly swooned!"

The Harmon Trophy was established to recognize the best aviators in the world. The 1960 title was given to three x-15 pilots, Scott Crossfield, Robert White, and Joseph Walker. On 28 November 1961 they were all invited to the award presentation by President John F. Kennedy at the White House in Washington DC. Bob White related the story: "I remember a time when

8. President John F. Kennedy admires an x-15 model presented to him by the x-15 pilots during the Harmon Trophy ceremony on 28 November 1961. To each side of the ever-grinning Joe Walker are Scott Crossfield and Robert White. Courtesy of the Robert White Collection.

we were in the White House and . . . I handed a model of the x-15 to the president. I have a picture of that at home. [In it] Scotty has a little grin on his face; he's enjoying it. I have a rather severe look on my face. I look as if I'm kind of worried whether or not we're doing the right thing in that exalted place." White saw Walker was different, even in front of the American president. "But Joe Walker, in that photo, I look at that, and there's that grin as wide as the Grand Canyon! I swear if the president had asked Joe a question at that time, I'm sure Joe would have said, 'Well, Mr. President, you've got a fine dining facility downstairs. Why don't we go there, have a beer, and discuss it.' That was Joe."

3. Gaining Speed

There's lots of unknowns, so the imperative is to equip
yourself to be the very best you can. That's when things
really start rolling.

Robert M. White

"He was about as opposite of [Joe] Walker as could be. I don't mean this in
a derogatory manner at all, but Kincheloe was a real salesman on himself.
He had me sold just as much as everybody else." Paul Bikle was talking of
Capt. Iven Carl Kincheloe Jr., who was to be the prime pilot for the x-15
program, chosen by the U.S. Air Force in September 1957.

By the time "Kinch," as he was best known, got to that point, he had already made a name for himself as a Silver Star recipient and Korean War
air ace. He earned that honor shooting down five MiG-15s in less than four
months in his North American F-86 Sabre. As flight commander of the 25th
Fighter Interceptor Squadron, Kinch then went on to destroy five more
North Korean fighters before his tour ended. Several years after the war,
he also garnered a world altitude record as a rocket pilot, achieving 126,282
feet in his fourth flight of the Bell x-2 on 27 September 1956. For this, the
press dubbed him "The First of the Spacemen."

Kincheloe was literally the air force's fair-haired poster boy, and he was
making the most of it. He once appeared on *The Arthur Godfrey Show* and
was asked, "Are you sure you haven't been selected by some Hollywood casting director to play the part of a test pilot?"

A native of Detroit, Michigan, born 2 July 1928, he graduated in 1949
from Purdue University, then went immediately into the air force, earning
his wings a year later. While in Korea, he survived 131 combat missions before rotating stateside. Soon after the war, Kincheloe received air force per-

mission to attend the Empire Test Pilots' School in Farnborough, England, graduating in January 1955. With all this behind him, and once he secured his reputation in the x-2 at Edwards, there was no doubt the x-15 was in his future. I asked Bikle if Kincheloe had actively sought out the x-15 assignment. Paul said, "I'm sure he did. Everybody did. Hell, that was what you were there for."

Walt Williams was running the x-15 program at the time Kincheloe came aboard. He said, "I invited Kinch on a weekend fishing trip with a bunch of bigwigs from NASA and North American Aviation. Kinch wanted to go at first and had even put up a deposit. He really enjoyed hobnobbing with the biggies. For some reason, he canceled out at the last minute. Don't know if maybe he felt he'd gotten in over his head or something." Whatever happened to make him change his mind, it was a fateful decision. Williams related what happened on Saturday, 26 July 1958: "That weekend, he pulled 'duty pilot' and was scheduled to do chase for a test flight on an F-100 mission out of Palmdale [for pilot Lou Schaik]. The flight was canceled, but the word didn't get back to Kinch fast enough."

Immediately after leaving the runway, a mechanical failure prevented the engine from gaining thrust, so he was unable to climb to safety. Kincheloe radioed, "Edwards, Mayday, 7-7-2 bailing out!" Because of the high T-tail on the F-104, the ejection seat was designed to fire downward. Understanding he couldn't get airspeed and altitude, Kinch rolled inverted, so as to not eject directly into the ground. The seat activated properly, but the plane had not gone completely upside down, and he was simply too low for his parachute to deploy fully before he hit. Less than three minutes after the 9:47 a.m. takeoff, Kincheloe died on impact, the seventh person killed by the F-104 in the span of four years since its initial flight.

"I found out when I returned from the fishing trip that he'd been killed," Williams said. "That was definitely a case where someone was killed flying when he should have been fishing." Butchart was one of those with Williams that weekend. He said, "If Kinch had gone fishing with us, like he was going to do, he'd be here today. . . . The future was all ahead of him. He really hadn't done much yet."

Bikle took over from Williams soon after the x-15 went into flight testing. Even after so many years, Bikle was still critical of Kincheloe, saying, "His flying experience was very limited. He couldn't have been as good as we all thought he was—although he was a very talented individual. . . . I often wondered if the x-15 program would have turned into . . . the 'Kincheloe Flying Circus.' Boy, did he lap up the publicity and attention. And that's as far from [Joe] Walker as you can get."

A week after the accident, Iven Kincheloe was buried with military honors at Arlington National Cemetery. His grave is now just a few feet from where President John F. Kennedy was laid to rest in 1963. At the time of Kincheloe's death, his wife, Dottie, was pregnant with their second child, a girl who would be named Jeannine. Iven III was just nineteen months old. President Dwight D. Eisenhower wrote a letter on behalf of the toddler, asking whomever was president when the boy grew up, to accept his appointment into the U.S. Air Force Academy. Iven instead chose to go into construction.

With a major void to fill after Kincheloe's death, less than three months prior to x-15 rollout, his backup pilot, Robert Michael White, stepped up into the number one slot.

New York City was Bob White's home. Born 6 July 1924, he said there wasn't much to tell until he finished high school and decided to go off to war. It was the summer of 1942, and Bob was not yet eighteen. "I wanted to go into the Aviation Cadet Training Program, [but] I needed my parents' consent. . . . Mother was obviously reluctant to do that, so I told her, 'When I turn eighteen I'll register for the draft and go and be a good infantry soldier.'" Laughing, Bob said, "So, of course, [my mother] relented."

Joining the U.S. Army Air Forces was a big change for a young, big-city kid. "Getting into something as exciting as flying was amazing. I'd never been off the ground before, and that had a tremendous impression. I fell in love with flying."

Following basic military training, he went on to flight school. By the time he earned his wings and arrived in Germany, the situation was not what Bob had imagined. It was July 1944, and he recalled, "I was disappointed I didn't get there when the great air battles were still going on. The Luftwaffe was already essentially defeated." As a second lieutenant, he flew bomber escort and low-level interdiction missions in the North American p-51 Mustang, occasionally going up against a German fighter.

White found out in February 1945 that no matter the state of the war effort, it was still a dangerous business. On his fifty-second mission, he recalled, "I'd just finished firing into an Me-109 [and] got hit. It was actually the third time I was hit [by anti-aircraft fire] when I went down [with] a direct hit to the engine. The oil covered my windshield, and I had to get rid of the canopy so I could see. My gauges were all going haywire, and I knew I was done. . . . I struggled for altitude to no avail." There was little choice but to jump for it. White laughed, "Errol Flynn did it, so why couldn't I? As I jumped out and pulled the ripcord, I got the opening shock of the parachute, then immediately went into the trees." His rigging lines and parachute tangled in the branches, bringing him to a sudden halt.

White relived the experience as if it were happening in real time—each thought was an abrupt staccato. "Hundreds of miles behind enemy lines; unhook; climb down from the tree; then quietly move. You want to get away. There's an old farmer coming up with a pitchfork. Take out my .45 [pistol]; get rid of the bullets. Field strip the weapon; throw it in the bushes—no use for it. I've got to move as far away as I can. Finally, come to a field where I'm at the end of a forest line. I slide down into a gully, and say, 'Okay, let's get our bearings here.'"

Gathering his thoughts, Bob paused for a moment, then: "I got out my little escape kit. . . . Now I have to come up with a plan. The odds are entirely against me . . . so I reconcile to myself, 'No regrets.' And from that day to this, I don't deal in regrets, and just accept it. At least, not until later, when I had time to think about it."

His evasion tactics proved fruitless, and Bob was captured by German soldiers. White arrived at a prisoner of war camp, but knew, this late in the game, he was lucky compared to those who came before. "There was no isolation or solitary confinement. There were such a lot of us that . . . you reinforced one another." There was some initial fear of retribution from the Germans. "We [had been] gunning their trains down and . . . blowing up anything that moved. . . . This is what war is all about." None of the feared retribution materialized.

The biggest problem became food. "For a couple weeks, we were down to very bare minimum rations . . . one piece of bread and a bean can of grass soup a day. Our compound was one large barracks building, no furniture except a kind of picnic table, a couple of benches, and a pot-bellied stove." It wasn't long after, "we heard the sound of trucks rolling into the camp.

They had Red Cross painted on the side . . . and had been driven up from Switzerland carrying prisoner-of-war food parcels. Indeed, that was a lifesaver."

This was the beginning of the end of the entire German prison camp system. White was there for a scant two months, compared to the years of ordeal some survived. "It wasn't too long afterward we evacuated the camp near Nuremberg and moved to one further south. . . . You could start to hear the sound of [Allied] artillery. . . . We knew we were going to win the battle. . . . We moved out and marched to a prison camp not far from Munich. It was there the Germans abandoned the camp when the gunfire started."

When World War II was over, there was a huge surplus of men in uniform. Most soldiers and airmen opted to return to civilian life, but White had designs on possibly staying in for a career. After he thought it through some more, he concluded, "If I opted to stay in the service, there was little opportunity to compete with the regular officers — the West Pointers and college graduates. I knew I had to get out and do what I wanted to do, which was go back to school." However, he did stay active in the reserves to keep his future options alive, serving at Mitchel AFB while attending New York University.

There was barely enough time between wars for Bob to graduate with his bachelor of science in electrical engineering in 1951. His unit was called to active duty for the Korean conflict, flying the Curtiss-Wright C-46 Commando transport with the 514th Troop Carrier Wing. "I applied for a regular commission [and figured] that, if I got it, I stayed; if I didn't, I have options." The commission came through. "So, I stayed on, and a career followed."

White was assigned overseas, at a base near Tokyo, Japan. With this transfer, he also left the C-46 squadron to get back into his primary love, fighters. The iconic image of air battles during this time were the famous dogfights between the F-86 Sabre and Soviet MiG-15, as happened with Iven Kincheloe. The frontline fighter preceding the Sabre was the Lockheed F-80 Shooting Star, the aircraft assigned to Bob White. The F-80 was the first to go head-to-head against the MiG-15 — and win. White became a flight commander on the Shooting Star, staying in that position until the war ended in 1953.

Moving up into a test pilot position gave Bob the best options for advancement in his field. White said, "With my intensity for how I felt about flight, I naturally gravitated to that thought. When I found out about the test pilots' school, I decided I'd like to try for that." His application was accepted as a member of Class 54B at the Experimental Test Pilot School at Edwards. He graduated first in his class in January 1955, the same time Kincheloe completed his course at the Empire Test Pilots' School in England.

Being first in class for White should have opened any door. Bob said, "The obvious place, 'The Place,' was the Flight Test Center at Edwards." But it didn't work out that way. Instead, he was given orders to report to Wright-Patterson AFB in Dayton, Ohio. That didn't sit well, and Bob's feelings became obvious. Bob recalled, "The commandant of the school came in, and he moseyed over to me and asked, 'Bob, how do you feel about your assignment?' I said, 'Sir, I'm crushed.'" The commandant told Bob that they would rate the class by standing. Since White finished first, he should have whatever assignment he wanted. With that, White was told to see the vice commandant of the school, Lt. Col. Harold G. Russell, about making the change. Once there, he was asked by Russell, "Why do you think you want to serve in Fighter Test?" to which Bob replied, "Because it would be a privilege to serve here." Russell was pleased with what he heard and told Bob, "That's the right answer! Go ahead and go up to Fighter Ops and meet your new boss."

Even though four years White's junior, when Kincheloe came to Edwards he eclipsed all others from the air force. White didn't mind, as he had plenty to keep him busy. He did not become involved in the early rocket planes like Kinch, but instead became proficient on what was called the "Century Series" of fighter aircraft. This came from a string of designations that began at F-100 and quickly continued to F-107.

White said one of the stand-outs, as far as he was concerned, was the F-105B. "I got a crack at the first formal air force test. That was very interesting, because I was the lead pilot. . . . It had the potential for being a good fighter-bomber, and it was going to be one that would carry an atomic weapon." Because of his evaluation and skills, they had to make a lot of changes, eventually coming out with the F-105D. "That was a very good airplane, and I got to fly it because I ended up in Europe [in October 1963 in the 22nd Tactical Fighter Squadron] as commander of an F-105D squadron. I could see all the things that were different and changed. It was a good feeling, because I knew that I had a part in helping to do that."

Bob's exceptional skills were acknowledged when he was chosen to participate in the x-15. Unlike Kincheloe, who craved the spotlight, White had no problem not being the center of attention. Being a research pilot on the program was what was important. Even though so different in temperament, Kinch and White were good friends, and Kinch first got Bob interested in the x-15. "Kinch started talking about it, so we were aware that it was coming up. . . . My career saw me at the right place and the right time to be able to participate in the x-15."

The situation changed with the f-104 accident and Kincheloe's untimely death. White stepped out of Kinch's shadow at that point and was recognized as the top professional test pilot in the U.S. Air Force. "When Iven lost his life, the air force pushed me into the catbird seat. . . . That's the way it went. It turned out to be a very good thing."

The template for all the various agencies to work together on the x-15 was set up early in the process, and Bob White was eager to move forward. He said, "We integrated right from the beginning. I got to know the NASA pilots, chatted with them, then the two primary engineers I worked with, Bob Hoey [with the air force] and Dick Day [with NASA]; they were good buddies. So we worked very well together." Bob decided to spend a week at the North American Aviation plant getting up to speed on all aspects of the x-15. "I wanted to talk personally with the engineers on each one of the systems. I wanted to get to really know all the systems in great detail, so I talked with the guys who designed them. That's where I started developing an intensity for the program."

Planning research flights meant making decisions on how fast to push things. That brought up the question of who was to fly each mission and be the first to get to any milestone. "There was no jockeying for position," said White. "If somebody thought there was, well, they could think that, [but] it wasn't the case. NASA was perfectly agreeable in seeing that Joe [Walker] and I would leapfrog in opening up the flight envelope." Bob laughed as he added, "Of course, Scott Crossfield felt it would be nice if he could do it all." White was anxious to get the air force into the action and finally had his opportunity with flight 1-4-9 on 13 April 1960.

White spoke about his flight and a special bet he made. "I looked forward to my first flight greatly. Joe [Walker] made the first government flight [but] Joe must not have taken it seriously because . . . he must have gone two miles past the point of [intended] touchdown." White wanted precision for his first mission and every one to follow. He said, "Working with Bob Hoey and Dick Day on the simulator one night, Hoey said, 'Bob, where are you going to land?' I said, 'You've got a little smoke thing going on,'" which referred to the smoke canisters lit off to give the pilot indication of wind direction at the surface on the lakebed. White said to Hoey, "'That's where I'm going to land. Give me plus or minus 1,000 feet.' So Dick Day said to me, 'Bob, bet you won't.' After we finished the simulator down at North American, [we went] out for dinner and a martini. Dick Day brought up the subject again, and said, 'I'll bet you a martini you won't do it.' I said, 'Dick, would you make it two?' He agreed." The bet was on.

The morning of 13 April was perfect. B-52 no. 003, with Capt. Jack Allavie and copilot Capt. Charles Kuyk, took off at 8:46 a.m., gaining altitude as they circled the local area to get into launch position. Three chase planes escorted the mothership. Both North American X-15 pilots were in an F-100 as lead chase. Al White was in the front seat, with Scott Crossfield riding in the back—a rare moment for Scott to not be in total control. Two F-104s completed the group, the first with NASA's Joe Walker, the second with Bob Rushworth, White's backup pilot for the X-15.

"The first flight," White related, "I was happy to get it going. I looked forward to it greatly, and without any fear or concern. . . . That's why we had this incremental approach [to envelope expansion flights]. You move forward a little bit at a time." White provided a prime example: "One time we had a little buckling of the wing. There was a little space [which was an expansion joint], and it created a hot spot and started buckling away. Could you imagine if we went another Mach number? It could have caused structural damage. . . . It's really problematic whether we could have saved the airplane or me."

South of Rosamond Dry Lake, the B-52's nose pointed east. White launched at 9:15 a.m. He explained in his pilot's report the events immediately following: "Chambers two and four on both engines were lighted immediately, and the airplane rotated to ten-degree angle of attack. Light buffet was noted, and I felt I would start uphill [to 48,000 feet]. . . . Deliberate pause was made between lighting the remaining [four] chambers, [and] waiting for a chamber pressure indication of 200 psi [pounds per square inch] before proceeding in like manner to each successive chamber. The lack of acceleration until all eight chambers were operating was a slight surprise."

Bob accelerated to Mach 1.90 as he followed the pattern counterclockwise around Edwards. "A tailwind put me directly over the main base with excessive altitude." This was a minor problem, easily corrected. He later stated he became frustrated by what he felt was inadequate aid from the NASA control room. Irritation is evident in his report: "It is hoped in the future that the ground control will be punctual in questioning calls that are misleading, erroneous, or questionable. This pilot requests and expects this assistance to help insure a successful mission." White's fuel readings were being incorrectly radioed to him and were even transposed, so it is understandable why he took umbrage at what occurred.

The smoke canisters were set off, and White approached the lakebed for touchdown. Bob armed, then jettisoned, the lower ventral fin. His only other complaint was something all pilots on the program understood could not be altered, saying, "Visibility from the cockpit is somewhat restricted during turns, but is considered adequate."

After all the preparations, even with the small hiccups, the flight went beautifully. What about White's bet with Dick Day? Bob said, "I did it, and I won the bet. You see, I practiced that landing [in the F-104], and I had it nailed." Dick bought White two martinis, officially ending the first mission for the U.S. Air Force on the rocket plane.

Roger Barniki, from the x-15 life support branch, said the martini tradition may have extended beyond the bet between Bob White and Dick Day. "The David Clark representative [the manufacturer of the x-15 MC-2 and later A/P-22S pressure suits], a guy named Norm Foster, started a tradition of a martini after the flight. . . . Norm had a briefcase that was labeled: 'Foster's Fine Medicines.'" Waiting in the suiting van, once the pilot was brought in to prepare to return to the base, Norm opened his case and mixed up the required "medicine." Roger continued, "[The pilot] just got out of the [pressure] suit [and he's] sitting there in his sweaty underwear, and Norm would say, 'Had a great flight; here you are,' as he handed him a martini. The pilot would be the only person that would have it. It wasn't even a heck of a lot of people who even knew that went on. We tried to keep it down."

On 6 May, as flight 1-6 came toward Rogers Dry Lake on final approach, White needed to make sure it was safe to jettison the lower ventral. That was a duty of the landing chase pilot flying close by the x-15, in this instance, Bob Rushworth. White asked, "Give me a call on this ventral will you?" Jauntily, Rushworth sent back, "Right-O!" With the steep x-15 glide slope, the ground was fast approaching. White prodded Rushworth: "How about here?" Rushworth saw they were passing over a base road, so he waited a moment to make sure White was clear. Once past the road, he radioed it was okay to jettison.

Inside the x-15, there was no indicator letting the pilot know the status of the ventral. He had to rely on the chase to tell him the situation. White fired the release charges, but he didn't feel it go. "Did it come off?" he asked. Rushworth had the bad news: "Negative, still on. Try again." White repeated the procedure, but still no joy. If the ventral refused to leave, the x-15 was about to be in deep trouble. As a last resort, a backup system was set to relieve the airplane of its burden when the gear extended. White popped the gear, and, thankfully, the ventral went, too. An engineer who worked on the x-15 once joked that if it didn't jettison, White would have made an awfully big furrow in the lakebed.

Bob White's next four flights, from 19 May 1960 through February of the next year, all went very well—just as White wanted. The first of these, flight 1-8, was only the second time in the program when the x-15 was launched away from the local Edwards area. It also was the first in a rocket plane to exceed 100,000 feet since the record set by Iven Kincheloe in 1956. When mission 1-10 came, on 12 August, the altitude increased to 136,500 feet, surpassing Kincheloe's mark by nearly 10,000 feet. White commented, "Kind of empty up here," as he floated in zero g on his ballistic arc, 26 miles above the earth.

Later, he was asked his impressions from that height, and said, "The flight today offered . . . no problems, nothing that could be considered a limitation as far as man's ability to fly an aircraft. What you see at this altitude impressed me as being the most dramatic point of flying [more than] 130,000 feet. [I saw] the very dark-blue sky, and the lighter band that was immediately surrounding the earth. . . . Looking to the future . . . I would particularly like to continue on in work that would take us to higher altitudes with manned aircraft."

This flight was the highest ever reached using the LR-11 interim rocket engines. Two flights later, on 7 February 1961, with mission 1-21, White also achieved the fastest velocity with the twin, four-chambered engines, touching Mach 3.50, or 2,275 miles an hour. It was the final X-15 flight to use these reliable rockets, the same that had first propelled Chuck Yeager to break the speed of sound in the XS-1 more than thirteen years previously.

The day following the landing, X-15 no. 1 was loaded on a flatbed truck to make the journey back to the NAA Los Angeles plant. While there, the aircraft was modified for installation of its first XLR99-RM-1, the big engine, which White was to use to go several Mach numbers faster than any pilot who had come before.

As highlighted by the work from Bob Rushworth, when White had his problem jettisoning the ventral, chase pilot duty was extremely important to the success of the X-15.

Usually, a minimum of two chase aircraft followed the B-52 mothership to the launch lake, providing information on the exterior of the X-15 as drop time neared. Their call signs were Chase 1 and Chase 2. Aboard the X-15, visibility was extremely limited, and the pilot only had instruments to tell him the condition of the rocket plane behind his cockpit. The chase pilots watched the jettison checks and verified the ballistic control jets test fired. Flight controls, such as the ailerons and rudder, were essential for safe flight, and chase pilots made certain all these were in the proper position for drop. Jim Robertson of North American Aviation said, "Usually, one second to launch, one chase would kick in the afterburner and get out there just as far as he could. The other chase would stay right by the bird through the initial drop, because that's the most critical time. The first guy would get way out there, but [the X-15 would] go past him so fast it was ridiculous!"

Depending on how far away from Edwards the flight was launched, other chase aircraft may have been assigned to critical spots. Sometimes there was a roving chase, or Chase 3, near an intermediate emergency lakebed in case of problems and one, usually designated Chase 4, stationed along the approach to Rogers Dry Lake for landing.

Joe Engle, who was to fly the X-15 starting in late 1963, discussed some of these duties: "The landing chase . . . was probably the one that was most interesting, because your job was to join up with the X-15 as it decelerated down into the pattern. . . . Making that join-up was probably the most demanding part. If radar was up and tracking both airplanes . . . they could vector you in. . . . You just kept looking up there until you finally picked him up."

If there was fuel left after the LR-99 engine burn, the pilot jettisoned it to bring the touchdown weight within limits, but also as a way for Chase 4 to find the X-15. "It'd make a nice, big, old white contrail that you could pick up," Engle said. Once in proximity, the chase pilot visually checked the X-15's exterior. "You looked for things like, if the airplane . . . had any pieces that had come off that weren't supposed to. [Basically] whatever was hanging out was supposed to be hanging out, and whatever wasn't supposed to be hanging out, wasn't." After the flare for landing, and as soon as the landing gear was released, Chase 4 confirmed the skids and nose wheel appeared down and locked. He then called out height above the ground, aiding the limited visibility of the research pilot inside the X-15.

In addition to the chase aircraft, the air force also had responsibility to supply and maintain both B-52 motherships. When it came to the X-15 itself, once the navy left the program, Paul Bikle said, "We pretty much split the piloting along agency lines [NASA and the air force]. We had a sort of understanding that carried throughout all of the x-airplanes, that the air force pilots would set the records. For instance, the official altitude record is still [Bob] White's, even though Walker went higher. We just never applied for any. It didn't make that much difference to us, but it did to the air force." Since the air force was footing the majority of the bill for the program, Bikle felt this was a pretty cheap price to pay to allow them the official glory.

Air force bias was evident in that only seventeen out of forty-eight chase pilots were from NASA or North American. Five different types of aircraft were used for chase duties at various times. A total of 741 chase sorties were logged during the X-15's 199 missions. The majority of those, 533, were flown by F-104 Starfighters. The T-38s and F-100s flew 129 and 67 sorties respectively, while the F-5D flew 11 times, and a lone F-4H just once.

The F-104 proved itself as a unique asset for the program in other ways, too. Looking at this aircraft next to an X-15 showed how much they resembled each other, and thus the F-104 became a perfect analog for pilots to gain experience in what it was like to fly the X-15, especially for high-speed, unpowered landings. White had proven the worth of the F-104 in this respect by winning his martinis from Dick Day.

Engle explained the setup required with the Starfighter to make it imitate the X-15: "We would set the F-104 with the speed brakes, landing flaps, and idle power, and we'd be able to closely duplicate the lift-to-drag characteristics of the X-15." Each research flight was expensive and time-consuming, so actual flight time for any pilot was minimal. With the F-104, all the pilots could do numerous landing practices before each mission. It proved to be such a great airplane that, when the chance arose, Bikle explained he bought a few for NASA as well. "They're really the only airplanes that NASA ever bought [up to that time]. They were bringing the F-104 production line to an end, and they had three airplanes on the tail end of that without any military commitment. [Lockheed] gave us a price of about $8 million, plus spares."

There did not appear to be any animosity against the air force for being the lead on the bulk of X-15 duties. As Bikle pointed out, he was more than happy to let them take this role, if it meant the research program remained funded. Engle said, "It was a really tremendous relationship. . . . It had been operating for a great number of years before the time I got in the program, and whatever problem areas may have existed — if there were any — were certainly ironed out by the time I got there. It was a totally [integrated] system. There didn't seem to be any jealousies or concerns about who was getting to do what."

One month after White flew the final LR-11 flight, he was ready to take to the air for the first time with the much larger LR-99, on mission 2-13.

At 10:28 a.m. on Tuesday, 7 March 1961, he activated the launch sequence. The X-15 rolled off slightly, with its right wing down. Bob said, "As you come off, you start flying . . . so you correct the roll and just shove [the throttle] right up. That's it, just 'Boom!'" He explained how the LR-99 compared to his previous flights. "A tremendous difference. . . . The LR-11s were nothing more than a pretty good fighter that's lit the afterburner. But the LR-99 at full throttle — zap, you're pinned to the back of your seat! The pressure continues to build the faster you go, until finally you're finding it difficult to breathe against all this pressure on your chest."

White started with 75 percent thrust, then rotated the nose upward. He pulled back to 50 percent until 70,000 feet, pushing over to level out, approximately eighty-five seconds after he left the B-52, at 77,450 feet. At that point, he brought the throttle back up to full, accelerating in twenty seconds from Mach 3 to Mach 4.43. By 127 seconds, it was time to shut down the engine. He said, "You're ready for it to quit. [When it does,] all of a sudden [that pressure] is relieved, and you can breathe again."

In those two minutes of powered flight, Bob White had become the first man to fly faster than Mach 4, surpassing the next highest speed attained in the x-15 by nearly a full Mach number. Yet there was still power to spare locked inside the LR-99 and the innovative design of the black, missile-like x-15.

It was unusual to make that large a leap in speed during the envelope expansion. The goal was for no more than half a Mach number at a time. As White had pointed out, they didn't want to overextend and find that a small problem had become too big to handle before anyone knew it even existed.

With flight 2-15 on 22 April, he jumped less than a quarter Mach number but did become the first to exceed 3,000 miles an hour. This was a good record, considering Bob almost had to abort soon after launch, when his engine refused to ignite. His post-flight report stated, "We went ahead and stroked the throttle right on up in one motion to 100 percent, and I got thrust. [But then] she quit." This was the first appellation of feminine characteristics toward the x-15 from a pilot on the program. Maybe it was the first time the aircraft had done something he didn't understand.

White continued with his speedy diagnosis of the engine problem. The indicator lights told him there was a malfunction—the engine was not ready to ignite. "I waited a few seconds, [then] went back and hit the reset again. And when I hit the prime switch and held it there, the 'igniter ready' light came on. I went back up with the throttle . . . delayed slightly, and went up to about 75 percent thrust. . . . I felt a good shove come in, then I pushed it on up." While Bob was busy figuring out his engine, and avoiding an emergency landing on the Hidden Hills lakebed, he had dropped 8,000 feet below the B-52 mothership and two chase planes.

With the unexpected problem, Bob got behind the flight plan. He started his on-board clock, the primary instrument that told him to do specific actions at specific times. Neil Armstrong, as NASA 1, called out the times and actions for him, and White found he was about eight to nine seconds off with his clock. He admitted his error, and said of Neil, "We were right on the second from what he was reading."

As Bob leveled out at 105,000 feet, there were a series of maneuvers to accomplish. He complained afterward he couldn't do them, because, "the airplane wallowed quite a bit up here, and . . . she just continued to oscillate in sideslip. The period was down low enough that it wasn't too exciting, but it just kept wandering back-and-forth. . . . You had to hang on to the control and work with the damn thing to keep up with it."

The x-15 came equipped with three different methods of directing its flight: a traditional center stick, as used in many fighter-type aircraft for aerodynamic control; a left-side hand controller, for using the ballistic rockets; and a right-side hand controller, which blended aerodynamics with the ballistic rockets as required. One of White's test objectives was to use the right-side hand controller to see if it was responsive and comfortable for the pilot. He said it was fine, but there was some getting used to it. "Prior to transitioning back to the center stick . . . , I felt a little bit tired, perhaps because I was working quite a bit anyway, jiggling this stick around. . . . It was just nice to get back to the center stick."

Two flights after breaking Mach 4, White became the first pilot to bring down a second Mach number. This time, the engine cooperated as he launched near Mud Dry Lake. "I came off the hooks, banged the throttle right up to 100 percent, and she just took off beautifully." Mission 2-17 on 23 June was off to a rousing start, accelerating to Mach 5.27 at 107,700 feet. As this was happening, the cockpit lost some pressure, and his suit made up the difference by inflating, thus becoming more rigid. White had to work much harder. He said afterward, "I was well loused up [because the suit inflated], but I didn't have any problem reaching over and pulling the throttle back and shutting [the engine] off." As soon as the thrust disappeared, Bob noted, "The deceleration on shutdown was quite noticeable. I was up against the straps. . . . The velocity holds up beautifully at the high altitude. . . . I had all the time in the world."

Movement with the inflated suit was restricted, prompting White to go back to the right-side hand controller he had tested the previous flight. "I felt as if I were happier with the side stick, so I moved this big, puffed-up arm over again and got back into position on the side stick, because I preferred to do it this way. . . . While my suit was inflated, I felt I could . . . use the sidearm control with greater accuracy." White worried somewhat about his flying ability because of the problems associated with having his pressure suit balloon up. He radioed to his controller: "Okay, heading here towards Mojave, make sure I don't go too far." NASA 1 recommended, "Yes, I'd turn back, I think, Bob." White explained that he would do so, "if I ever get this suit un-inflated, I can't twist my head now." He worked his head to the right position, and the suit deflated as he rapidly descended to the ground.

Bob prided himself on pinpoint touchdowns. Coming from 20 miles high, at more than 3,600 miles an hour, White hit within a hundred feet of his intended spot. He said, "My primary concern was to try to get as close to the spot as possible. I am interested in how the records look . . . and hoped to arrive and touch down on the point." With his record Bob could have won martinis from Dick Day after nearly every mission.

Launch that afternoon had occurred within five seconds of the stroke of 2:00 p.m. Two days into summer, afternoons in the California desert were often hot. This was especially true within the cramped cockpit of a black aircraft, cooking away inside an inflated silver pressure suit.

Originally, the attempt was made to get the B-52 off the runway early in the morning, but problems cropped up, forcing White to get back out of the x-15 and into the suiting van to literally cool his heels — and the rest of his body — as he waited for the technical problems to clear. Afternoon rolled around and the heat rose, then it was time to fly. White commented to his debriefers of the situation: "Now with these temperatures out there, believe me, it is insufferably hot. You can take it just by gritting your teeth, but I think it does an awful lot as far as effecting efficiency."

Bob lamented the loss of the morning launch opportunity: "You could just feel the difference in the outside temperature compared to what it was in the morning." He thought rapid preparations were the way to go in the hot weather. "It was a real fast go. . . . I like this, and wonder why we can't always do this, with the pilot getting in the cockpit . . . say, ten minutes to engine start. You start by slamming the face plate closed, close the hatch, and [the B-52] engines are turning over. This is pretty nice."

The second taxi led to a smooth, but very warm, initial takeoff at 1:09 p.m. "It was just miserably hot. It was good to get off the ground and start getting upstairs where it cooled down a little bit." As he launched and flew the 3,603-mile-an-hour mission, the heat became secondary, especially as White prepared for touchdown. Coming into the lakebed proved to elevate the pilot's heart rate as his entire system went into overdrive during one of the most critical mission requirements: a safe landing. "Even if you are tired or hot," White said, "you become stimulated again when you finally get down to the landing phase." Over the radio network, the life support engineer told everyone involved with the flight that day: "They are getting a real tired-looking pilot out of the cockpit!"

Heat was one thing, but coping with the reactions of the test pilot's own body to stress was something else. High heart rates went hand in hand with these situations, but, until the x-15 came along, no one had been aware of the phenomenon.

The rocket plane program was a compartmentalized one, especially after the orbital program took the spotlight. When it came to Project Mercury, the entire world seemed to be watching. An early test of the Redstone rocket and Mercury space capsule used a chimpanzee named Ham on a sixteen-minute suborbital jaunt. When telemetry showed his heart rate skyrocketed, there was worry a human astronaut might suffer seizures, or even death, if subjected to the same situation.

Walt Williams knew the rates were not a significant problem, because he'd seen them in the first tests of the x-15, before transferring to oversee the project of putting Americans into orbit. He spoke about the situation, saying, "They went so far as to propose stopping the [Mercury] program because of these heart rates. Then enters the x-15 biomedical data that showed it was natural to get these high rates at peak points in the trajectory. They had just never seen what was happening before in high performance jets, because they had never looked, and these rates were not abnormal at all." Walt said that once the data became known, at first it did not persuade the doctors to clear astronauts to fly but rather caused a momentary near calamity. Walt exclaimed, "I thought they were going to cancel the x-15 instead of clearing us to fly Project Mercury!"

Charles Donlan, from NASA's Langley Research Center, talked about the conundrum at an x-15 symposium. When the question came up in a meeting, he sent for medical data to compare to early flights by Scott Crossfield in the x-15, to reassure everyone the astronauts would survive. A return message from North American Aviation's chief flight surgeon, Toby Freedman, stated, "I went to the School of Aviation Medicine looking for heart rates. I found some that were higher than [Crossfield's], but these occurred during copulation. However, not many people have died from that."

The air force's second pilot after Bob White was Robert Rushworth, who explained what was done on x-15 flights with regard to gathering medical data. "Blood pressure and heart rate and respiration were the primary things we were interested in. Burt Rowen was in charge of the Bioastronautics Branch, and he was the one that was primarily interested in getting the medical monitoring in the x-15. They did get it in, and it worked. . . . Of course, if the guy's heart stopped, all you can say is, 'He just died.' There isn't much you're going to do about it."

White knew he was in prime condition, and his heart was working just fine. He never had a qualm about flying faster and higher in the x-15. He had already knocked over Mach 4 and Mach 5. Design specifications for the rocket plane and LR-99 engine stated they could easily get at least one more of these Mach numbers laid to rest, and Bob was more than willing to make the flight to do so.

The 11 October flight by White was a break from the speed increases. Instead, he flew the first time to more than 200,000 feet, experienced two minutes at zero gravity, and had the outer pane of his left windshield shatter during reentry. He told his debriefers, "I had come off the high g [and] was level at about 60,000 feet . . . between 1.6 and 1.8 on the . . . Mach meter. [The windshield] cracked all at once. I could notice little slivers of glass falling away once in a while."

Each windshield had an inner and outer pane. The rectangular shape created hot spots as the frame expanded during high heat portions of the flight — usually above Mach 5.5 — and contracted as it slowed and cooled. This deformity cracked the pane, but luckily it never occurred with an inner and outer pane simultaneously. If that happened, the pilot may have lost the aircraft, as superheated boundary-layer air blasted into the cockpit. As it was, the loss of a pane meant the pilot lost his ability to use both windows for visibility, compromising his depth perception. White said after a later flight when the same thing happened, "It's pretty obvious now that both windshield panels are required, because I had the impulse to move over to use both eyes through the one clear panel. But, of course, with the restricted headroom, you can't move over. . . . I still had both eyes operating, even though one windshield was opaque."

Eventually, the problem was solved after the no. 2 x-15 was rebuilt with a modification of the windshield into an elliptical shape. The no. 1 and no. 3 aircrafts were not modified at the same time, because they were no longer to be used for flights into the speed range where the shattering occurred. However, there were still incidents where the other aircraft also suffered from cracked panes.

White tried to take each difficulty in stride, but made a comment concerning what might happen if there were a cascade of failures. "I think emotional stress levels will go up by a factor of [the number of] things that fail in the airplane. There is nothing really to get excited about. You are just . . . wrestling with the little tasks as they show up . . . but, if things start failing, or the suit blows up, or [the stability augmentation system] goes out . . . , then I think you might talk about emotional stress levels."

On 27 October Bob was disappointed when his flight was aborted upon arrival in the vicinity of Mud Dry Lake. Overcast had moved in since the morning check flight, obscuring the ground. Two more delays from mechanical problems, on 2 and 3 November, again brought frustration, the second of these especially so, since the countdown reached ten seconds before being called off when the engine igniter did not function. Of all 199 missions accomplished by the x-15, nearly one-third, 63 in all, flew on a Thursday. This was the case with flight 2-21, which finally and successfully launched on Thursday, 9 November.

White had a habit of not using a standard countdown as launch neared. He never said why he was the only x-15 pilot to avoid this convention. Maybe he thought it was too corny. He described his launch that day at 9:57 a.m. high above Mud, by saying, "At about five seconds, I turned on the launch light, then gave another quick look across the board that everything was good. It was time to go, so I threw the switch."

With the low winter sun angle and a south heading in flight, White was set up for lighting problems. "As soon as I rotated, the sun was right in my eyes, I actually was flying the airplane with my hand up, blocking out the sun to get a good look at the instruments. . . . As the acceleration increased, I finally gave up and . . . dropped my hand down."

This flight was aiming for a maximum speed run for the x-15, and another Mach number was about to fall. White stated in his post-flight report, "When I leveled off at about 101,000 feet, I made a little downward pressure [on the control stick], because I didn't want to be climbing. I remember . . . going along watching that [Mach] meter reading roughly 6,000 feet per second, [and] saying to myself, 'Go, go, go, go!' We did just crack it, because we knew that bringing all the proper things together, we could or should get just about Mach 6."

In order to achieve the goal, the flight plan called for pushing the LR-99 to the point of exhaustion instead of manually shutting down the engine at an arbitrary point. White said, "The shutdown seemed to be a little bit different this time, compared with a shutdown by closing the throttle. It seemed to occur over a longer time interval." His speed at burnout was Mach 6.04, equating to 4,093 miles an hour.

Deceleration was immediate, as drag overcame the now depleted rocket thrust. The altitude and Mach meters both started to wind down. As he passed through Mach 2.7, the outer pane on his right windshield broke. Instantaneously, it was like looking through a window that had been covered in white soap. At first, Bob hadn't even noticed it because he had been concentrating out the left window, the direction that he was turning into the Edwards landing pattern. "I could see out the left windshield panel fairly well, and the lake was just off to the left." Then he discovered the problem out the right-hand window. "As I got down lower, I realized I couldn't see out of the right side. For all intents and purposes, the visibility . . . was nonexistent. I asked the chase plane to stay in close, thinking right after it happened that it might also happen on the left side."

If both had gone opaque, there was no way to see outside to set up for a safe landing. White immediately had an idea if this were to happen: "In that event, I considered going to high faceplate heat when I got subsonic, jettison the canopy, and see what happened from there. . . . I was quite surprised at what a compromise it offered, being able to see out of only one windshield." Touchdown occurred at 10:06 a.m., with no further problems from uncooperative windshield panels.

In eight months, the speed of manned aircraft had doubled from Mach 3 to Mach 6, an amazing feat for the x-15 program and the pilot who accomplished the task.

Sitting on the enclosed patio of his home in Florida during our interview, Bob looked at me thoughtfully for a moment, then said, "You know, I never thought about it until later on — and that just shows you how this program was working — I didn't realize until afterward that I was the first [Mach] 4, 5, and 6." He counted off each on his left fingertips as he said the numbers. "Now, if some NASA guy had said, 'Hey, wait a minute, we want Joe [Walker] to break one of these here, too,' I'm not sure what would have happened." Then he laughed and said, "Unless they did it for me intentionally, because, if they were going to lose a guy, better me than Joe!"

In the next six years, the x-15 surpassed White's speed only three more times, adding a scant 427 miles an hour to the rocket plane's top end. The Mach 1 through 6 records now read: Charles Yeager, Scott Crossfield, Mel Apt, Robert White, Robert White, and Robert White. All six of these records were accomplished in just over fourteen years. No one has added a new Mach number to human flight since that day in 1961.

It seemed White had several flights when it took many tries to get underway. This was certainly true of mission 2-23, with two times in late April, then two more a month later, before the x-15 was released on the fifth attempt. It was 1 June 1962, and this flight became the one hundredth time one of the three rocket planes left the ground attached to the B-52, but only the fifty-fifth time it actually was launched, making for a ratio of close to one launch for every two attempts. Those numbers were to improve over the next few years but dropped again toward the end. White fell right in the middle, eventually making thirty-two flight attempts, but launching only sixteen times, for a ratio of exactly two to one.

Missions 2-23, 3-5, and 3-6 all went without a hitch, and all were accomplished within three weeks in June, thus becoming the most successful month during Bob's tenure on the program. The last of this series, 3-6, on 21 June, also was his highest altitude to date of 246,700 feet. Yet even this heady height was just a precursor to his penultimate mission in the x-15 and his ultimate as far as altitude was concerned.

It started on 10 July with a mission that had to be aborted soon after takeoff because the mothership had landing gear trouble. The B-52 has an unusual arrangement, with four landing gear all mounted inside the fuselage, two fore and two aft, down the centerline of the bomber. When they are extended, the two pairs each rotate outward, one to the left and one right. In this instance, the aft left gear refused to retract after liftoff, so the B-52 and mated X-15 were forced to turn around and head right back to Edwards.

The following day the B-52 gear worked fine, but the X-15 auxiliary power unit pressure regulator ruptured, so, once again, it was a return home. On 12 July it rained at the primary launch lake, Smith Ranch, which brought up the idea of changing to a different lake, in this case, Delamar. South and east of Smith Ranch, Delamar was unaffected by the wet weather. All personnel stationed remotely in case of emergency had to be shifted to the new position, taking away another day. This brought the launch to 14 July, but then a "Small Boy" got in everyone's way.

Flying from Delamar to Edwards meant the X-15 and all associated aircraft were going to pass over the southwest corner of an area of restricted airspace in Nevada called Frenchman Flat. This was a lot of square miles owned and operated by the Atomic Energy Commission, then in charge of running tests of America's atomic bomb arsenal. Four atmospheric tests, as part of Operation Sunbeam, were scheduled to be exploded during July, and one of them, nicknamed Small Boy, was to be lit off right under what was the X-15's flight path on 14 July. The AEC requested the X-15 team delay their flight to another day. Under the circumstances, there was no hesitation from the air force or NASA in doing so.

Suspended ten feet above the desert floor, at 10:30 a.m. the tiny bomb (by atomic standards) was commanded to explode. This was called a tactical nuclear weapon, in that it was set to a very small yield. In a theater of operations, it would hopefully endanger only enemy troops and not nearby friendly ones. The casing for this weapon was barely fifteen inches in diameter, weighing just sixty-four pounds. When it went off, it had the power of 1.65 kilotons of dynamite, less than 10 percent of what was used in World War II at Hiroshima. This was the final bomb from the Frenchman Flat area of the Nevada Test Site range to be ignited in the open atmosphere. The other bombs in this series also had innocuous nicknames: Little Feller I and II and Johnnie Boy, all three of which were testing what was called the Davy Crockett warhead, a dubious legacy for the American frontiersman.

The X-15 crew gave it two days for Small Boy's mushroom cloud—and its associated radioactivity — to fade away before flying uprange for another attempt on 16 July. This, too, failed, as an umbilical connector inside the pylon holding the rocket plane to the B-52 came loose because it was too short. One more day, and White was finally able to launch at 9:31 a.m. on 17 July. Rain, numerous mechanical problems, and even an atomic bomb couldn't keep the X-15 from making its first mission into space.

Operations engineer John McTigue gave a synopsis of what happened leading up to the flight: "The two flights that [White] had previous to that, he did in nine days, so we were hoping to do that on this one. Well, twenty-five days later, we flew. We had all kinds of things. We had an intercom problem; we had three aborts. I had to change an engine, and then the engine gave more thrust than before, which actually helped Bob get to altitude."

The intent of the flight was to make 282,000 feet, which was to be the first above 50 miles, and thus the first winged vehicle to officially — according to American standards — fly into space and back under pilot control.

Designating a specific boundary for spaceflight has always been arbitrary, since the atmosphere does not just suddenly stop and a vacuum begin. In the United States during the X-15 era of the 1960s, the line was officially drawn at 50 miles. The Federation Aeronautique Internationale, which governs aerospace records, uses the metric system, and they decided on 100 kilometers, equating to approximately 62 miles. The American delineation of space at 50 miles actually has good scientific rationale in that this is the boundary between the mesosphere and thermosphere, whereas the FAI line does not match any specific atmospheric change. It does, however, correspond to the esoteric "Karman Line," where atmospheric control of a vehicle is theoretically no longer possible.

At the time, only four other Americans had been into space: Alan Shepard and Gus Grissom on suborbital lobs down the Atlantic Test Range, and John Glenn and Scott Carpenter on orbital missions, all aboard Mercury capsules that had no aerodynamic controls for either leaving or returning into the atmosphere. The test pilots at Edwards, especially those working on the X-15, were of the mind that Mercury was simply a "man-in-a-can" scenario, with little to no value toward making spaceflight accessible, routine, and cost-effective over the long term. Bob White was poised to prove the viability of wings in space for the first time, and because of the finely tuned LR-99 installed by the team in the rocket shop, it all went much better than expected.

The X-15 was mated to the mothership, and final checks were made after fueling and servicing before the pair was set to head down the taxiway. McTigue spoke of how a small leak helped lead to the big change in upcoming altitude. "We got ready for flight that day, and we had a peroxide jettison valve that was leaking just a little bit. We repositioned the B-52 so I could properly vent the peroxide, to see if it was still leaking. We raised one end of the B-52, and the leak sealed. In the process, we were able to tilt the [B-52] enough that I was able to add more ammonia to [the X-15 fuel tank]. Well, that gave Bob a little . . . more time at thrust." The entire crew understood this flight, if successful, was to be a momentous one. McTigue said, "The two things I did were to give him a higher thrust engine, and more fuel, because [White] was very sensitive for wanting to make altitude after he missed it on the flight before."

After Capt. Jack Allavie and Sq. Ldr. Harry Archer took B-52 no. 003 to the launch coordinates near Delamar Dry Lake, White proceeded through his checklist. With less than a minute to go, the stability augmentation system dropped off line. White quickly dove in to correct the problem. He remembered, "In the early days, there was still the maturing process as far as the systems were concerned. . . . This was the third attempt. I guess it was so near launch that I instinctively reacted by just resetting the flight control system. It reset properly." Uncharacteristically, Bob counted down, "Three . . . two . . . one . . . launch." He dropped from the mothership's pylon at 9:31 a.m. on Tuesday, 17 July 1962.

Flying off Bob's wing, in an F-100 as Chase 1, was Chicago-native Maj. James A. McDivitt. Jim was in line to become the next officially selected air force X-15 pilot, so this was part of his orientation. As the LR-99 came up on power and the rocket started to pull away, McDivitt told White: "Everything looks real good, Bob." White pointed the nose upward, and it took exactly eighty-two seconds to get to the spot where the engine provided enough energy to push the X-15 upward more than 50 miles. Bob shut down past five times the speed of sound. With the unexpectedly high engine thrust and the extra amount of fuel squeezed into the tank, White was heading 34,000 feet higher than his aim point. He passed through 300,000 feet, another first for a manned aircraft — now spacecraft. The X-15 briefly touched 59.6 miles—314,750 feet—then started its unpowered drift down-ward toward dense air.

Robert M. White had just become America's fifth astronaut. He spoke about the experience: "I was up roughly over the California-Nevada bor-der, in the area of Las Vegas, because we were just beyond Vegas when we launched. When I described it, I said, 'It looked as though I could spit into the Gulf of Baja and toss a dime into San Francisco Bay.' Indeed, there was no question, Columbus was right, the Earth is round. It was dramatically impressive. It's short term, but you take it all in. Then, pretty soon, I had to start dialing in my angle of attack and prepare for reentry. The reentry was very dynamic." White explained further, "I have to say, that was the most dramatically impressive reentry, with the forces that are throwing you for-ward and down. As I was pulling the debriefing, I was splotched here [in-dicates a large bruise over his right shoulder and upper arm]. All the blood vessels burst. . . . I had an achy feeling, then, in a week or so, it went away. No problem."

Coming back down, the aircraft was heading too far from its intend-ed ground track. White had to make a tight right-hand turn to get back where he wanted. "I overshot Edwards. I passed over the field at Mach 3.5 at 80,000 feet, so I had to make the largest 360 in history getting back."

White touched down ten minutes and thirty-two seconds after launch, changing the idea forever of piloted reentry. Walker radioed to his friend, "This is your Happy Controller going off the air."

Once safely back on the lakebed, White had no time to relax. "We were scheduled to go to Washington DC, to get the Collier Trophy, and didn't have a real post-flight debriefing. An engineer came out to the van when I got out of the pressure suit, and he did a quick little debriefing." Bob was now in a hurry to catch a flight east with fellow X-15 test pilots Joe Walk-er from NASA, Scott Crossfield from North American Aviation, and For-rest Petersen representing the U.S. Navy. The morning of 18 July, they were all scheduled to be presented with the Robert J. Collier Trophy by Presi-dent John F. Kennedy, "For invaluable technological contributions to the advancement of flight, and for great skill and courage as test pilots of the X-15."

Outside the White House that morning, the pilots and other represen-tatives from NASA and the military assembled in front of a contingent of

9. Capt. Jack Allavie and Sq. Ldr. Harry Archer bring B-52 no. 003 in for a low sweeping pass above x-15 no. 3 at the end of Robert White's record altitude flight of 314,750 feet on 17 July 1962. Courtesy of the Armstrong Flight Research Center.

press covering the event. Walker and Crossfield were dressed in business suits, with White in his air force dress blues. Petersen looked especially dapper with his high-collared navy dress whites. After some opening remarks, Kennedy stepped to the microphone:

This large and distinguished trophy has been held by some of the most distinguished pioneers in the field of aviation and science, and some of them are with us today. All of them have won it by being willing to extend the horizon of either knowledge or of human endeavor, and particularly human endeavor, which requires not only great courage, but also the highest kind of talent and coordination of science and personal qualities.

We are very proud of these four young men, who are among the finest we have. We have welcomed the astronauts who also occupy this same domain, and now these flyers, who [took] a manned aircraft into space yesterday to over 300,000 feet, and at over 4,000 miles per hour. . . .

So we are very proud to have them here, not only for what they have done in space, but also because they represent the kind of Americans whom we want this country to be identified with. We hope that their example will encourage others to do something that requires reaching beyond what we now know and what we now do.

Bob White took Kennedy's place at the microphone, giving recognition to everyone involved with the research program: "Thank you Mr. President. On behalf of the x-15 pilots, we are very happy to accept this very wonderful and great aviation award. . . . I am sure the gentlemen with me today would agree that we accept this trophy on behalf of all those people who are not here present today, but who we think did a magnificent job in allowing us to get on with this program and accomplish some of the objectives which we have achieved. So, on behalf of all of these people, we gratefully accept the trophy, Mr. President. Thank you."

There was a big smile from White as he related a special observation of that day with Kennedy. "There was an air force [public relations] guy that met us when we came into the airport, and said, 'God, Major White, if we had dreamed this up, if we had scripted it, we couldn't have done it better!' He's going on and on; he's a crazy man. He said, 'Here it is, you're coming to the White House, the newspaper's got headlines, you broke this record, the President's going to see you. . . . The timing couldn't have been better.'

And he was right, you know, just by accident, we couldn't have written a script any more perfect."

Within two months of the astronaut qualification flight by White, the new air force x-15 pilot who had flown Chase 1 that day, Jim McDivitt, got a different offer. This was from NASA's astronaut office in Houston to join their ranks. McDivitt became the first man to walk away from an x-15 assignment, although Jim went on to fly two highly successful space missions with *Gemini* 4 and *Apollo* 9.

With White ready to leave the x-15 himself and rejoin the military, a replacement had to be found for McDivitt. Robert Rushworth was already there, but they needed to have at least two from each organization. Bob White wasn't too worried. He had accomplished everything he had envisioned on the x-15, and so much more. Only one pilot in history will ever have bragging rights to three Mach numbers and the first winged spaceflight.

Five months after the White House ceremony, there was another mission, White's sixteenth and final one in the x-15. Flight 3-12 on 14 December was perfect. He was back to saying simply "Launched!" as he fell from the B-52. Bob zoomed up to a mere 141,400 feet and 3,742 miles an hour. He again complained about having to hold his hand up to block the winter sun as he powered south to Edwards. At 10:53 a.m. his rear landing gear touched Rogers Dry Lake. He said, "As soon as I felt the skids skipping along, I jammed the stick right up against the stop and held it there. The nose impact, my impression was that it came down pretty hard, but I suppose this is probably kidding myself because I'm helping shove it down. . . . It's been five months since I felt it, so you always have this impression that it's a pretty hard impact."

Talking with his debriefers, White said, "I was past the aim touchdown point a little bit, but not too awfully far, I don't think." Not far at all. Bob landed his final flight in the x-15 within fifteen feet of his intended spot, another perfect martini touchdown.

He didn't leave Edwards immediately. As anyone who has been in the military can attest, the catch phrase is "hurry up and wait." Before his orders arrived, he stayed somewhat active with the x-15, performing six chase flights up through 3-18 on 29 May 1963. During his last flight associated with the research program, Joe Walker's inner left-hand windshield shat-

10. Following a flight Bob White stands up in the x-15 cockpit and prepares to use the ladder to step down to the lakebed. Notice his A/P-22S pressure suit has been zipped open and a portable air conditioner unit has been connected to the inner suit. Courtesy of the Edwards AFB History Office.

tered, something Bob could certainly identify with. Walker made it home safely, and White transferred away.

With outstanding credentials and name recognition from his years on the x-15, White could go anywhere. Two major aerospace companies wanted him for their own — a route taken by many famous astronauts. Instead, the only thing he wanted was to get back to being a military pilot. "I was a professional officer, and that was the way I was going to go. I wear the uniform. I was dedicated to staying [and] some things broke my way. . . . Gen. [Bernard A.] Schriever had plans for me to go to the Cape [in Florida] and work my way up to being a launch director on the Titan II missile." Unmanned ballistic rockets were not what Bob had in mind, so he told his boss he wanted back into tactical aviation. That meant fighters, such as the Republic F-105 Thunderchief, which White had wrung out while in flight test at Edwards. "My boss finally agreed," White said. "Bitburg, Germany, was a premiere fighter outfit then, [and] that's where I went." On his way there, Bob stopped by Nellis AFB, Nevada, to get checked out in the aircraft, then to Germany as an operations officer. "I hit the place running when I got there," he said, working his way up to commander of the 53rd Tactical Fighter Squadron in July 1964.

Bob's list of educational and military accomplishments is a long one. He considered the x-15 very special, and it opened many doors for him, but there was so much more to his career. By May 1967 one of his goals was met by requesting an assignment to the Vietnam war zone. He thus became the only x-15 pilot to serve in three armed conflicts. At first he flew F-105s, often affectionately called the "Thud," out of Takhli Royal Thai AFB, then later from Tan Sun Nhut Air Base in Vietnam. White racked up seventy missions against North Vietnam, earning the Air Force Cross, but also losing squadron members in the process. "Any loss rate in South Vietnam was practically nil compared to what was going on in the north," he said. "Because there, we were facing heavy anti-aircraft fire, the 88-millimeters, and anti-aircraft missiles." Bitterness was in his voice when he recalled, "Some out of my group . . . were shot down. I always flew when we went north. . . . I knew what would have been at the end of the trail had I gone down. I had some of my comrades tell me when they came back, 'Oh, yeah, they would have loved to have seen you.'"

Returning to stateside duty once his combat rotation finished, he became one of the first pilots to fly the McDonnell Douglas F-15 Eagle at Wright-Patterson AFB, Ohio. In mid-1970 he returned to California and the environs where he had spent so many productive years on the x-15, assuming duty as commander of the Air Force Flight Test Center on 31 July. He served in that capacity until 17 October 1972.

Bob moved through the ranks, eventually peaking as a two-star, or major general. This seemed to be the rank that most air force x-15 pilots attained before retirement, which Bob White decided to do in February 1981. He talked about the lack of will that appears to have infested the aviation research world since he left: "I did a thesis when I was at the Industrial College of the Armed Forces [for one year, starting August 1965] on the need for an advanced manned research aircraft. . . . It pointed out that there has to be a separate, independent research program. If you postulate, as I was doing, the scramjet technology being developed, maybe we could get to a . . . 13 and 14 Mach range. There would have to be a number of development programs going on coincidently. . . . Then, if the technology was developed, you could come together and have a new research vehicle."

Unfortunately, the scramjet program moved at a snail's pace, and a flight engine was never built. White went on, saying, "There had to be a technological revolution. Look at what the hell has happened in the electronic world. I went through this with vacuum-tube technology, coils and resistors. . . . I had a professor who said that one day engineers will be able to do almost anything. He was prophetic." But first you had to point the engineers in the right direction. "It's hard to get people excited about developing a technology that's such a complex subject. . . . You can excite their imagination when you talk about being on Mars, but there's a hell of a lot between Earth and Mars before you're ever going to get there. A heck of a lot has to happen."

Talking of his love of a good martini, Bob wanted to make sure it was understood that, "I wouldn't touch a drop of alcohol at least three days in advance of the day I was going to fly. I took things seriously. . . . This was new-frontier stuff." Then his jovial mood turned reflective as he recalled two friends associated with the x-15, now long gone. "I've often thought back that I was real fortunate to be at a place and time where I could do that—and survive. Makes me think of [Iven] Kincheloe and other guys, like [Joe] Walker, who didn't."

The x-15 was a big adventure midway through Bob White's air force career. "We realized we were going somewhere we hadn't been before." Bob explained what it was like at the first moment when he joined this experimental program, which was to take him into space and push the boundaries of speed. "They used to have a little glass-fronted shack out alongside the [Edwards] runway, and you'd watch as airplanes came in, checking them through field glasses to make sure their gear was down. It was kind of [a] boring duty. My boss came out one day and said, 'You know you've been designated [as an X-15 pilot]. Maybe you want to go home and talk to your wife, and you want to think about it.' I said, 'No, if I have to think about it, they've got the wrong man.'"

4. Naval Engagement

You didn't have time to look out the window to see what
the scenery looked like. You were busy every second.

Forrest S. Petersen

The U.S. Navy saved the x-15 program. The ambitious goals set forth required sustainable funding over many years. At times, that money was difficult to find, and was more so when the navy had plans of their own re-
garding high-altitude hypersonic flight.

Interservice competition was evident in many programs, such as in late
1953 when Scott Crossfield and Chuck Yeager each made a bid to become
the first man to break Mach 2. Crossfield used the D-558, Phase 2, and naval funds to first reach the magic number, over Yeager's attempt with the
x-1A from the air force. This small-scale rivalry was a model for what later became the space race of the 1960s, albeit between two branches of the
American military instead of the United States and Soviet Union. Whatever the scale, the idea was the same: the other side must not be allowed to
capture any record, because to do so meant embarrassment.

As the government flight research agency, NACA wasn't part of the equation. Their job was not to set records but to answer fundamental questions
about flight. The military services tolerated brainiac scientists and engineer
pilots but never saw them as competitors.

Along came the x-15, and the situation had to change. The idea was born
out of NACA studies and designs, but it was understood their small agency
could never secure the money needed to bring it to fruition. The air force
had the political clout in Washington DC, which NACA lacked, so Walt Williams spearheaded the idea of partnering with the military service to secure
funding for the rocket plane. Even then, the question was asked, would this

be enough? No other research aircraft had the cost potential of the x-15. At the same time, the navy was looking higher and faster than the Mach 2 and Mach 3 aircraft in service or envisioned.

Space was the ultimate goal, as military doctrine was always to secure the high ground. There was nothing higher than space. The air force and NACA thought the x-15 was the best route, while the navy started research for a follow-on D-558 craft, the Phase 3. Their goal was an altitude of 1 million feet, or nearly 200 miles. Based on the knowledge and materials available, this goal was unattainable using a winged and piloted aircraft. However, in the mid-1950s, just about anything appeared at first to be possible, so the studies moved forward.

Once reality crept into the navy proposal, it became evident they had set their sights too high. With the military budget tightening during the administration of President Dwight D. Eisenhower, the idea was proposed to merge the competitors' programs. The x-15 accomplished what no one thought possible, bringing together two branches of the military to work alongside each other, while welcoming the input of NACA. Walt Williams said, "Navy funding, although small and rather limited, was enough to keep the program going in some hard times before completion of the first aircraft. The air force and NACA were maxed out, and the navy was able to put it over the top."

Navy lieutenant commander Forrest Silas "Pete" Petersen completed the U.S. Naval Test Pilot School at Patuxent River, Maryland, in 1956. Pilots always referred to the facility more simply as "Pax River." Outstanding performance in the classroom and in the air earned Pete the rating of first in class. After graduation, he decided to stay on at Pax River to train other pilots. This kept him close to the world of naval flight testing when it came time, two years later, to select pilots to fly the x-15.

It was rumored that another navy pilot was originally selected but turned down the offer because he didn't feel it worthy of his time while pursuing a proper military career. Petersen had no trouble accepting the assignment. When I sat with Pete in his kitchen one Saturday morning to discuss his role, he said, "I was there when they decided to send a navy guy out to the x-15 program. . . . The program was formulated and the money had been appropriated, and they were constructing the three aircraft, when the navy got serious about sending somebody out to the desert. Having decided that,

I think the people at Pax River felt that it probably ought to be one of the people from there to represent the navy."

Petersen confirmed the rumor of other pilots, saying, "There was no formal selection process. They considered several guys that were down there at that time, including me. Some of them weren't terribly interested in the program. I was. They insisted I go home and talk to my wife about it. She didn't seem to have any objections, so I guess you'd say I volunteered. I reported [to Edwards AFB] in August of 1958." His orders were open-ended, meaning the length of his assignment was to be determined by U.S. Navy necessity, and how important they believed the X-15 was to the continuation of their overall mission.

Behind the scenes, the navy almost immediately began to question that necessity. Of what use was a rocket plane for blue-water sailors? This radical design certainly was no analog for a future aircraft carrier–based fighter or bomber. However, the money was either committed or already spent, so Pete's initial role was secure—to a point. Sheri McKay-Lowe, daughter of X-15 research pilot Jack McKay, recalled her uncle talking of Forrest's arrival at NACA. "The [other pilots] clued him in that he wasn't going to be having that many flights. Pete was stunned. The navy had put so little money into the program compared to the air force, for instance. That's just the way the cards played out. [Petersen] was very disappointed to hear that. He asked, 'Hey, I came all this way, and I'm not going to be doing that much on this program?'"

What Pete was there to do was his job. He still felt it was an important one, and that's what mattered most. Any revelations such as these were political background noise, something in which he preferred not to involve himself.

Like the majority of X-15 pilots, Forrest Petersen came from a small community. In his case, it was Holdrege, Nebraska. Located in the south-central part of the state, about two-thirds of the way from Omaha to North Platte, the town is six miles west of Funk. The area reportedly sits atop one of the world's largest water reservoirs. Even in the early twenty-first century, the community still fell short of 6,000 residents. Forrest arrived on Tuesday, 16 May 1922, to parents Elmer and Stella. The family soon moved about forty miles northeast, to an even smaller town, Gibbon, which originally sprang from a Union Pacific Railroad storage area in the 1860s, as new tracks cut across the Great Plains.

Petersen spent most of his childhood through the late 1930s in Gibbon and the Platte River Valley, an area that had once been at the bottom of a vast, shallow sea during the Cretaceous period, a hundred million years previously.

After finishing high school, Pete headed to the University of Nebraska at Lincoln but attended only two years before being accepted in 1941 as a midshipman into the U.S. Naval Academy at Annapolis, Maryland. Any acceptance at a military academy is a hard-won honor, and Petersen never regretted his decision to make this his career. With World War II waging, there were ample opportunities for an ambitious young man.

His Annapolis graduation on 7 June 1944, with a degree in electrical engineering, was the day after the D-Day invasion on Normandy beaches by Allied forces. Petersen, as a member of the class of 1945, was part of a three-year accelerated academic schedule to commission academy graduates quickly in wartime.

The newly minted Ensign Petersen was assigned to the Pacific theater of operations, aboard the Fletcher-class destroyer USS *Caperton* (DD-650). This destroyer class was named for Adm. Frank Friday Fletcher, who earned a Medal of Honor in 1914 at the battle of Veracruz, Mexico. In World War II there were 175 of these warships commissioned, the most of any type destroyer. The purpose of the destroyers was to move fast. The Fletcher-class ships were capable of a crisp thirty-eight knots.

The *Caperton* fought as part of the campaign to return Allied troops to the Philippines and roust the Japanese. Pete arrived soon after the Battle of the Philippine Sea, which had decimated the Japanese fleet under the command of Vice Adm. Jisaburo Ozawa. After securing the Mariana Islands in the battle, the fleet moved westward toward the Philippines. Petersen saw action in several of these battles, his destroyer often serving as a screen to protect other ships crippled in attacks until they could reach safe waters. He was a fire control officer, meaning it was Pete's job to determine the range and direction of where to point the ship's guns against the enemy.

As part of Task Force 38, under the command of Adm. John S. McCain Sr. and Third Fleet's commander, five-star fleet admiral William F. "Bull" Halsey Jr., the *Caperton* survived Typhoon Cobra in December 1944. The destroyer was attempting to refuel when operations ceased, and the fleet was inadvertently sent directly into the storm's path, due to inaccurate weather predictions. Petersen's ship went unscathed by nature, as it did with all future enemy encounters at Okinawa, Formosa, and the Battle of Leyte Gulf. This last was the action that destroyed the Japanese ability to wage war with its fleet. From that point, they relied primarily on land-based assets and began last-ditch efforts through kamikaze attacks on American naval forces. By October the liberation of the Philippines was in full swing, the Japanese military fighting valiantly but losing the war they had started against the United States in December 1941.

Following the war Petersen decided aviation was a better career path than gunnery and applied for flight training at the Naval Air Station in Pensacola, Florida. Before this, he had another important goal, marrying his sweetheart, June Berkshire. She accepted and they wed on 2 February 1946, six months after World War II came to an end.

Pete completed training in Florida, attaining his Naval Aviator wings on 14 June 1947. Several years later, he applied for Navy Postgraduate School, returning to Annapolis in August 1950. Two years after that, Pete earned his bachelor of science in aeronautical engineering. Attending college through these years took him away from the conflict in Korea. He eventually obtained his master's at Princeton University, New Jersey, in June 1953. Also sharing the campus during this time at Princeton was Albert Einstein, who ran the Institute for Advanced Study until his death in 1955. It is unknown whether Petersen and Einstein ever crossed paths, but considering his course of study, there is a good chance he met the famous scientist.

After graduation Pete returned to active duty, serving with the VF-51 fighter squadron for three years. Interestingly, this was the same squadron where Neil Armstrong had flown a couple years previously.

In 1956 Pete left VF-51 for test pilot school, attending the same class as Alan Shepard, who became America's first astronaut four years later. Pete flew various fighters, such as the Grumman F11F-1 Tiger, a favorite naval aircraft that served the Blue Angels aerial demonstration team from 1957 to 1969. Petersen and Shepard shared an office as instructors at Pax River at the time Pete accepted his new assignment with the X-15. Eight months later Shepard also left the naval school when he was chosen for the Mercury orbital spaceflight program.

"I went to Edwards and got settled into an office at the NACA High Speed Flight Station," Pete said. "It was only a few months after I got out there that NACA became the nucleus for NASA. . . . I went out and got acquainted with people, and immediately became involved in the program. . . . I would say that I was a full-fledged member of the pilot team and took part in as many things as I wanted to."

The fuselage, wings, and tail of the x-15 are covered with various markings. These identify everything about the aircraft, from the serial number to how to rescue a pilot from the cockpit in case of an emergency. Along each side of the fuselage are large white letters: "U.S. Air Force," and once the three aircraft were turned over from North American Aviation, the tail was emblazoned with a wide yellow band, outlining the black letters "NASA." Considering the nature of the joint program, which included the U.S. Navy, the stencil denoting that association is conspicuously absent. Petersen reflected on that fact, saying, "That was one of the things they used to say, you want to give us a couple hundred million bucks we'll put 'Navy' on there."

The cost of the x-15 was substantial, although not as high as Petersen mentioned. When considered next to other research aircraft that came before, it was indeed expensive, but was a small investment compared to the space program that led to the Apollo lunar landings a decade after the x-15 began to fly.

Initial estimates in 1955 put the cost of the entire program at $12 million. This included both the airframe and the LR-99 development costs. By the time of first flight, the airframe cost alone ballooned to $74.5 million, with the rocket engine adding another $59 million. Of this, the navy contributed $6.4 million, with an attempt made to squeeze an extra $1 million from the service in 1960. NASA provided facilities and testing. Other than that, all costs were on the shoulders of the air force budget.

Besides the share of funding and the inclusion of Petersen on the short list of pilots, the navy also supplied one other critical element: the centrifuge at the Johnsville Naval Air Development Center in Pennsylvania. This facility, more formally named the Aviation Medical Acceleration Laboratory, became important, first for supporting the x-15, then later in training America's astronauts.

The gondola was mounted at the end of a fifty-foot arm, which was then rotated at high speeds, simulating g forces for pilots to gain safe experience on what it was like flying new supersonic fighters then coming into service. For the x-15 something more was required. A replica cockpit was inserted into the gondola, and instead of having the "flight" controlled by an operator outside the centrifuge, the x-15 pilot had control of the profile. Petersen and the other pilots accurately simulated all aspects of flight, from launch, acceleration to hypersonic velocities, then the crushing g forces of reentry into the atmosphere. The idea proved very successful and opened the avenue for astronauts in the Mercury, Gemini, and Apollo programs to better gain knowledge of how to fly their spacecraft.

Precise control was a necessary element to prove the x-15 research plane could do what it was designed to do and be safe for the pilot throughout the envelope. Wendell H. Stillwell, in his book x-15 *Research Results*, wrote, "The simulation contributed materially to the development and verification of the pilot's restraint-support system, instrument display, and side-located controller. The x-15 work proved that, with proper provision, a pilot could control to high acceleration levels." These simulations were crucial for verification of piloted and winged reentry techniques from space. Before the first actual launch on 8 June 1959, more than four hundred simulated x-15 missions had already paved the way at Johnsville.

It took two years after his arrival before Petersen was ready to take his first x-15 research flight. Being with the initial team of pilots meant waiting for the systems to be wrung out by Scott Crossfield, then having the no. 1 aircraft turned over to the government, where both Joe Walker and Bob White had first dibs. By the time it came around to Pete, Crossfield had completed eleven flights, and Walker and White had four each. None of this bothered Pete. "The whole two years was [spent in] preparing for that flight. Engineering problems, investigations, committees, all those things, which all the pilots were involved in, were preparations for my first flight."

By August 1960 Pete knew his time was coming soon. "Probably six or eight weeks before, you firmed up what you thought you were going to do on that flight and would begin perfecting your technique in the simulator to do what was expected." He worked closely with the operations engineers and flight planners, making sure there was nothing being asked that he didn't think was possible due to time constraints or other factors. "All of those things were normal for almost every flight," he said. "On my first flight, I wasn't going into any area that anybody else had not already been a number of times."

After Bob White completed his first mission above Mach 3, with flight 1-12 on 10 September 1960, the aircraft was run through post-flight checks, then turned over to the support crew and Petersen to prepare for his own launch. He made his first attempt ten days later, but an auxiliary power unit failed to start after the B-52 was in position to make the drop.

That Friday, 23 September, the no. 1 X-15 was mated to B-52 no. 008 and ready to go again. Pete explained his standard procedure upon arriving at the rocket plane: "Normally, you were there in your pressure suit, and planned to get strapped in and squared away, and start working with the people on the checkout procedures, about an hour before scheduled taxi. There were a lot of checkouts accomplished in that hour. Frequently, there were delays of some kind. . . . One time I spent about four hours in [the cockpit] before taxi."

For flight 1-13-25, preparations went quickly. Capt. Jack Allavie and Maj. Fitz Fulton ran up the B-52's engines and pulled out of the mating area. Pete said this was the worst part of the entire flight. "The only rough part of the ride was when you first started to taxi. Those damn wheels had flat spots on them and it felt like you were taxiing over two-by-fours!" By 9:10 a.m. the mated pair were at the end of the Edwards runway and cleared for takeoff. Once in the air and on their way southwest to the launch point over the Palmdale area, the ride smoothed out. "Even in turbulent air," Pete said, "the old B-52 had springy enough wings that it wasn't a rough ride. . . . They had some real pros flying the B-52."

As a simple, local-area familiarization flight, Petersen was ready to go to only Mach 2 at 50,000 feet. At 9:52 a.m. the shackles released, and he was flying free for the first time, ready to light the eight chambers of the two LR-11 engines. He noted later that as he lit the engines, "the rocket chambers were started slower than optimum, but all eight chambers started normally." By the time everything was running, he had dropped to 36,500 feet. He chided himself, saying, "The slow start undoubtedly contributed to this low value."

Pete pulled the nose up to regain lost altitude. At 42,500 feet he started to push the nose back down to come level at the planned 50,000 feet, where he continued to accelerate toward maximum velocity. "No control peculiarities were noted during the pushover, and the acceleration from [Mach] 1.2 to 1.6 was made at 50,400 feet."

Heading northeast, he then made a left turn to change direction to the west. As he was in the turn, experiencing about 2.5 gs, he "noticed the chamber pressures and manifold pressures on the upper engine dropping to zero." In other words, the upper engine was dead. On the radio a surprised Petersen called, "Something happened to my engines."

The malfunction actually went further than this, but he was initially too busy working the problem and failed to notice as other things went haywire. His post-flight debriefing shows how centered he was on being precise in the descriptions of what occurred, to the point of speaking in the third person. "At this time, an overspeed light was noticed on the upper engine. While keeping the airplane in a left turn, two attempts were made to start the upper engine. . . . The pilot believed, erroneously, that the lower engine was still running. . . . Inability to hold altitude and airspeed variations from values expected for single engine operation forced the pilot to the inevitable conclusion that both engines were shut down."

Once the full situation was recognized, Pete set up for landing a little earlier than expected, at the north end of Rogers Dry Lake. He set the remaining fuel to jettison but wasn't sure when it was completed. He asked Joe Walker, in the chase plane, "Lost my jettison?" Joe responded, "It stopped dribbling out a while ago." Pete was lofting about 1,500 feet on approach. From his chase plane, Walker radioed his suggestion to lower the flaps and extend the landing gear early to increase drag and lower the x-15's altitude. Pete said of the final moments of flight, "The approach, flare, and landing felt comfortable. . . . The jolt resulting from nose wheel contact was less than anticipated." Pete summarized the flight, saying, "Nothing during the flight surprised the pilot with the exception of early engine shutdown."

Rocket engines can be touchy beasts, as Pete found out on his first flight. The LR-11s, and later the LR-99s, required a special group of people to prepare them for flight and to service and repair them as needed, to make sure these volatile pieces of high-powered machinery were safe enough in which to entrust a pilot's life. These men called their office the "Rocket Shop" and every one was extremely proud to have worked there.

Billy Furr started out at Edwards as a mechanic on the B-29 mothership for the x-1 series of rocket planes. Eventually, his bosses, Clyde Bailey and Mac Hamilton, asked Billy to enter the x-15 rocket shop. He stayed there a decade, from five weeks prior to rollout until the program came to an end in 1968. He talked about his job, saying, "We did all the servicing, propellants, and pre-checks out in the hangar. We supplied nitrogen, helium, whatever, when they were running their tests."

The nuts and bolts of what he did was fine, but it was the camaraderie of working with the people in the shop that he truly enjoyed. "The guys were great. I was known as the guy who would come in with a full lunch pail. I started eating when I came to work, and I was still eating out of my lunch pail when I left to go home. Sy Burnett was always on me about that. I did like to eat." This conversation was especially appropriate at that moment, since Billy and I were sharing a pizza as we talked at a small hangout in the mountain town of Squaw Valley, about sixteen miles east of Fresno, California.

Like all the people in the shop, Billy attended a special two-week course called the Rocket Engine and Pump Assembly School, which provided the basics of operation and maintenance. In the long run it was their affinity for the complicated maze of tubing and wires that gave the maintenance technicians the special touch required of the recalcitrant rockets. With the LR-99 a hefty punch of nearly 60,000 pounds was packed into a space less than six feet long and four feet in diameter. Dexterous fingers were a necessity.

For the LR-11 the fuels were liquid oxygen and ethyl alcohol, the latter now better know as ethanol. When the change was made to the LR-99, the alcohol was replaced with a much nastier fuel, anhydrous ammonia. During the development difficulties early in the program, the Rocketdyne division of North American Aviation made an attempt to usurp the LR-99 with a rocket of their own. Harrison Storms remembered it would have upped the ante even more if it had been chosen for production. "I ordered a special, hydrazine-fueled engine. . . . It was the highest output you could get from a rocket. It was probably about thirty percent more power. That fuel would have been a kicker!" The bottom line was, "you had to keep it away from people. As long as you know it's toxic, you can protect yourself against it. It's when you don't, that you have a problem." In the end, even with the improved performance, it is understandable why this fuel was not pursued.

When the rocket program was first initiated, the idea was for Reaction Motors to modify their XLR-30 engine, but soon it became apparent the modifications were a lot more than envisioned, so a new production number was called for. Within RMI the designation for the project was TR-139. On 21 February 1956 the official designation of XLR99-RM-1 was provided through the air force's Power Plant Laboratory. Around this same time, someone from Reaction Motors made the prescient comment that development of the engine would be a difficult task, which certainly proved to be the case.

The most complicated part of the design was to make the rocket throttle-able. No one had done this before, although many in the industry thought it would be a necessity for future spaceflight. On the X-15 engine, the lowest setting was to be 30 percent, equating to 17,100 pounds thrust, while maximum delivery was 57,000 pounds. The range was eventually limited at the low end to 50 percent, while the maximum increased to 60,000 pounds. To solve the throttle problem, feeding the fuel and oxidizer by changing turbopump speed made the LR-99 possible.

A white-handled throttle arm was installed at the front left-side cockpit pilot's panel to control the percentage. The arm was set all the way to the rear of the slot while at idle, then moved forward and pulled slightly to the right, like a car gearshift, to place the engine at minimum setting. This action was referred to as moving the throttle "out of detent." To increase thrust the pilot slid the arm forward, until reaching the end of the slot for 100 percent.

Rocketdyne lost on making a viable alternative to the LR-99 with their hydrazine-powered entry but was asked in March 1958 to instead construct an alternate thrust chamber and injector nozzle for the LR-99 itself, in hopes of speeding delivery. They were reluctant to get involved further, although they finally decided to take the contract—and the money. Finally, the more mature Reaction Motors design passed important milestones, negating any need of further input from Rocketdyne.

By 18 April 1959 acceptance tests on the Preliminary Flight Rating Test Engine were finally completed at the Reaction Motors stands outside Denville, New Jersey. Another year was required to bring the flight version to fruition and delivery to Edwards. That first engine was lost in the ground explosion on 8 June 1960, through no fault of the LR-99. It was not until

11. North American Aviation technicians install the first XLR99-RM-1 rocket engine into X-15 no. 3 in April 1960. This engine was destroyed in the test stand explosion on 8 June 1960. Courtesy of North American Aviation.

15 November that Crossfield was able to make a successful first flight with the big engine as part of the North American Aviation contract.

On 7 March 1961 the government made their first flight using the LR-99. Bob White was at the controls of the X-15 that day on mission 2-13. Forrest Petersen was close by in an F-4H Phantom II chase plane. It would be the only time this aircraft was used in conjunction with X-15 operations.

Although both the LR-11 and LR-99 were considered good engines, it was impossible to get a consensus about which was considered more dependable. Harrison Storms said, "The LR-11 was one of the most reliable little engines in the world. Just flip the toggle switches and go!" Then he qualified the statement when he joked, "Except when it blew up!" Wade Martin, who worked for the air force in quality control, believed the LR-99 was better technology. "That small engine, well, they didn't have the materials, the welding technology, and the inspection capability we had for the LR-99." Wade explained further, saying, "On the big engine we did a lot of X-

12. X-15 no. 1 is secured in the Propulsion System Test Stand for an engine run. The pilot is in the cockpit and several technicians check out the various connections to the aircraft. Note the condensation on the outer skin from the very cold liquid oxygen. Courtesy of the Dave Stoddard Collection.

rays. But on the LR-11 we hardly ever X-rayed the [combustion] chambers. We did that a few times, then we stopped doing it. . . . Reaction Motors looked at them, and [found the] heads were all warping, so we just stopped watching them! It scared us because we were clearing them for flight anyway." Whichever was considered the more reliable, throughout the program the rocket engines caused more flight delays from a mechanical standpoint than any other component. The outside factor of weather was the only one that beat that record.

Nature also affected the engines on the ground, or at least the personnel who worked on them. In the high desert at Edwards, it could get very cold in winter and extremely hot in summer. At the Propulsion System Test Stand, sometimes referred to as the test cell, unless inside the blockhouse for an engine run, the men were exposed to the elements. Wade said they had a solution, but it didn't always work the way it was supposed to. "We had a shelter to put over the PSTS. It was on railroad rails, and we pushed it over to keep some of the wind and cold out, but it was usually so much trouble, we said, 'The heck with it!' I don't recall it ever being so cold or snowy we couldn't work on the cell. We had heaters in there that could melt you! We also had some coolers that could freeze your butt off!"

Working for the rocket shop always presented a hazard with corrosive and poisonous fuels, and so many other ways to get hurt. The guys who worked on the engines each day took it all in stride. Operations engineer John McTigue talked about the differences between then and now. "Today, you couldn't meet the safety requirements, let alone anything else. We took risks. They don't allow you to take risks nowadays." McTigue went on to give an example: "On the x-15, we had turbines to charge pumps for the oxygen and ammonia. We used hydrogen peroxide as a catalyst, which gave us superheated steam and free oxygen. Well, that great big turbine was probably about that big in diameter [gestures a circle of approximately eighteen inches], and it had a seal all the way around. That seal leaked every once in a while, and you had to catch it when it was just starting to leak, or you would lose a lot of your pressure and not be able to make a flight."

Then the tricky part began as the leak had to be tracked down. "You can't see superheated steam. It's invisible—and it's hot! Our inspectors went in there with a mirror and ran the mirror around [to see if] the steam would condense on the glass. . . . On the side of the airplane there was a hole about [twelve inches in diameter], and . . . I'd take [the inspector's] mirror and stick my head inside that hole while we were at igniter idle, which means we were running a few thousand pounds of pressure out the back end. They wouldn't allow me to do that nowadays."

The time of the x-15 and earlier rocket planes was unique. "Today, we could do it differently by putting sensors around it, go in with cameras, and so forth. Back then, we didn't have that option. You stuck your head in and you looked."

Meryl DeGeer, also an operations engineer with NASA, continued with another example: "One of the guys found he could stop a leak by just spraying some water on it, because it was all so cold! Everything was cryogenic. If you could see . . . gas coming out of a seam, you'd spray a little water and it'd freeze up solid and stop the leak!"

The men there didn't like having NASA Headquarters breathing down their necks. Meryl continued, "We had Paul Bikle, and . . . he decided what Headquarters found out about. A lot of it was settled right there, [and] it was done quietly. A great example is, we had the wooden lifting body [the M2-F1] ready to fly before Headquarters even knew we were working on it." McTigue added, "Back then, people took responsibility for what they were doing. You raised your hand and you went and did it."

There were actually two different types of engine runs conducted at the PSTS. For the first type of run, an x-15 was prepared for flight at the NASA facility, including installation of the LR-99. Once completed, it was brought to the test cell and locked down. Then the LR-99 was run through a short ignition and firing sequence to verify all systems were operational. For the second method, they hooked up an engine alone to what was called the PSTA, or Propulsion System Test Article. This was basically a skeletal version of the x-15, consisting of the tanks, plumbing, and wiring, all connected to the engine. There was no cockpit and thus no pilot; all control was handled inside the PSTS blockhouse.

One of the men who ran the engine under these circumstances was Jim Townsend. He considered his job very special to have such power at his command. Jim explained that when the engine was connected to the PSTA, it came with its own dedicated control box, which he had the pleasure to operate. "What happens is, the engines are burned into the control box itself. They were matched to the engines, so, wherever the engine went, its throttle box went." He expressed pride when he said, "I'm going to make a statement that's going to piss a lot of folks off: I have more time than all other throttle pushers [pilots] combined because of what I did."

Jim was very sick when we were able to speak. His son, Daryl, heard about my research on the x-15, explaining his father didn't have long to live, but maybe he would be okay to sit for a while. Jim was surrounded that afternoon by his son and another rocket shop friend from his time on the x-15, Dean Bryan. Daryl said later this was the brightest and most animated he had seen his father in a long time. The high point for Jim was when he showed off his most prized memento: a throttle box, presented to him when he retired from NASA. Jim said, "They found one they didn't have a mate for, so they gave it to me. . . . I bawled like a baby when that happened."

Daryl also shared a special memory, which had been possible because of his father. "He took me out on the last engine run on the test stand. It was 1968. . . . I was in the blockhouse. I was a young, chubby kid then, maybe twelve. The Mach diamonds were what I was most impressed with. . . . The noise was incredible. The whole earth was rumbling. It was pretty impressive."

That sound—or in some cases an absolute lack of it—was one of the more impressive aspects of being around an LR-99 rocket. NASA instrumentation technician Glynn Smith said, "That X-15 engine, when it fired up, it was just beautiful. It was awesome. When we'd shut down, it would go like it was sucking all the flame right back up in it, like running a movie backward. This old LR-99 had a sound all its own. Fire that thing up, throttle up to 100 percent. . . . Holy cow, it was loud!" As he talked of this, Glynn's recollections were punctuated by the sound of the howling wind outside his house, the doors and windows rattling with the overpressure. It was an eerie, yet appropriate, experience.

Wade Martin added to the mystique. "The engine overpowered anything I have ever heard. It's just a phenomenal thing that anybody can control such an explosion. . . . It runs like heck!"

Wade related that "the LR-99 engine had a really harsh, high pitch." Then he shared a bizarre side of the physics of rocket engines when he talked about what was called the "cone of silence." He said, "We would wander around the lakebed sometimes looking for artifacts, because the test cell was located right on the lakebed. If you happened to be back there and the engine started, you wouldn't know it [because] walking directly behind it you can't hear it. That was my first experience with the cone of silence. I was just flabbergasted as to why I couldn't hear that." A similar phenomenon happens sometimes in places like a concert hall where the sound waves can literally cancel each other out, creating a dead zone. Wade said, "If somebody new [came to the shop], we'd take them back there when it was running and — nothing. . . . I bet there must have been an area of about fifty feet where you could walk back and forth and not hear a thing. Every time I went back there, I was amazed."

All the men of the rocket shop were special, but none more so than Jim Townsend. Within just a few weeks of our meeting, Daryl phoned me to let me know Jim had passed away.

In the first two years of the program, the LR-11 was used for thirty flights. Twenty-one of these were in the no. 1 aircraft and nine in no. 2. X-15 no. 3 never used the small engine.

Modification of no. 1 to accept the LR-99 rocket engine took six months at the North American facility. Bob White had flown last with the LR-11 in this aircraft before it was trucked south for LR-99 installation on 8 February 1961. Once there, the two LR-11s were disconnected and pulled out. In the cockpit, the eight toggle switches controlling the individual chambers of each engine went away, to be replaced by the throttle control slot and handle for the big engine. The LR-99 was mounted, and all electrical connections were repeatedly checked and verified.

Once buttoned up and ready for transport, the flatbed truck pulled up outside the NAA plant, the giant "Home of X-15" sign sticking up from the roof. The X-15 was loaded aboard, then headed north through Los Angeles toward Edwards. The trip passed west of the city on the coastal plain of the Los Angeles Basin, then up over the Sepulveda Pass into the San Fernando Valley. Crossing through the expansive valley, the truck and its precious cargo climbed into the San Gabriel Mountains on a winding traverse that took it to the southern end of the Mojave Desert, through the towns of Palmdale and Lancaster, then across the flat to Edwards and the NASA Flight Research Center. The slow trek was made during darkness so as to not interfere with the notorious daytime traffic encountered throughout Los Angeles and its environs.

The thirty-ninth flight of the X-15 was the first in the no. 1 aircraft to use the LR-99. It fell to Petersen to run this checkout on his third mission in the program, flight 1-22. This was also his first outside the local Edwards area—in this instance, Silver Dry Lake, about a hundred miles east.

For any situation that required an emergency landing, the pilot had to know what to do immediately. There was no time for improvisation. That was where the F-104 Starfighter came in as the perfect vehicle to simulate the high-speed, high-drag X-15 landing characteristics. Pete said, "You practiced landing in an F-104. It had exactly the same lift, drag, and everything as the X-15. The only difference was the visibility was considerably different out the F-104's bubble canopy. . . . You could practice in an F-104 until you felt at home, [and] as a result, we never had any difficulty with landing." Smiling at the recollection, Forrest said, "Even a navy guy could get an airplane from the air force to fly anytime he wanted to—if he was scheduled for the X-15."

On this first flight with the LR-99 for Petersen, everything went fine, but he said it was nice to have the confidence F-104 simulated landings provided. "I think that's one of the reasons the program was so damn successful. You had a simulator [where] you could really perfect the techniques. . . .

Then you practiced landing under emergency or normal flight termination circumstances with the F-104 to your heart's content. It wasn't uncommon to make seventy-five or a hundred landings with an F-104 in the week or so before you'd go out to fly the X-15."

At 10:27 a.m. on 10 August 1961, mission 1-22 got underway, or at least that was the plan. Pete discussed the problem after the flight with B-52 pilot Jack Allavie: "[I] pushed the drop button, and nothing happened. I leaned on it some more and nothing happened. . . . I had just about given up and was reaching for the mic button to tell you to drop me off manually and 'klunk.'" Allavie said, "I could tell that something was wrong. . . . I was waiting for old Pete to say, 'Launch me' when . . . away he went."

After the few moments of confusion, the X-15 dropped away, but instead of grabbing the LR-99 throttle, Pete latched onto the speed brake handle just behind it. He joked, "It surprised me when it came off [the shackles] because I had given up. So I immediately grabbed the speed brakes and tried to light the engine!" Once he had hold of the correct control, the flight started to proceed normally. "I finally got a hold on the throttle, and it lit real nicely. It seems like less delay than there is down on the test stand for a ground run. . . . It came up just as steady as a rock."

With the LR-99 operating properly, the rocket plane leapt forward. In less than two minutes, Pete went from the speed of sound to more than Mach 4, at 2,735 miles an hour. This was the first time past this milestone for both Pete and aircraft no. 1. It was more than twice his previous high speed in the X-15 and showed the potential power of the LR-99.

Following burnout, an oscillation began. Pete figured out the problem quickly—somehow, he was feeding the motion into the airplane. In the post-flight comments, again using the third person, Pete said, "During the deceleration from [Mach] 4 to 3, it became apparent that the oscillation was being fed by airplane motions through the pilot control loop. He removed his hand from the stick, and the next two or three cycles were obviously damped. When he took hold of the stick again the oscillation built back up." Pete complained, "The simulators are quite steady and do not fly at all like the X-15. . . . The major discrepancy uncovered by this flight is the fact that X-15 oscillations in flight are not simulated by the X-15 simulator."

Also in the descent, Petersen lost pressure in the cockpit, and his suit inflated to compensate. He joked as he radioed, "Cabin has blown up!" NASA

I deadpanned back, "Rog, understand cabin blown up."

Petersen entered the landing pattern and prepared for touchdown. He said, "I let loose the landing gear handle after I dropped the gear and it flipped back and busted the tank pressure gauge. Also, I was a little bit late getting rid of the ventral. I imagine there is a groove in the runway out there."

The skids slid onto the dirt, and the nose gear slammed down. After a mile the x-15 came to rest with a total flight time of nine minutes, twenty-four seconds. Sixteen minutes later, as Pete was getting his land legs back, Jack Allavie and Harry Archer brought B-52 no. 003 back to home base.

Petersen's fourth flight, 2-19, went out of Hidden Hills Dry Lake, about fifty miles north from his last launch at Silver. The guys on the crew made it a special occasion for Pete before he entered the x-15 at Edwards since he had just been promoted to commander in the U.S. Navy. The NASA life support supervisor, Roger Barniki, talked of his plan: "I went to the illustrator and had a U.S. Navy insignia made up that we put on the side of the x-15. We then found a guy who was a bosun's mate from the navy, one of the mechanics at NASA. He brought his bosun whistle in on the day of the flight."

As Pete was preparing in the suiting van, a three-foot-wide red carpet was laid on the tarmac leading to the portable stairs that he had to climb to get into the cockpit. Roger picked up as Pete exited the van: "We all lined up, piped him aboard, and went through that whole ritual, bringing him on board the x-15." Petersen saw the carpet in front of him, showed his surprise and humbleness at the honor. All around him were smiling and clapping crew members. Pete was helped up the stairs in his bulky pressure suit as the bosun's whistle blew. Inside the cockpit, he appeared especially proud. The white shield was on the right side of the x-15's fuselage, just forward of the canopy. The naval aviator wings were in the center with "U.S. Navy" below, and his name above, reflecting his new rank.

To receive the promotion, Petersen was technically supposed to have sea duty in the navy, which was hard to come by while assigned to a base in the desert. A way was found around the requirement by assigning him to a naval squadron just north of Edwards at the China Lake Naval Ordnance Test Station. Although China Lake was just as land-locked as Rogers Lake, the navy still considered it sea duty. Pete said, "All they did was assign me to that squadron up at China Lake, and I stayed at the same office [at Edwards] where I'd been. After a year, I had accumulated the sea duty, and they promoted me to commander."

Years later, when I spoke with Pete, he was still moved by the honor displayed to him that day. He smiled, and I saw him hold back a moment, but he declined to elaborate on his feelings. Pete gathered himself, then went back to his military style, giving me his impressions as he prepared for the mission: "The launch is a crescendo of checkout procedures and situations [before] you were ready to go. At that point, you could launch yourself, or the b-52 pilot could launch you, or the guy in the back end monitoring the top-off of the fuel systems had a console from which he could actuate the release mechanism. We agreed how it was going to be done before a particular flight. I think I launched myself on all but my first flight."

A primary goal of the x-15 program was to achieve high temperatures through aerodynamic heating on the Inconel-X skin. Pete's initial mission on aircraft no. 2 had heat as its goal. On 28 September he launched flight 2-19 and pushed the aircraft to Mach 5.30, attaining temperatures above 1,000 degrees Fahrenheit for the first time. Petersen said, "They packed the maximum amount of data acquisition, and, therefore, pilot activities, into a flight. . . . You didn't have time to look out the window and see what the scenery looked like. You were busy every second. . . . That's what the taxpayers paid for."

Petersen's peak altitude was 101,800 feet. This, combined with his high speed, marked the only time a pilot recorded his highest Mach number and altitude on the same flight.

Pete was much happier with the correlation between the simulator and the way the aircraft handled this time out. His push had definitely paid off: "On this flight, the airplane responded very closely to the motions predicted by the fixed-base simulator."

Even though the simulator was providing a better pre-flight environment, Pete admitted, "There were some characteristics and sensations that were a little surprising the first time. That big engine hit you like an eighty-second catapult shot [off an aircraft carrier]." It was the job of the pilot to understand the forces at work on him and the aircraft and to compensate appropriately. "A lot of people would strain against those forces. That would be about one and a half gs initially, but pretty close to four gs at burnout. . . . I think we learned very quickly just to relax against the support systems

13. After flight 2-19 on 28 September 1961, Forrest Petersen removes his pressure suit helmet and gloves. The portable air conditioner sits by his left foot. Courtesy of NASA Headquarters.

. . . and could make the control inputs a lot more accurately that way than we could fighting against this force."

Forrest Petersen's last flight didn't go nearly as well as his first. It was January 1962, three and a half years after he was assigned to Edwards to begin his work on the x-15. Pete decided he had been away from his naval career too long and was leaving soon, but there was still flight 1-25 to accomplish.

"The profile dictated how far [uprange] you would proceed with the flight." Pete said. "You wanted to end up at Edwards, so the profile would dictate where you were going to drop the airplane. We always required, that if the engine didn't run at all, or it ran for one second . . . you had a safe place you could put the aircraft." The series of lakebeds along the High Range corridor provided that safety. In the case of Pete's final flight, the launch was set to occur near Mud Dry Lake, the seventh time this area was used for a long-distance x-15 research mission.

Flight planning, simulations, emergency contingencies — the pilot was intimately involved in every aspect. Pete explained the difficulties in getting ready for a flight. "Sometimes, the way you would get a particular piece of data was to get the airplane going through some kind of oscillation and let it record its motions at that particular condition. . . . But how it was accomplished, what control inputs you used to get that condition, was something that the pilots always won on if there was an argument."

Prior to getting the pilot involved, the engineers had first crack. Pete related, "Engineers got on the simulator and . . . became very proficient at it. That doesn't necessarily mean we could get in the airplane and accomplish the same thing. We sometimes had differences of agreement as to what we ought to do under different circumstances. . . . I don't want to imply they were too serious in any way. They were friendly differences. Reasonable men can disagree on almost anything, [but] pilots always won."

Scheduled for Mach 5.7 at 117,000 feet, this flight was to expand on Petersen's experience. Again flying with Captain Allavie in the left seat of B-52 no. 003 and Major Bement in the right, the mated x-15 and mothership headed north at 11:30 a.m. Approximately fifty minutes later, they passed east of Mud Lake, and Allavie started a large-radius 180-degree turn to come down the west side of the dry lake. By 12:28 p.m. they were in position. Allavie called, "Ten seconds." Pete wasn't quite ready, replying, "Okay, I'll need that few seconds right here. Hold her up." He finished his last checks, then said, "Everything looks good. Three . . . two . . . one . . . launch!" Pete later commented, "Everything up to launch was as perfect as I have ever seen it."

The LR-99 appeared to catch, and Maj. Walter F. Daniel, in Chase 1 confirmed, "Got a light." Within two seconds, Pete said, "Malfunction shutdown."

Pete remembered, "On my last flight, the engine didn't run. When they dropped me at 45,000 feet, we're doing about Mach 0.82, and when you drop off of that B-52, you're going to hit the ground someplace within three minutes if you don't get an engine running. So you've got three minutes max." The LR-99 normally took about three seconds to be up and running. It was a much quicker sequence than on the LR-11s, where the pilot had to individually throw eight toggle switches to get full thrust from all chambers. For Pete, on flight 1-25, "I dropped off of there, the engine lit and ran about two seconds, then turned itself off." He tried for a restart, but it didn't fire. Pete confirmed the news: "Malfunction shutdown. Going to jettison." With just a few seconds of thrust, Petersen never reached Mach 1.

"The jettison procedure," Pete explained, "could take about two minutes, and when you had only three minutes until you hit the desert, there was only enough time to try one restart. If that didn't work, then the fuel had to be jettisoned, and hope it all got out before you had to commit for landing. Since the rule was to eject if you couldn't dump fuel, you didn't screw around with restarts."

The x-15 was headed south at launch. Mud was east of this position, to Pete's left. Even though officially classified as an emergency, the process of getting down to the lakebed was smooth, and he actually found there was excess altitude, so he flew farther south beyond Mud before making a turn to head back north to land across the widest section of the lakebed.

The runway was marked with eight-foot-wide black stripes, placed three hundred feet apart. Small perpendicular marks at one-mile intervals showed him each of the four miles he had to work with as he slowed and descended. Pete's aim point was at the one-mile mark from the south end. After he flared, he figured his touchdown point was smack in the middle, at the two-mile spot. Landing was right where Pete wanted it to be. Allavie informed everyone on the radio network: "All stations, B-52 over Mud. x-15 appears down, rolling very nicely [on the] center lane on the north/south runway."

Mud Dry Lake was about ten miles south of the Tonopah Airport, and many people in the area watched for x-15 launches, possibly listening in on the radio frequencies. The B-52 and its formation of chase planes were hard to miss in the Nevada sky above the town. Someone at the airport decided to check out the landed rocket plane but was quickly seen by sharp-eyed Allavie. He sent a warning: "Chase 1 and 2, there's a light plane circling Mud—you might watch it." Joe Walker, in Chase 2, called, "Chase that guy out of there." It didn't take long for the civilian pilot to be intimidated by Daniel's F-100 and Walker's F-104, and he disappeared back to the airport.

Pete spoke about post-landing on the lakebed: "There were some shut-down procedures we went through. We had to shut down the auxiliary power units, turn things off in general, but that was a relatively simple check-off list. You always had help. I guess, you could get out of the airplane, unstrap yourself, and open the canopy, but it would be pretty difficult. We always had people at any anticipated landing point to help you get out of the airplane."

A C-130 Hercules orbited the primary launch site, so they were nearby for just such an emergency. Pete went on, saying, "You could raise [the canopy] from the inside, but it was pretty heavy. That flightsuit was built in a sitting position, and you had to be careful to get yourself unstrapped and make sure you didn't hit a bunch of switches that might turn something on instead of off. . . . Your helmet was fastened to oxygen hoses, and the suit and ventilation systems were attached to nitrogen hoses at various places. Getting all those things disconnected, so you could step out of there and crawl over the side, you could do it yourself, but it was nice to have help."

Considering the short amount of time available from the moment of launch until the aircraft landed—if the engine did not light—Petersen could make only one restart attempt before having to initiate jettison. He admitted to getting very annoyed that someone obviously did not understand the constraints he was under in these circumstances. "After landing on this flight, one of the engineers came up and asked me, 'How come you didn't try a restart?' I should've hit him!"

The first emergency landing at a lakebed away from the Edwards area had been successfully accomplished. This was only the second emergency that occurred, after Scott Crossfield had an engine fire on the third powered flight in November 1959, but that landed in the local area at nearby Rosamond Dry Lake. There were ten more such events still to happen as the program progressed.

As exciting and productive as the X-15 was, in the end Petersen knew he had to leave, saying, "The guys in charge of the Atlantic and Pacific fleets were not terribly interested in contributing to [the X-15], or, if you did that long enough, they would have serious doubts about your usefulness. . . . So, it was my decision [to return to the navy]. I wouldn't have even been able to stay that long with the x-15 and still get back on active duty, had they not gone to bat for me a number of times, making sure my name was considered for a squadron while I was on detached duty with NASA."

His decision was made in late 1961, prior to the emergency landing in January 1962. Pete's orders said he was to leave by the end of the month, and there was an attempt to get one additional flight before that deadline, but events conspired against him. Evaluation of data from his emergency revealed unusual landing gear loads. Engineers wanted to understand what was happening before again clearing no. 1 for flight. Only one mission, by Neil Armstrong, got off the ground between Petersen's landing at Mud and the first week in April. On 30 January 1962 Pete left Edwards for Naval Fighter Squadron 124. After several months of training, he assumed command in July.

There was definite melancholy about leaving behind the people Pete enjoyed working with, but there was little choice. "Four years is a long time to take out of a young naval officer's career. I feel I probably could have stayed there as a research pilot with NASA if I had desired, but that would have required getting out of the navy. . . . It was not an easy choice to make."

Bill Dana, who flew the x-15 nearly four years after Pete, said, "Petersen was a very knowledgeable individual, a lot of smarts and very talented. He had his own personal goals, and I think when he saw the next possible step up for himself, he took it."

I asked Pete why the navy decided to not put another pilot into the program. He surprised me with his answer: "They did. They sent a guy named Hoover out there. . . . He went out and came back within a month. No indications whatsoever that it was anything because of his qualifications or so forth. He came from Pax River, just like I did."

At that point, Petersen became reluctant to go on record about exactly what transpired with Hoover when he reported to Edwards. Pete eventually decided to provide background by admitting his replacement was sent with no prior warning from the navy. Pete said Joe Walker called and asked, "Who the hell is this guy?" The air force and NASA got the impression that it was the navy's intention to rotate pilots "out west to rack up a few records in a rocket plane," then send them home to put in another guy to do the same thing, what Pete called "the revolving door effect." Apparently, this is similar to what they had done on the D-558 series of rocket planes in the 1950s. With that program, the navy was in charge and could do whatever they wanted. With so little navy funding in the X-15, the air force and NASA did not sit still to allow it to happen on their watch.

Bob Hoey, a U.S. Air Force flight planner on the X-15, explained this wasn't something to just jump in and fly. "It was the kind of airplane where you spent a lot of time training the guy for that very first flight, and you didn't want to be doing that all the time. If we're going to just do checkout flights, that's not a very efficient way to use the airplane."

Two X-15 pilots, Bob Rushworth and Milt Thompson, spoke about the situation. Bob said, "They sent a representative to monitor the program, but he was not invited to participate in the flying. . . . If you couldn't fund it, you couldn't buy a seat. Just that simple." Milt expanded on that, saying, "Yeah, there was another guy that showed up out here and said, 'Okay, I'm Commander Pete's replacement.'" Milt laughed and said, "Everybody just kind of ignored him. . . . He showed up here pretty much unannounced, and he didn't have a hell of a lot of backing within the navy, either. We took the position that, if the navy pressured us, we might do something, but if not, he wasn't about to fly. He went away." Milt was adamant when he added, "They really got way more, piloting-wise, than they deserved, as far as the money they spent."

Paul Bikle ran the entire program and had the ultimate responsibility at Edwards. He said of navy participation in the X-15: "The air force attitude was, 'Why are those guys in there, anyway? They didn't pay their share.'" Paul then echoed Rushworth: "If you didn't buy the ticket, you shouldn't go for the ride." However, when it came to Petersen personally, Bikle was very positive: "Of course, Pete was already in the thing when I moved over to NASA. . . . He did a lot of good work when he was in there. He got very little credit for it because of our not being too anxious to antagonize the air force. We supported Pete as much as we could."

In the end, it was a matter of protocol. Pete said, "In my case, they knew I was coming. I had been out there and made a courtesy call on all of them and had talked to the boss." When Hoover arrived, no one had "greased the skids," according to Petersen. At that moment, the U.S. Navy was out of the business of manned hypersonic flight.

It wasn't long after he got back into the regular navy when Pete starting moving up the ladder. "I had a wonderful time in the navy once I came back. I got a squadron but wasn't there very long [before] they pulled me out to go to nuclear power school. Then Adm. [Hyman G.] Rickover sent me to the uss *Enterprise* [cvn-65]." His two-year assignment as executive officer aboard America's first nuclear-powered aircraft carrier included deployment to Southeast Asia, with planes from his ship taking part in raids against the Viet Cong.

By November 1967 Petersen was given his first sea command with the amphibious assault ship, the uss *Bexar* (apa-237). During his year at this post, the *Bexar* supported incursions by marine battalions into the Mekong River Delta region of Vietnam.

In January 1969 the *Enterprise* suffered a terrible accident when a rocket exploded aboard an f-4 fighter aircraft on the flight deck, killing twenty-seven sailors during the ensuing fire. At the time, Captain Petersen had been reassigned to the Atomic Energy Commission school for advanced training. Soon after he finished in June, he received command of the *Enterprise*, following repairs from the incident. During his time in charge, he oversaw a refit of the ship with a new nuclear reactor core before he and the aircraft carrier returned to combat duty in the seas around Vietnam.

Petersen continued his upward momentum. As a rear admiral he oversaw Carrier Division 2, as commander Carrier Strike Forces of the Sixth Fleet. He then received the three-star rating as vice admiral, the highest military rank achieved by any pilot associated with the x-15. His effective date of rank was 1 May 1976, which led to his final assignment on 29 October as commander Naval Air Systems Command.

As he moved higher in the structure of the military, Petersen also moved further away from flying. He said, "I'm one of those guys who didn't have any trouble transitioning to a desk when the time came. . . . Even the last several years I was on active duty, I didn't do much flying. What that would do was take flying hours from young guys who needed it a hell of a lot more than myself. We are already not providing enough time to most of our young pilots to keep them as ready as I think we should, so it would have been a selfish thing for me to do." However, Pete didn't turn down flights when they were available. "Oh, I flew an f-18 [Hornet], and a [av-8b] Harrier, and all those things the last year I was on active duty, but those were with somebody else. . . . Some of those pilots have to fly in order to stay sane or something. But that didn't bother me."

After retiring in 1980 Pete started a one-man consulting firm. "I try to keep involved in one major aircraft company, a helicopter company, an engine company, and a missile company. Those vary from time to time, but I've generally been able to do that, and I plan to do it for a few more years."

He had two sons, Nels Christian and Forrest Dean, and a daughter, Lynn Elizabeth, from his first marriage. Pete was proud that both sons followed him into the navy. At the time we spoke, he said, "I have a son that flies in the back seat of F-14s [Nels], and I've got another one that's over at the Naval Academy now [Forrest, academy class of 1986]." Pete's wife, June, died in 1977. He remarried thirteen months later to Jean Baldwin.

Forrest Petersen had one of the most successful careers of any pilot in the x-15 program. He made career decisions that sent him on the proper path toward a powerful position within the navy. By the time he retired, he had spent three years at the U.S. Naval Academy and thirty-six more as a commissioned officer, for a total of thirty-nine years wearing the uniform of his country. Out of that, he spent less than three and a half years at Edwards working on the research program and flying the x-15. Pete said those were some of the best times in his life. "One of the things I am most grateful for is the fact I made so many associations when I was out there during those years, and a great many of them have stood the test of time." He went on to say, "I just thought it was a wonderful opportunity, a wonderful program. I was extremely fortunate to be able to take part in it."

Pete did have one regret that came from his years of being a pilot, and especially from the x-15, saying, "We ought to have more national emphasis on aeronautical things. Research aircraft fall into that category. . . . I think the things that the NACA used to do on a national level — which were so much a vital part of our preeminence in aircraft design — are not being done today the way they should be done."

Being part of the x-15 early in his career brought Petersen into the rarified air of being seen as one of America's spacemen. At rollout in 1958 the country expected the x-15 to be first into space. No one had any idea ballistic missiles and tiny capsules would forego wings and pilot control. Pete had no interest in pursuing the path opened by NASA with the Mercury program. "I had my hands full, and I enjoyed what I was on." His path might have taken a sharp turn away from the navy if just one small physical trait of his was different. Forrest Petersen explained, "I never really considered getting out of the x-15 program for the astronaut program. One of the reasons was that I was about an inch and a half taller than their specifications were at that time." At slightly more than six feet, Pete simply could not fit into a capsule.

5. Changing Course

I believe that the top of my helmet hit the lakebed
first, although at the time it did not appear to be
a severe blow.

John B. "Jack" McKay

Eons of hot, dry sun baked the ancient lake bottom to near-asphalt hardness, the color bleached to a dirty white. Close up, the surface was crisscrossed with fissures, giving it the look of dragon scales. Seasonal rain repeatedly scoured the top layer smooth. Roughly circular, the dry lake was more than five miles across, a perfect oasis of level ground among the scattered hills surrounding the area. Southeast of the silver boomtown from the early 1900s of Tonopah, Nevada, the lake was given the inauspicious name of Mud.

No vegetation marred the pristine surface. Maybe an occasional lizard, snake, jack rabbit, or coyote might dart across the outer edges, but there was little reason for any living thing to trek across the vastness of the silent interior. All these inhospitable qualities were perfect for one special purpose. On the morning of 9 November 1962 that purpose was fulfilled, as the quiet was shattered by a sonic boom. Overhead, several specks came into view, one of them black. This one quickly grew in size until it slammed down onto the lakebed.

The x-15, with pilot Jack McKay at the controls, had not achieved full thrust after launch, so he needed a place to land. Mud Dry Lake was directly to the east of his flight track. He turned and made for the ground. As he neared touchdown, McKay set the flaps, but they did not respond. His landing was going to be very fast. The rear skids slapped down, followed quickly by the nose gear. Bouncing slightly, the left rear skid failed under the strain. A chain reaction began as the now-crippled aircraft went out of control. With the skid gone, and thus no support on that side, the left wing and rear stabilizer scraped along the ground, soon ripping the stabilizer from its mount, taking the landing skid with it. The wing continued to slide, turning the aircraft, before the right wing suddenly caught, flipping x-15 no. 2 onto its back.

As the accident unfolded for McKay, time slowed as adrenaline kicked in, heightening his test pilot reflexes. Jack sensed the beginning of the flip, instinctively reaching for the canopy jettison switch. He knew if the aircraft went over, he would be trapped inside if it remained in place. Forward momentum stopped, the aircraft coming to rest on its back as McKay had feared. The canopy landed upright a hundred feet farther on. Besides being upside down, the x-15's nose was facing almost back along its path. With the canopy gone, the weight of the aircraft transferred to Jack's helmet, compressing his neck. Quickly, the twin-bladed Piasecki H-21 Workhorse helicopter, on standby nearby, moved in and used its downwash to keep the area clear of fumes from the toxic fuels aboard the stricken rocket plane.

The Norfolk Naval Shipyard in Portsmouth, Virginia, was once known as the Gosport Navy Yard. In 1861 during the first months of the American Civil War, the burned-out hulk of the uss *Merrimack* was salvaged by Confederate shipbuilders to be turned into the world's first ironclad warship, the css *Virginia*. North of Portsmouth, the Elizabeth River leads into the confluence with the Nansemond and James Rivers. The waterway is known as Hampton Roads, before it empties into the Chesapeake Bay, on its way to the Atlantic Ocean. On 9 March 1862 the Union ironclad uss *Monitor* met the css *Virginia* in a three-hour battle that ended in a standoff. The engagement, however, changed the course of naval history. From that point forward, shipbuilding turned away from wooden vessels to the more formidable material of steel. Sixty years after the historic battle, on 8 December 1922, Jack McKay joined the world in Portsmouth, with his twin brother, Jim.

The two boys shared many adventures in their early years. Their father, Milton, was a U.S. Navy photographer in St. Thomas in 1928, where Jack and Jim first saw Charles Lindbergh and caught the aviation bug. Three years later the family was living in Haiti, and Jack and Jim were hard at work building model airplanes. It was during this time when tragedy struck.

Jack's daughter Sheri recounted how it began, saying, "They were into model airplanes and they got bigger and bigger . . . making one large enough they could sit in." She explained that in the meantime, "Their father, my grandfather, was a hunter. He loved to hunt wild boar, geese, wild turkeys, just all this stuff. He had been off with some of his friends hunting in Haiti, and they found [an] ammunition shell about this big [indicating eight inches]. . . . My grandfather brought it home and put it on the mantle over the fireplace." Sheri's grandmother told her husband to get rid of it, but he procrastinated. "My grandmother gave it to a servant and said, 'Okay, you get rid of this.' So the servant took it out the back door of the kitchen and went through the backyard into some tall grass."

This accomplished nothing except to put it back outside to be found again. Sheri continued, "A short time later, Jack and Jim were hammering away, building their plane, [but] there was only one hammer . . . so Jim went looking for a rock or something to pound nails with. He found this shell in the grass and brought it over and was pounding with it, and the thing just blew up!" The accident cost her uncle his right arm below the elbow, and his leg was amputated just above the knee. He also lost a finger on his left hand. Their sister was close by. She'd broken her arm previously, and she was thrown back against a fence. The explosive shock rebroke her arm.

Sheri talked of the irony of the situation, saying, "Here they are making an airplane, just doing what they loved, and my uncle was so horribly injured. . . . My uncle was courageous to go on. [After he grew up] he had his own private plane and became an aeronautical engineer at Edwards. Of course he would have wanted to be a test pilot or an astronaut, but the door was slammed shut that day when he was nine years old."

In the carnage her father, Jack, was amazingly unscathed. "Other people might have stopped playing with airplanes . . . or their parents would have said, 'No more of this.' But that didn't happen. They both went on into careers in aviation."

The day prior to Jack's nineteenth birthday, the Japanese attacked Pearl Harbor and declared war on the United States. On Jack's birthday, President Franklin D. Roosevelt addressed a joint session of Congress, saying in part,

14. Jack McKay in the cockpit of his F6F Hellcat on the deck of an aircraft carrier in the Pacific during World War II. Courtesy of the Sheri McKay-Lowe Collection.

"Hostilities exist. There is no blinking at the fact that our people, our territory, and our interests are in grave danger." At 4:10 p.m. Eastern Time, the formal declaration was issued and the United States entered World War II. Three days later the writ was extended to include Germany and Italy. The world was at war, and Jack McKay, like most other young men his age, was eager to get involved.

After completing his U.S. Navy pilot training, he was assigned duty aboard an aircraft carrier (most likely the USS *Yorktown*, CV-10) in the Pacific Theater, flying primarily the Chance Vought F4U Corsair. His flying skills earned him an Air Medal, signifying, "Meritorious achievement while participating in aerial flight." He also received an oak leaf cluster, the equivalent to the award of a second Air Medal. Jack's son, John, said of his father's war experience: "As I remember, Dad . . . flew in support of bombing raids in China. . . . The only time he ever mentioned his navy time was, he told me they were flying back from China, and there was radio silence and a darkened ship. He mentioned it was a bitch to get back. He didn't talk about the war much, but I know he loved the navy."

When the war ended McKay took advantage of the GI Bill to pay for his college education. He chose Virginia Polytechnic Institute's aeronautical engineering program, graduating in 1950 from the seventy-eight-year-old school in Blacksburg. The school motto was "That I may serve," appropriate considering his naval, then civilian, service for his country.

Jack met Shirley Mae Londeree from nearby Lynchburg prior to the war. From all indications he was smitten from the beginning, but there was a four-year difference in their ages. With a worldwide conflict requiring his service as a pilot, Jack didn't have any intentions of a serious relationship until he was home from military service. Shirley had grown up and was working for the Atlantic Coastline Railroad. They married on 2 August 1947. Jack was twenty-five and Shirley twenty-one.

Soon after graduation McKay landed a job with NACA at the High Speed Flight Research Station at Edwards AFB, reporting for work on 8 February 1951. The former Muroc Field had been renamed the previous year, in honor of Capt. Glen W. Edwards, who died on 5 June 1948, in the crash of the massive YB-49 jet-powered flying wing prototype.

Jack's arrival coincided with that of the Bell X-5, which tested variable sweep wings. Already in its test phase was the Northrop X-4 Bantam. McKay eventually flew both aircraft. But this had to wait, as his initial job with NACA was as an engineer, not as a pilot. He set his sights on achieving his flight status with the agency, which occurred on 11 July 1952. He joined Scott Crossfield and Joe Walker in test piloting duties. Getting into the airplanes was sometimes an interesting task, as McKay was a stocky man with short arms and a bulky build. Some described him as a bear.

The move from Virginia to the Mojave Desert was tough on Shirley. The couple's daughter, Sheri, said, "My poor mother. I remember her . . . sweeping out all the dirt, and there was just one scrawny tree in the front yard. Here she'd been raised with all this green." Shirley took a job in the Contract and Purchasing office at Edwards. Being on the base meant she saw and heard the all-too-frequent accidents. Sheri said of her mother, "She told me in the last week of her life that, in the beginning, she had thought about leaving Dad and going back to Virginia, that it was all too much for her. But then, she settled in, more children arrived, and she went on to have an interesting life."

Jack's first rocket plane flight came on 27 April 1955, aboard the Douglas D-558, Phase 2, Skyrocket. It was the third D-558-2 built, a hybrid with both jet and rocket engines. He flew using the jet engine alone the previous December but then joined the exclusive fraternity of rocket pilots for his second flight in this aircraft.

On 22 March 1956 McKay was ready to make his eighteenth rocket flight in the D-558-2. It didn't look good for a drop that day due to a fuel tank pressure relief valve. In the cockpit of the P2B-1S mothership, Stan Butchart and Neil Armstrong were dealing with other problems. The right-hand outboard engine quit, but the propeller refused to stop windmilling. This was causing a dangerous overspeed, giving Butchart little choice but to make the decision that McKay and the Skyrocket would have to go. He yelled back to Jack, who reportedly responded, "Hey Butch, you can't drop me!" Stan made the command decision: he had no choice but to go ahead. It took three attempts before the D-558-2 fell away. As McKay dropped, struggling to control an airplane not ready to fly, the mothership's propeller spun furiously and came off its hub. One of the blades came through the bomb bay where McKay had been, exiting the fuselage and striking a port-side engine.

McKay was furious with Butchart, even though the drop had possibly saved his life. After jettisoning his fuel, since the pressurization valve failure precluded lighting the rocket engine, he was able to glide back to a safe landing on Rogers Dry Lake. Because of the serial damage to the P2B caused by the lost propeller, it had only one of its four engines left running to make a long, slow, circling descent. Butchart's controls did not respond, so Armstrong made the careful and safe landing.

After the incident McKay went home that night. Like many test pilots, he didn't bring his work with him and didn't share a lot in that regard with Shirley. Jack's son, John, told me what happened later that evening: "I think some people came over and started talking [about the incident], and my mom almost crapped her pants because my dad hadn't said a word to her."

For whatever reason, McKay's attitude changed toward Butchart. Under the circumstances there hadn't been much Stan could have done except to get McKay away from the failing bomber. But it was also said it may have been the action of dropping Jack that precipitated the prop coming off the hub. If he and Neil had instead gotten the aircraft slowed down enough, the windmilling might also have subsided, and the accident could have been much less severe. John told me, "My dad was not a fan of Butchart's at all. . . . I know a few times he came home and he was pissed off at him." McKay's daughter, Sheri, always liked Butchart a lot, but admitted it was different for her father. "My dad didn't like Stan Butchart, and I don't know why. . . . Stan was such a nice guy. . . . If it was the situation where he had to drop him, [my father] should have liked him very much after that for saving his life. . . . I think he had a great deal of respect for [most pilots], and probably Stan Butchart, too. I just remember that if there was anything negative, it had to do with Stan."

Although the D-558-2 program lasted another nine months, other new rocket planes were also coming into NACA's possession, such as the X-1B. This aircraft was first flown for the air force by Jack Ridley, but he accomplished only two glide flights before other military pilots took over. Even that program was short-lived with just eight more flights, then it was turned over to NACA. Jack McKay became the only pilot to explore its limits for nearly a year, from 15 August 1956 to 8 August 1957. The nose gear failed his first time airborne. The fault was quickly corrected, and he went on to fly a total of thirteen flights.

One of the primary purposes of the X-1B was to accomplish the first experiments with a set of reaction control system rockets. These small rockets, later called the ballistic control system, or BCS, on the X-15, were to push the aircraft about in pitch, roll, and yaw and were fueled by hydrogen peroxide over a silver catalyst bed. McKay's last two missions in the X-1B were to test the characteristics of an extended left wingtip, which was a mockup of the actual system, installed as he left the program. Neil Armstrong completed four more flights, three with the actual rocket system in place, before the aircraft was retired from the inventory.

McKay worked on other NACA, then NASA, research aircraft. By the fall of 1960, he was preparing to be dropped on his first X-15 flight. Like all pilots on the program, he used aircraft no. 1 for his familiarization mission. On Friday, 28 October, four seconds before 9:44 a.m., Jack fell away from B-52 no. 008 to begin flight 1-15. Still using the smaller LR-11 engines, his flight was planned for Mach 2, cruising for several minutes at 50,000 feet. With workmanlike precision he hit Mach 2.02 and was slightly high on his altitude by a few hundred feet. A local area flight, he circled around to the east and north of Rogers from the Rosamond Dry Lake launch point. The only problem on the flight was the lower ventral parachute failed to deploy after being dropped during final approach for landing. This was a problem that dogged the program until the jettisonable portion of the lower fin was deemed unnecessary.

Other duties with NASA programs took McKay away from the x-15 for nearly seventeen months. He returned to fly aircraft no. 2 the first time with mission 2-24, on Friday, 29 June 1962. This mission was launched near Hidden Hills Dry Lake, which was on a line northeast of Edwards, on the way to Las Vegas. He said of the area after his flight, "I could never actually see Hidden Hills from where I was sitting, although I must have been fairly close to it. If I hadn't gotten an engine light, I believe I would have pushed down and headed for a spot I couldn't see, and that would have probably been it."

This was also McKay's first ride using the big LR-99 rocket engine. It was a huge change. His scheduled speed of Mach 4.2 was more than twice what he had accomplished before. The g forces he experienced after launch and engine light were what Jack called a "good boot." The LR-99 ran so smoothly it was deceiving. McKay said during debriefing, "The engine was like the ground run, I didn't get a bit of vibration, I was impressed by the thrust."

The LR-99 ran for more than 112 seconds, but this was less than half the normal run time using the LR-11s. His previous experience with the smaller engines went against hitting this mission's targets. He went well past his planned speed, maxing out at Mach 4.95, or 3,280 miles an hour. Jack's instincts from flying with the LR-11s told him he didn't have enough energy to get all the way back to Edwards. "When I burned out, I expected to get a call from NASA 1 saying I had better head for Cuddeback, but I looked over my nose and there was Edwards." At that moment, it hit home how much power he commanded. "It looked to me like I was heading for Pt. Mugu," eighty miles farther southwest.

Vince Capasso and Bill Albrecht, both operations engineers on the program, spoke about lighting off canisters of colored smoke to aid x-15 landings. Bill said, "A routine landing required that you dropped a smoke canister so the pilot could easily see which way the wind was blowing." Vince went on more specifically with what they had to do: "Meryl [DeGeer] and I were out on the lakebed, throwing the smoke flares to mark the landing spot . . . then took off perpendicular to the runway to get away from it, [in case] the pilot touched down early. . . . Jack McKay was notorious for being a little short [on landing]. You had to watch him," he joked. McKay noted in his post-landing report: "I no more than touched down when I started to drift to the left, and this is why I cut across the runway to the left of the smoke bomb. I don't know where the smoke bomb was placed, but I came to a stop just about 200 yards ahead of it."

Once this flight was completed, McKay became more integrated with the X-15 program. Instead of the seventeen months between his second and third missions, his fourth flight was three weeks later, on 19 July. Again flying aircraft no. 2, Jack exceeded Mach 5 this time around, although he stayed relatively low at 85,250 feet.

The purpose of these high speeds and low altitudes were to expose the airframe to maximum heat loads. The X-15 expanded knowledge on these effects to aid in the design of future winged vehicles when returning from orbit. Besides learning the techniques required to return safely from high altitude, these heating experiments were one of the primary purposes in building the X-15 in the first place.

As with McKay's second flight, the ventral parachute failed after it was jettisoned. Six flights later, just before Jack's next mission, the ventral was removed and not reinstalled again until a small series of tests more than three years later. He continued with two more flights in the no. 2 aircraft. Missions 2-29 and 2-30 were accomplished in less than two weeks, the first racking up the longest burn time of the LR-99 to date. The second was from a launch point approximately a hundred miles north of Las Vegas, near Delamar Dry Lake. This was the farthest distance he had yet launched away from Edwards, making a total run of 235.5 miles at touchdown.

Jack McKay and his wife Shirley had a large family. Sheri led off, born in Virginia before her dad first brought the family west. Sheri shared hazy memories of the move when she was less than three years old. "I kind of remember the car ride out from Virginia a little bit. A lot of ups and downs over the desert and that sort of thing." Her brother, Jim, was slightly more than a year old during this westward trip. Six brothers and sisters joined the family in California over the next several years: Joanne, John, Milt, Mark, Charlie, and Susan. There was also baby Shirley, who was stillborn.

Sheri said of her life: "It was exciting. I think one of the things that has

15. Sheri and Jim McKay with their father, Jack, exploring the desert at Red Rock Canyon, now a California State Park. Courtesy of the Sheri McKay-Lowe Collection.

always intrigued me . . . was the amount of space, just physical space, of the Mojave Desert. That was such a big thing to be able to play in the desert. . . . I think the desert, and being bonded to the land, is probably a very primal memory for me." She recalled her father as "being kind of fun and lighthearted, and kind of a joker. [He was] ready to have a good time, but do serious work too. He'd come home and give us all big hugs." Family trips were commonplace. "Red Rock Canyon wasn't too far away. My parents loved to go on picnics. My dad would go down to Mexico, deep-sea fishing with friends. . . . We'd go out for drives in the desert. He would take my mother out target shooting. We'd go on picnics to various places and over to Bakersfield for lunch on Sunday afternoon. He had a favorite Chinese restaurant called The Rice Bowl. It's still there, I hear."

When Sheri was five, the family moved from Mojave to a home in Lancaster, a favorite spot to settle for many from the air force and NACA at Edwards. The community had barely 2,500 people at the time. "My mother took some ceramics classes, making little figurines, and my father took a wood-working class. He made a beautiful coffee table and a book case and some other things. Everyone then were really happy campers. We had a growing family. My mother had been an only child for a long time and wanted lots of children. . . . It was that generation, too. They tended, after World War II, to have quite a few children. I mean, it was the baby boom."

A favorite trip for McKay was to hike up Mt. Whitney in California's Sierra Nevada Range. As the highest mountain in the contiguous United States, it was a challenging exercise, but one that he accomplished on several occasions, sometimes taking his two eldest sons, Jim and John. John remembered that he was nine years old the first time he went up with his dad, which placed the trip in 1962. "Nowadays you just do a day hike and fly up to the top if you're a marathon maniac. We didn't climb it and then come back; we climbed it and then went back on the other side, then spent three or four days fishing. And there's nobody there—at least there wasn't, except for the F-104s that would fly by! If [Joe] Walker was up there, he would know where you were and would always fly through Owen's Valley and then back over [Whitney]." Things didn't always go the way they were planned, however. "Both me and my brother got altitude sickness. We were probably all of two hours from the top, and we started barfing up Spam like you wouldn't believe. And that was [Dad's] vacation. . . . He must have had the patience of Job."

Mark, the second youngest son, said, "I can remember we got into the 4-H Club. I think [my father] wanted us to get involved in a lot of these outdoor projects with animals. We also had a large garden. We had horses and chickens, things of that nature." The next oldest, Milt (who prefers "Mac"), said, "[Dad] wanted a simple life, with a garden, which is kind of weird. Here he's a test pilot, and he's out there with a hose watering the garden." Other than some of these types of activities, Jack wasn't a typical dad. As John said, "I only remember him playing catch with me once, and that probably lasted all of five minutes. He got bored easily."

The purpose of flight 2-31 was to verify the x-15's stability with the lower half of the ventral fin removed and to investigate the turbulent boundary layer air passing along the aircraft skin at hypersonic velocities. Speed was planned for Mach 5.5, with an altitude of 125,000 feet. This was going to be Jack McKay's seventh flight in the x-15, but his fifty-fourth in a rocket plane since he first joined NACA in 1951. With the exception of Scott Crossfield, McKay was the most experienced rocket pilot who flew on the x-15 program.

The first attempt on 7 November 1962 was canceled after the B-52 and x-15 were serviced and ready for takeoff, because of a faulty ammonia safety valve on the rocket plane. After replacement, the second try was again thwarted prior to leaving Edwards when the scouting planes reported heavy cloud cover over the Mud Dry Lake launch area. On Friday, 9 November, at 9:29 a.m., the B-52 was finally able to leave with its two precious pieces of cargo, x-15 no. 2 and Jack McKay.

Maj. Russell P. Bement commanded B-52 no. 008, with Maj. Kenneth K. Lewis Jr. in the right seat. Chase pilots Joe Walker in an F-104D, and Bob White in a T-38, escorted the mothership north, each with a photographer in the back seats of their planes. They all crossed over the Nevada state line, passing east of Mud Dry Lake. Bement made a left turn, circling north of Tonopah, before heading south to line up for the x-15's launch on the west side of Tonopah and Mud.

Radio chatter was by the book, verifying all systems were operating normally. Both auxiliary power units worked properly, the flaps deflected as planned, cabin pressure switched to internal power, the emergency fire extinguisher was set to automatic, and the inertial navigation system was reading within specifications. Radio communications, both sending and receiving, were perfect, as Jack confirmed, "5 square." The B-52/x-15 combination was aloft fifty-four minutes out of Edwards when the final countdown began. At the one minute to drop point, McKay set the LR-99 engine to igniter prime, the equivalent of inserting a key in a car's ignition and turning it to the position before the starter engages. Twenty seconds to launch, he put the rocket into its idle position, then called "Ready to launch." At 10:23 a.m. Jack counted backward from five, then confirmed, "Launch."

"It's a good light, Jack," he was told by White from his chase plane, but Bob spoke too soon. Within seconds, it became evident the engine was not producing enough thrust, refusing to throttle up past 30 percent. McKay tried again to push the throttle to 100 percent. The engine refused to advance, his acceleration barely pushing him to Mach 1.49 at 53,950 feet, a climb of less than 8,000 feet from the drop point. Jack confirmed he was shutting down the engine and proceeding to the emergency lakebed at Mud. NASA 1 passed on directions: "Start your left turn and look for the lake. You're right opposite it at about ten miles."

Inside the cockpit, Jack glanced left and saw the large dry lake. He later stated in his post-flight report, "The runway markings were just visible from this altitude; however, they were sufficient to help program position and other cues essential for satisfactory approach and landing." He started the jettison sequence to rid the aircraft of excess fuel. With a full load just after launch, this was a dangerous time to land if it could not be fully offloaded through the pipes at the rear of the fuselage fairings on either side of the engine. Three years previously, Scott Crossfield, in this same aircraft, had broken the x-15's back in a similar situation on landing. That had been the program's first emergency; this by McKay was to be its third.

McKay extended the dive brakes to slow his descent, while bringing the nose of the aircraft up thirty-three degrees above the horizon. As he passed below sonic velocity, Jack had to cut the fuel jettison sequence and pressurize the tanks for landing. He informed control he would be coming in due east. White gave him an update on the conditions ahead. Jack would have some crosswind hitting the right side of his aircraft as he touched down.

Inside the x-15 the motor clutch release controlling the flaps had been improperly torqued. It was impossible to know for certain afterward, but this may have been the root problem that started the cascade of events over the next several seconds. When McKay sent the signal to lower the flaps, which would have slowed the x-15 to a safe landing velocity of approximately 220 to 230 miles an hour, the flaps did not respond. Without the larger airfoil the flaps created after extension, the wings would stall at a much higher speed.

Six and a half minutes after launch, and four seconds prior to touchdown, McKay released the landing gear. As Jack was about to hit the lakebed, he was still traveling at 292 miles an hour, 46 miles an hour faster than any previous landing. The skids contacted, then, less than two seconds later, and 330 feet farther down the lakebed, the nose gear slammed down. Unable to bear the loads, the left skid appeared to have failed at the moment of nose gear contact. The skid collapse caused the fixed portion of the lower ventral to gouge into the dirt, tearing it to bits. This allowed the left horizontal stabilizer to dig in, twisting about its axis and ripping from its mount. As it departed, it struck the already failed left skid, taking it off as well. With the loss of the skid and stabilizer, the nose wheels started to lose stability, soon coming off completely.

The accident report detailed the final sequence of events: "After the failure of the nose wheels, the aircraft slid about 1,400 feet on the nose gear strut, the right main-gear skid, the remains of the ventral fin, and the left wingtip. It gradually turned broadside with the nose to the left. . . . Finally, the airplane rolled to the right, dug the right wing into the ground, and flipped over on its back."

The report detailed how the x-15 was situated on the ground: "The airplane came to a rest upside down on the upper vertical tail, the right wingtip, and the fuselage nose. The pilot had jettisoned the canopy just prior to rollover, and the headrest was forced a few inches into the ground." McKay's helmet dug into the dirt with the headrest, too much weight literally resting on his neck and shoulders. McKay described the situation from his position within the cockpit:

After the nose gear came down, the airplane seemed to list to the right just slightly, and then back to the left, and then it seemed to be banked very heavily to the left. The airplane then began to diverge to the left and I thought about getting rid of the canopy because it looked like the airplane was going to spin. At about the ninety degree point, when the airplane appeared to be going sideways, I felt the canopy blow, although I do not actually remember ejecting the canopy. . . . [The airplane] appeared to be slowing down at a pretty good deceleration, then it rotated on its back at a pretty fast rate. At this point the airplane came to a stop. I believe that the top of my helmet hit the lakebed first, although at the time it did not appear to be a severe blow. After the airplane came to a standstill, which was almost immediately, I was trying to grab for the knife to cut myself loose, but in this position this was a very difficult maneuver to make. I gave up the idea when the rescue crew came on the scene and got me out of the airplane.

Emergency personnel descended on the lakebed quickly. Early that morning, prior to the b-52 departing Edwards, a c-130 cargo plane deposited an air force fire truck and a NASA jeep for just such a contingency. Also on site, the H-21 helicopter, piloted by Capt. Paul J. Balfe, lifted off from its position and was the first to get to the crash scene. Balfe dropped off Dr. Lynne B. Rowe, Sgt. Charles L. Manes, and Airmen Larry Hough and Rob-

16. Aftermath of Jack McKay's rollover accident on 9 November 1962. X-15 no. 2 is pointing nearly back along its flight path, and the shadow of the H-21 helicopter is on the right. The H-21 used its downwash to protect McKay from fumes when he was still trapped upside down in the cockpit. Courtesy of the Armstrong Flight Research Center.

ert Mills to get to McKay as quickly as possible. He then took off to hover over the aircraft. Fitz Fulton related, "They were trying to get him out and the ammonia fumes were really bad. . . . A pilot [Balfe] ended up hovering his helicopter overhead to keep the fumes away while they dug Jack out. . . . I'm sure it was thirty [to] forty minutes at least."

The fire truck started to hose down the X-15 and any dangerous liquids. The C-130 landed with an additional jeep and crew members, one of whom was NASA technician, Ray White. He recalled that he arrived about the time McKay was dug out from the cockpit, saying, "I noticed that the APUs were still running, [and] here's the airplane sitting upside down. So, I crawled up under there, and it was a little confusing for a second, but I turned the switches off to shut down the APUs. Engineering said that if they'd have run another thirty seconds, they would have blown up."

Back at Edwards, the situation took a while to become clear. Possibly because of the low altitude and distance from Edwards, Bob White, in the lead chase aircraft, was not getting through to NASA 1. As events unfolded,

Capt. Mervin P. Evenson, in a chase aircraft closer to base, relayed White's initial report on the condition of the aircraft. The Edwards control tower had been waiting for the x-15 to land so other traffic could be cleared in the area. NASA 1 asked, "Eddy Tower, have you been reading the transmissions?" The tower responded, "Roger." The NASA controller explained, "The x-15 will not be coming back here to land." Back at Mud, the people coming to McKay's aid radioed to reassure everyone on the network, "[Jack] is conscious and they are trying to get his suit off him to make a further determination. . . . He seems to have some internal pain. . . . He does not seem to appear to be in terribly bad shape, though."

Once they verified he was stable, McKay was loaded into the C-130, and immediately flown back to Edwards for hospitalization. He stayed there four days, with injuries rated as "minor." The medical report delineated the problems: "Patient sustained minor abrasions [to the] back of [his] neck and right leg. A laceration of left leg (4 stitches). Generalized muscular bruising of [his] back, especially trapezius and lateral intercostals." The doctor praised McKay's performance during the emergency, saying, "Patient recognized [the] situation and reacted quickly, possibly preventing serious consequence."

Was McKay's action in jettisoning the canopy the right thing to do, especially considering he was initially trapped under the aircraft with his helmet dug into the dirt, compressing his spine? If he had not done this, there is a good chance he would have died at the crash site. Estimates vary in how long it took to release him from under the aircraft, but the pilot's oxygen supply is limited and may not have lasted long enough if the canopy was still secured. Additionally, the design of the cockpit had the area full of pure nitrogen to prevent fire, which was deadly if inhaled for more than a few breaths. If his pressure suit was compromised in any way around the helmet seal, his oxygen might have been gone in moments. There would have been no way to save him. Bill Albrecht said, "Jack McKay saved his own life. As soon as he saw it was going to tumble over, he jettisoned the canopy. . . . He was smart enough to think, 'If I go over, and if I'm not breathing air, by the time they get here to try to rescue me, I'll be dead.'"

Once McKay was safely on his way to the hospital and the x-15's toxicity was stabilized, the ground crew then spent three days to get the aircraft righted, picked up all the pieces scattered along the landing path, loaded everything onto a flatbed truck, and moved back to Edwards for examination. From that point, until 6 December, x-15 no. 2—now termed as a salvaged aircraft—was gone through thoroughly to find everything possible that may have gone wrong.

On Friday, 7 December, the x-15 was trucked south to the North American plant for the contractor to assess the damage and complete a feasibility study on the cost and extent of repairs. From the time of the incident, it took only six weeks to complete the report of the malfunctions and emergency landing. The final version was released in December, before the Christmas holiday break. Donald R. Bellman chaired the investigation board, the same position he held five years later for the accident involving x-15 no. 3 and Mike Adams.

The primary cause behind the engine throttle difficulty, and thus the first link in the long line that led to the rollover, was traced to a burned-out diode and transformer in the electrical controls, but how that occurred baffled the team and was never fully explained. More worrisome was the left skid failure. Structural analysis showed the gear sustained only a force equal to 80 percent of its design limits. In other words, it should not have collapsed, even under the severe load of higher-than-normal landing weight and speed. Add in the extra burden of not having flaps, however, and that equation changed.

The investigation report stated their reasoning behind what happened: "The most probable cause of the ground crash was the excessive landing gear loads imposed by the lack of wing flaps. The cause of the wing flap failure could not be determined."

Bill Albrecht, the NASA operations engineer on aircraft no. 2 at the time of the accident, shared his thoughts and admiration for the x-15: "My first view of it was from the window of the [c-47] Gooney Bird, which took a bunch of us up to Mud Lake. When I saw that thing upside down, I said, 'That's the end of that airplane.' But, a steel airplane is something else. It's tough stuff."

Like the doctor's report after McKay left the hospital, Bellman's report went on to praise McKay's performance, saying, "Evidence indicates that the pilot acted in a skillful and acceptable manner throughout the entire flight."

The relatively minor medical problems noted during McKay's time in the hospital under air force care did not take into account several other factors that developed into major difficulties for Jack over the coming years. His son, John, saw the damage done to his father: "I had come home for lunch and my mom said [my dad had] been in an accident, but that he was okay. . . . I remember his shins had knocked up underneath the instrument panel and were gouged about a quarter-inch deep. If I remember right, the x-15 had an f-104-type setup where you had stirrups that held your legs. I remember it hurt looking at it. . . . And his back was killing him." Sheri remembered, "I'll never forget when he came home. It was shocking. He looked like someone who had really been through something. The look in his eyes—I was fourteen and I knew he was a marked person. I said to myself, 'This guy's not going to live long.' He did recover, but there was a lot of damage."

The biggest problem was three vertebrae in Jack's upper spine were compressed to the point where he lost approximately an inch of his height. The debilitating pain, initially covered up by McKay due to the adversarial nature most pilots have with doctors for fear of losing their flying status, became so severe it was about to change his entire personality—and eventually cost him his life.

The rollover accident was initially seen as a severe setback. With the no. 2 aircraft in tatters, the scope of the research program was questioned, and it was unknown for several months whether this x-15 was to be rebuilt or scrapped. Very quickly the idea took hold to not only return the aircraft to service but seize the opportunity to expand beyond the scope of the original design. Since the overall program was still maturing, there was a lot of support to not be satisfied with the status quo.

By January 1963 North American Aviation, working alongside NASA engineers, delivered their proposal. It would transform the vehicle into a testbed for new concepts in hypersonic aerodynamics, including the ability to fly the x-15 to Mach 8, two Mach numbers above its initial design limit. To accomplish this, provisions for two external tanks were included, which radically changed the overall look of the aircraft for the few flights on which they were eventually carried. Historically, this image of the heavily modified x-15 has now become the iconic one for the overall program.

With the speed afforded by the extra fuel, the exotic idea was put forth of mounting a small test version of a supersonic combustion ramjet engine, more simply called a scramjet, to the underside of the aircraft. This new type of engine required hypersonic velocity to even begin to provide propulsion.

The lower ventral was modified to be removable so a mounting pylon could take its place whenever the scramjet became available for flight. The scramjet required a liquid hydrogen fuel source, so two new tanks, giving a total of ninety gallons of propellant, were installed in the x-15's midsection. To accomplish this, the aircraft was literally cut in two, just aft of the forward edge of the wings. A twenty-nine-inch fuselage plug was constructed between the liquid oxygen and anhydrous ammonia tanks, extending the overall length of the aircraft. Vince Capasso said the extension, although an exciting idea, "got to be twenty-nine inches of trouble, because that created a few problems due to splicing all the wiring and rigging between the two halves." A similar difficulty was experienced with the no. 3 aircraft, where the rear half of that airframe was decimated, and subsequently reconstructed, after the test stand explosion in June 1960.

Several additional upgrades were proposed and accepted, including the installation of a star tracker to test the equipment required for navigating the Apollo lunar spacecraft on its way to the moon. The system was placed inside the x-15's equipment bay behind the cockpit. This bay already had an elevator for easy access to components, and a special "flip-top" door was added to the access hatch so the star tracker was exposed to space during the minutes at peak altitude during ballistic flight.

The x-15's outboard right-hand wing was replaced with a removable section, allowing new instrumentation to be installed, as well as the possibility of testing new types of structures, although that part of the idea never went beyond the proposal stage. The side fairings were extended to the end of the LR-99 engine tail cone, affording room for forty additional gallons of hydrogen peroxide propellant.

Another visible change in the look of the aircraft was replacing the rectangular cockpit windows with elliptical ones. There were many instances where the inner or outer panes on the windshield shattered during high heat loads. The stress points created by glass corners were the problem. Since the modified aircraft was to fly at higher velocity and experience higher temperatures, the distinctive oval windows eliminated this flaw.

Modifications also were required on the motherships, so both B-52s were equipped with heavyweight shackles and hooks on the x-15 pylon, as well as strengthened wing structure to handle the added weight of the advanced x-15 and external tanks.

By 10 May 1963, six months after nearly being destroyed, x-15 no. 2 was officially sanctioned by NASA to be rebuilt by North American. NAA's Howard A. Evans was the director of the operation and was tasked with its completion in less than eleven months. As the operations engineer at NASA in charge of aircraft no. 2, Bill Albrecht also went to North American. "I spent a lot of time down there with Howard Evans," he said. "I got involved with defining the configuration of the airplane as it was at the time of the crash, which is significantly different from the configuration as [originally] delivered by North American. They had all their documentation indicating how it was at delivery [on 10 April 1959], and we had reams of work orders where things were changed."

Some of the tankage and plumbing were not included by the time the vehicle was ready for redelivery to NASA and the U.S. Air Force, as was also true of the large external tanks. These would be added later, as the program required. On 15 February 1964, three weeks ahead of schedule, the newly modified and redesignated x-15A-2 was rolled out of the Los Angeles facility. Three days later the A-2 was delivered to the government's NASA Flight Research Center at Edwards and officially accepted in a ceremony open to the media on Monday, 24 February.

While all this was going on with respect to the aircraft, McKay continued his medical recovery, taking much less time to return to flight status than x-15 no. 2. Five months and two weeks after the accident, 25 April 1963, Jack was ready to go again, this time on a Mach 5.32 flight at 105,500 feet. It was his eighth mission, and the thirty-fourth for aircraft no. 1. The drop from B-52 no. 008 occurred at 2:03 p.m., rather late in the day for a flight, but not unheard of. Jack said at his debriefing, "What is there to be said about a flight that went according to schedule and flight plan?"

Sheri McKay-Lowe explained why Jack returned to the airplane that almost killed him: "[My father] just loved what he did. For someone to explore the unknown, that's really something. Most of us would rather be comfortable, but I don't think he felt he had any other choice but to go back into the cockpit. That's what he loved to do—ever since he was five years old."

Sheri went on to say, "When my father and his twin brother were growing up, other kids made fun of them because they weren't out doing sports and all that. They were in their rooms making models, [or] at the park flying remote control airplanes. It was their obsession—their passion. Just what he loved to do."

The landing gear and associated equipment caused many difficulties on the X-15A-2 once it was flying again, but these problems were not always distinct to the no. 2 aircraft. In the case of McKay's second mission after returning to flight status, 1-35 on 15 May, the launch and acceleration past Mach 5.5 were perfect—until the nose gear ram scoop door opened. McKay said afterward, "There was a sharp noise . . . accompanied by airplane vibration. The aircraft immediately went into a left yaw and roll, and considerable aileron was necessary to maintain level flight. . . . It was a real effort to get it turned around and get it back in the direction of the lakebed."

Jack continued to fight the X-15 for stability all the way back to Edwards, not knowing exactly what had gone wrong. As he approached Rogers Dry Lake and extended the landing gear, the chase pilot called, "Your nose wheel looks cocked." McKay said later, "The touchdown felt pretty soft, and I could feel the skids touch or skip along the lakebed for a short period before the nose came down. I got the call from the chase about my nose gear being skewed [about twenty degrees to the right], and felt quite a bit of roughness, especially during the early part of the runout. Then I began to shed rubber. Large pieces were flying by the windshield and didn't quit until at least half of the runout was complete. . . . The structural failure of the nose gear door was a new experience."

The left tire failed the moment it touched the lakebed, transferring the entire load to the right tire. It failed, and the remainder of the runout was on the rims. A worried NASA 1 controller asked, "Okay, you're in good shape though?" To which McKay replied, "Roger." Then he smiled and said, "I kept the canopy on this time, incidentally."

The rough landing may have shaken things up a bit much for Jack. It was ten months before he returned to the X-15 cockpit. During that time, his condition deteriorated, but he was able to maintain his status as a research program pilot.

Jack made special trips away from Edwards to get medical help. His son,

Mac, remembered, "After it happened, my dad had to go to New Mexico, [to the] Lovelace Clinic. I didn't even know until later that he popped a couple discs out of his back, and all they gave him was this sling to rest his neck in. As I got older, I thought, 'That's not gonna do crap!' He just had to suffer with the pain." It was a tough situation for Mac to watch his father go through. "I remember as a kid, one night he came home, and it took him probably half an hour to get from the front door to the bedroom. I was sleeping in the den at the time, and it woke me up. He just stopped and grabbed his back. It was the first time I ever heard him in pain. . . . He wasn't going to show Paul Bikle that it was killing him, because that would have dragged him off flight status."

Jack Sheri noticed the toll on the family, especially her father. "After his accident, his personality changed. He was more removed, more remote, stayed out more. . . . I read that happens when people are in heavy-duty car accidents, or they have some sort of head injury, that's just kind of the thing that happens. He was less jovial, more serious, and seemed to be somewhat fearful at times." Sheri's brother, John, agreed, "Yeah, he changed. And I don't think it was any good. I think the drinking got heavier. I mean, all of these guys [tended to drink]. I know [Bill] Dana, he quit drinking. I know Milt [Thompson] did later on. All these guys had to. I guess my dad didn't quite get it. . . . Back then, you couldn't admit you were an alcoholic; it's an unmanly thing to do. . . . I think he was in a lot of pain. And I think that's how he took care of it."

Jack McKay probably saw it as a choice between alcohol and medication. In the culture of the 1960s, taking pills was frowned upon, whereas having a drink was expected. A certain amount of medication was okay, since he had been in the accident, but only to a point. John continued, "As I understand it, the [painkiller] he had was pretty potent stuff. I think his choice was to medicate himself to where he wasn't in pain, then he could stand it for a certain length of time. . . . I think they had some pretty heavy shit back then. I did see a change in him. But it wasn't until years later I admitted that change."

When Jack returned to the x-15, he was slated for flight 3-27, his first in aircraft no. 3. On Friday, 13 March 1964 at 9:46 a.m., McKay prepared to launch from b-52 no. 003, cruising at 45,000 feet near Hidden Hills Dry Lake. He was a little overanxious, saying, "On the drop I made a lunge for the drop switch and missed it, and that's why I had that two-second [launch] delay." Once away, the flight proceeded by the book to Mach 5.11 at 76,000 feet.

This x-15 was equipped with a different stability system than the other two aircraft. It smoothly blended the aerodynamic surfaces with the exoatmospheric ballistic control system rockets, to make the airplane easier to fly. McKay noted that tasks performed consciously on the other two aircraft were intuitive on no. 3. "The approach and the turn into final, and everything right down to the flare, was simpler with this system than it was with airplanes no. 1 and no. 2. And with a little more practice, as far as the actual touchdown phase, I think it might improve in the future."

McKay picked up the pace of his flights. During his first twenty-nine months on the program, he completed nine missions. During the next twelve months he flew eight more.

Mission 3-33, on 26 August, was planned as a Mach 5 flight, but the engine actually over performed, and McKay achieved his highest speed on the x-15 at Mach 5.65, or 3,863 miles an hour.

On 15 October Jack made the first test flight with new pods attached to the wingtips of the x-15. The left pod had an experiment, which at high altitude, opened to catch dust from micrometeorites. Like many flights with this experiment, it caused electrical transients that were difficult, if not impossible, to trace. As McKay was returning to base, the experiment extended itself into the airstream as he was decelerating through Mach 1, causing the aircraft to buffet and roll in the direction of the pod. Jack compensated and landed without incident.

Since its reconstruction and delivery back to Edwards in February 1964, the x-15A-2 had been undergoing a rigorous checkout in preparation for its return to flight status. Robert Rushworth did the first three missions in June, August, and September. He encountered a warped right-side horizontal stabilizer on the first, and landing gear difficulties on the second and third.

On 6 November, and again on 16 November, McKay went aloft in the A-2 for two captive flights, 2-C-58 and 2-C-59. These were the first times he entered the cockpit of the rebuilt no. 2, the same aircraft that had nearly taken his life when it rolled on top of him two years previously. As captive tests the x-15 stayed firmly attached to the B-52 pylon while landing gear modifications were checked out following Rushworth's troubles on the previous two missions.

All went as planned, and two weeks later at 12:09 p.m. on Monday, 30 November, McKay dropped away to start flight 2-35-60. There is no record of McKay ever speaking his thoughts during this mission, how he felt being in the same physical space that had nearly proven fatal the last time. Jack McKay did as his test piloting skills dictated: he flew the mission and brought back the data.

McKay accomplished five more flights in the new A-2 aircraft through mission 2-42 on 2 September 1965. He praised the performance of the rebuilt aircraft and compared it to the unmodified X-15, saying, "Once you made up your mind you wanted to do something to make that airplane do it, it'd do it." This flight was the second mission for McKay above 200,000 feet and, at 239,800 feet, was less than 10,000 feet below the highest altitude the no. 2 aircraft ever flew. In his post-flight report, he talked about how the X-15 operated at these altitudes, providing some contradiction to his previous statement. "I think the most adverse piloting task was the point just before burnout, where I was trying to reach for the throttle. I did not have any real aerodynamics on the airplane, and it was beginning to get a little peculiar, but reaching for the throttle, I couldn't control the airplane through [the ballistic control system], so the airplane was wandering around at will."

One month later, on Tuesday, 28 September, McKay passed the demarcation line of 50 miles altitude, qualifying as the fifth astronaut on the X-15 program with mission 3-49. His peak was well across the 50-mile boundary, at 295,600 feet, although the plan had been to be much closer to the mark of 264,000 feet. Jack McKay's feat was not in doubt, although his status as an astronaut was not officially recognized by NASA until nearly forty years later.

Accomplishing twenty-five missions was an impressive achievement on the X-15. McKay became the third to reach this milestone with flight 1-63 on 6 May 1966. Joe Walker ended his stay on the program at this number on 22 August 1963, and Bob Rushworth made it on 17 February 1965. Jack was the last to do so. No other pilot ever came close, with a maximum of sixteen by four others: Bob White, Joe Engle, Pete Knight, and Bill Dana.

On 30 November McKay was scheduled for the first of what became six unsuccessful attempts at this mission. Repeated problems, primarily with winter weather and wet lakebeds, forced delays for tries in December, January, March, and twice in April. By the time 6 May rolled around, no x-15 mission had flown for more than six months. Bob Rushworth was also thwarted on three occasions to get the A-2 airborne. Several failures of the stability augmentation system kept him from launching over the previous three weeks. McKay was finally allowed to go first, with 1-63.

Heading uprange toward Delamar Dry Lake, the B-52/x-15 and the chase planes flew a few miles west of Las Vegas. About seventy-five miles farther north, Fitz Fulton made a sweeping left turn to get the mothership headed back toward Edwards and the safe haven of Rogers Dry Lake. As they approached east of Delamar, Jack called, "Launch!" Pete Knight, who had just two x-15 flights under his belt, was the lead chase pilot. He radioed to McKay, "You got a good light, Jack. Good heading." Very quickly, that changed.

McKay pushed the LR-99 throttle forward, accelerating past Mach 2 and 68,000 feet, when a low fuel warning light illuminated on his instrument panel. Jack informed everyone on the radio network: "Fuel line low." Bill Dana was serving as the NASA 1 controller that day, ready to help McKay through the emergency. At less than thirty-two seconds into the planned eighty-second burn, Jack shut the LR-99 off, then radioed, "Where's the field?" Dana asked, "Are you going back in?" McKay's answer was lost over momentary static.

West of the ongoing emergency was the Beatty, Nevada, High Range tracking station. T. D. Barnes was the site manager at this remote location. He recalled that the radio beacon emitted by the x-15 was strong enough that it was a rarity when they lost track. This would usually be only if the aircraft went over their heads, since the radar dish could not be directed straight up. This was not a problem for the flight that day. "We actually kept track of it all the way down to the lakebed," he said. "Edwards probably had it if we didn't, but the beacon was so strong it wasn't a problem. . . . We didn't really think too much about it. Just part of the mission. . . . Once you locked onto the airplane, we'd very seldom lose it."

Beatty and Edwards, as well as the third station farther northeast at Ely, Nevada, kept watch on aircraft no. 1 and McKay, as they descended to Delamar Dry Lake.

17. Running off the lakebed at Delamar on 6 May 1966, McKay jettisoned the canopy in case of a second rollover. Courtesy of the Armstrong Flight Research Center.

In his post-flight report Jack described his approach to the lakebed: "Visibility was good and there was no difficulty in seeing Delamar Lake in setting up for the downwind [leg] for a landing to the south. However, doing the base leg, and while going through about 18,000 feet, the lakebed became obscured by several layers of clouds at the north end of the field, and a turn to final was made earlier than normally planned, to prevent overshooting." The slight cloud coverage deflected Jack's approach just enough that when he touched down, he was over a mile from the edge of the lake, halfway through what was already a relatively small emergency site less than three miles wide.

The skid marks showed a very smooth landing, with just the barest disturbance of the hard-packed surface before the full weight of the x-15 bore down. The nose wheel touched soon after, with its usual jarring suddenness. The aircraft continued forward, veering slightly to the right as it slowed. After 6,132 feet on the ground, McKay was still traveling at roughly 90 miles an hour. Seeing he was not going to stop in time, Jack pulled the handle that ejected the canopy, again wanting to ensure he would not be trapped inside, whichever way the x-15 might be facing when it finished its slide. About twenty feet later, he reached the edge of the lakebed—and kept going. There was a slight dip as the aircraft transitioned to the rougher surface, then a low ridge and another dip, which momentarily popped the nose wheel off the ground. After that, the x-15 plowed on through the stunted sagebrush surrounding the ancient lake surface. Loose soil gave way under the landing gear. After 265 additional feet, and less than six seconds off the lakebed, drag brought the vehicle to a full stop.

The emergency personnel nearby got to McKay and helped him from the x-15. At first glance, the airplane appeared a mess, but this was primarily superficial dust, dirt, and scratches. The landing gear remained intact, even under the strain imposed by the rough surface outside the boundaries of the lakebed. The cockpit canopy came to rest upright, still on the lake surface. Blow-out panels on the fuselage fairings beside the engine were gone, probably from the moment the engine malfunctioned.

Somewhere during the rollout, the micrometeorite experiment in the nose of the left-wing pod opened without being commanded. As part of the fixes after no. 1 was returned to Edwards, a special control circuit was installed, allowing the pilot to manually extend and retract the mechanism. It didn't solve the overall problem, and this caused consternation well after the Delamar incident.

Behind the cockpit bulge, sticking out of the top of the fuselage, was a piece of thin, twisted, silver pipe. It appeared as if someone had taken a wire coat hanger and jammed it in as a makeshift radio antenna. At launch from the b-52, three feet of liquid nitrogen line and the plug attaching it to the x-15 had been pulled away as McKay dropped. There was no connection between this and the engine problem. That turned out to be the casing of the lr-99 fuel pump, which gave way. With no pump, the engine was starved of fuel and could no longer operate. It was the third such failure, but the other two had occurred during test stand runs and had not endangered the aircraft.

Dean Bryan, of the nasa rocket shop, was part of the crew sent to Delamar in the aftermath of McKay's second emergency landing. "We got there, and here's the airplane sitting out in the sand and sagebrush. It was all intact—the airplane wasn't damaged or anything. Tough airplane." Dean re-

18. Coming back after an emergency landing on Mud Lake in January 1962, the convoy stops for a break at a gas station. The locals from Beatty, Nevada, have turned out to see the rocket plane strapped to the back of a flatbed truck. Courtesy of the Armstrong Flight Research Center.

lated, "We went in and got rid of the fuel and purged the engine and auxiliary power units, things like that, just to make sure the airplane was safe, and for the crew to be able to get in and load it on a truck. It would take us about four or five hours to have the airplane all purged out, then however long it took them to get cranes in there to pick it up, then haul it home."

Cleanup of aircraft no. 1 allowed it to take to the air less than a month later, with McKay again piloting. Two attempts on 2 and 10 June were aborted due to inertial system malfunctions. Pete Knight then took charge of this airplane and made a successful flight on 12 July. Jack came back two weeks later to make a precision flight. This was in preparation for a future experiment, where the x-15 would need to meet a specific schedule to fly and be in the proper position for a missile launch from the Western Test Range at Vandenberg AFB, on the central California coast. McKay's excellent flight skills proved the concept, but the actual experiment, carried out after he left the program, never achieved its goal.

Two more flights, 1-66 and 1-67, also went perfectly for Jack on 11 and 25

August. By this time, it was becoming evident to McKay that he could not continue to hide his growing pain. Exactly when he made the decision to leave the x-15 program is unknown, but his twenty-ninth and final launch in the rocket plane also proved frustrating and disappointing.

Smith Ranch Dry Lake lies almost 300 miles directly north of Edwards AFB and 125 miles east of Reno, Nevada. Another 150 miles farther east is the Ely radar site of the x-15 High Range. Smith Ranch was used for the longest flights during the research program, and Jack McKay flew the first mission from this launch area on 8 September 1966. Flight 1-68 was planned for Mach 5.4 at 243,000 feet.

Maj. Charles J. Doryland commanded B-52 no. 008, with Col. Joseph P. Cotton in the right seat. Cotton was the first pilot to fly the XB-70 Valkyrie triple-sonic bomber on 21 September 1964, along with Al White, from North American Aviation. White had been Scott Crossfield's backup on the x-15 program. Three months previous to this x-15 flight, on 8 June, Cotton was in the rear seat of a T-38 Talon as part of the formation flight that ended in tragedy for Joe Walker and Maj. Carl Cross, when Walker's F-104 collided with the no. 2 Valkyrie.

Controlling McKay's x-15 flight on 8 September was Maj. Michael J. Adams, who was anxious for his first rocket plane flight, scheduled for later in September.

After an hour of flight, all x-15 systems checked out fine. Doryland had the B-52 headed back south, and McKay launched himself from the bomber's wing pylon at 10:39 a.m. The initial calls all showed excellent readings on the rocket plane, then déjà vu must have kicked in for McKay as he saw the same readings as his last emergency flight, occurring at nearly the same time and altitude following launch. He took a few moments to assess the situation, and thirty-two seconds into thrust, he retarded the throttle to 50 percent. The warning light refused to go out, so thirteen seconds later, he initiated the shutdown sequence.

Jack radioed, "Fuel line low light, going back." Unsure if Adams heard his radio call at Edwards, McKay said, "Shutdown. Do you read, NASA 1?" Adams verified, "Rog, going into Smith." Mike then advised McKay: "Going to be a hard turn to the right, Jack." A chase pilot encouraged the tight turn, saying, "Bend it on around, Jack."

McKay was bleeding off airspeed and altitude as he approached the lake, then explained his plan: "Okay, I'll be landing from northeast to southwest." This was the long axis of the seven-mile-wide Smith Ranch lakebed. Jack's light touch on landing was again evident, as noted in the operations report: "A normal touchdown with very low main landing gear loads was made." Soon after the nose gear came down, out of the vast expanse of the approximately eighteen square miles of lakebed, a surveyor's nail was lying exposed on the dirt. The left-hand tire ran directly across it, causing a puncture. The aircraft seemed to not even notice the problem, with the remainder of the runout relatively straight and uneventful. Although it was Jack McKay's third emergency landing—his second in only four months—this was the one where he experienced the least amount of problems.

Doryland, from the B-52, relayed the information to Mike Adams at NASA control: "Good touchdown, no problems." Mike sounded relieved. "Rog, tell them I appreciate that, I'm supposed to fly that airplane!"

Initial feeling was that a fuel leak had occurred. McKay said, "Before I left the bird and got on the C-130, I looked around [but] I could not see any evidence of fuel leaking out. . . . I just did not feel like I wanted to come home with this thing staring me in the face. The bird was not injured."

The fuel tank regulator was deemed the next most logical cause and was removed immediately from no. 1 and flown to Edwards. The aircraft was prepped for transport with the upper vertical tail and the two horizontal stabilizers removed before it was lifted onto the back of the flatbed truck the next day for the return to base. The regulator was later confirmed as the culprit.

Writing his post-flight report that afternoon back in his office at the NASA Flight Research Center, McKay noted, "The landing appeared to be very normal. If anything, the rotation after touchdown did not appear to be as backbreaking as some. . . . I think we can get the bird back and turn it around and fly it." Since the no. 1 aircraft was always used for pilot familiarization flights, Mike Adams had nothing to worry about with regard to his fast-approaching first mission.

The worst year for X-15 emergencies proved to be 1966. McKay's 6 May flight was the first, followed less than two months later by Bob Rushworth in the first attempted flight of the X-15A-2 with the external fuel tanks. Then McKay again into Smith Ranch, then one month later, Adams got the flight he had been worried might be delayed. The fuel tank bulkhead in the airplane ruptured, causing his first mission to become an emergency, landing at Cuddeback Dry Lake.

Flight 1-68 was the final one for McKay in the x-15. Perhaps he figured he was pushing his limits time and again, and it was best to leave while he could. The people he worked for, Stan Butchart and Paul Bikle, both also had input into Jack's decision. The last time McKay took to the air in support of the x-15 research program was as a landing chase pilot for flight 2-50 on 18 November 1966.

Stan Butchart said it was a tough decision with regard to McKay. "After I became director of Flight Operations, I picked Jack as the chief pilot. But he just didn't seem to want to do anything. . . . Nothing was happening; he wasn't organizing anything; he wasn't giving any directions; he wasn't doing anything at all. Finally, I had to move him out. I put [Donald L.] Mallick in and moved Jack into the regular pilot's room, and then it just kept getting worse and worse with his drinking and all."

When Stan spoke, it was easy to see he was still upset after all the intervening years. Whatever divide had occurred between him and Jack, Stan still tried to consider McKay his friend. That became hard, as he explained, "Way back when he was flying the x-15, Jack showed up out at Joe Walker's house one night, and he was drunk [even though] he had a flight the next morning. Joe told me about it some time later, and he didn't know what to do. . . . We couldn't let him go fly the next day." Reluctantly, Stan admitted, "Finally, we had to move him out of flying completely. It was a difficult task. When I talked to him, I was kind of surprised that he almost seemed relieved somebody was taking this step. We moved him up front, and he acted as our safety officer for NASA. I thought he might be upset, but he wasn't."

After that Jack McKay flew only a desk until 5 October 1971, when he retired from the agency after nearly twenty-one years. His career on the x-15 spanned more than six years, longer than any other pilot. He flew twenty-nine research missions, five less than Bob Rushworth's record, and one of those made it past the boundary of space, qualifying McKay as an astronaut.

"There was brain damage, lung damage, a lot of spinal cord damage," Sheri sadly remembered. "Years later, I talked to a couple experts in the field, and they said, 'The fact that he survived is amazing.' Most of the time, when there is an accident with a test pilot, they are killed outright. They aren't hanging around in the halls [at Edwards] as a reminder of what might happen. I think my father felt more alienated from everyone and everything. I don't think he could handle the brain changes, and he was in a tremendous amount of pain."

The toll got worse as the years passed. Once Jack retired from NASA, his wife, Shirley, decided it was time to take a job to help support the family. Her son Mark said, "Mom took care of us. She went to work at Rockwell around 1971. . . . She had eight children to take care of. . . . I respect Mom for what she went through." Mark's older brother, Mac, explained his dad did not take well to the concept of a working wife: "I remember the day Mom told Dad she was going to take a test for a job. She goes, 'I'm going to go work at Rockwell.' And Dad goes, 'No you're not!'" In the end, Shirley took the job, working on various programs, including the B-1 bomber, then doing reentry heat tile work on the Space Shuttle orbiters when they were under construction at the Palmdale plant.

Sheri spoke more about the repercussions, saying, "There was a big fallout for the family. My father started drinking more . . . yet everyone tried to carry on as best they could. The accident was so severe there were definite personality changes that became more apparent over time, like when my father would take utility bills and put them in his briefcase and take them to work. Then suddenly the lights would go off."

It was around this time when Sheri left to finish her schooling. "In the end, I wasn't there. . . . I put myself through college [at the University of California, Santa Barbara]. My father had always wanted me to go to [the College of] William & Mary, but he was just losing it, and I could tell he was losing it. I think it really scared my mother. . . . She was an amazing trooper."

The memories, bad and good, were reconciled when I spoke with McKay's children. The good ones didn't take long for them to find. A special one from Mark involved his father's hair. Over the years, McKay allowed his signature flat-top haircut to grow longer, but he always kept it slicked back in the style of the day. Mark said, "I remember he used like a gallon of Vitalis in his hair! So, every time we'd walk by when he would lie on the couch . . . he would whip his comb out and have one of us sit at the head of the couch and comb his hair for him. We didn't know how long we were going to be there! And if Mac or Charlie walked by, it's like, 'Here ya go,' and he'd hand them the comb, too."

The slightly longer cut was okay for their dad, but when it came to the boys, Jack had a different standard. Mark recalled, "He'd march us all down there to Homer's Barber Shop off Lancaster Boulevard. We always had to get a crew cut, every week, practically. So we were always shaved." Mac jokingly shuddered and recalled, "It was time for me to get a haircut. I must have been bawling my head off, because I knew what was going to happen. I was going to get a flat-top or just a butch. So, Sheri comes in, bless her heart, and goes, 'Well, Dad, Jesus had long hair.' Dad said, kind of sarcastically, 'Thanks Sheri, thank you.' Vroomp! Vroomp! and my hair was gone! Dad said, 'Come summertime, you're going to thank me for doing this.'" Charlie jumped in, saying, "I remember his motto was, 'Go around again until you run out of hair!'"

Mac spoke of another fond memory: "One thing I remember happened after Joe Walker's accident. We used to go up to Tropico Gold Mine in [Rosamond]. They had an annual gold panning contest and chili cook-off. During the event Dad presented an x-15 trophy on behalf of Joe Walker. . . . [His son], little Joey Walker, was there. In fact, Joey went up with us, and Dad let him present [the trophy]. It was bigger than he was!" This event most likely occurred in October 1966, attracting people from all over southern California and beyond. Local dignitaries, and even the reigning Miss Rosamond, Patricia Wombacher, attended the western-style event. Cowboy hats and scraggly miner's beards were everywhere. Mac's recollection is especially poignant in that I attended this event myself as a child and have wonderful memories of meeting Jack McKay and remembering how much I wished I could have the beautiful trophy with the x-15 model on top. When I told Mac of my own connection to the event, he laughed and said, "You probably saw us running around and getting in trouble!"

In the late 1960s and early 1970s the United States was changing. The war in Vietnam divided the nation, as nothing had since the Civil War. Protests, sit-ins, and marches occurred across the country. Sheri came of age as all this happened around her, and her stance was at odds with the conservative values of her father. It wasn't that Sheri wanted to distance herself from her parents and family as much as it was following her own path. She spoke of the turbulent times: "I was going to UC Santa Barbara. . . . There were some riots there and the bank burned, plus there was just a lot of radicalization. . . . I protested the war in Vietnam. I hung out at the women's center . . . and just a whole host of other things. It was very, very exciting. So, it's not that I was against my parents, but I was *for* something else and something more. . . . That was my era. I was born at a time to be involved in that, and I still carry a lot of those values."

As much as she embraced the changes of those times, she also regrets the isolation this caused with her father. "I don't feel like I was rebelling against my father; however, he certainly noticed. Before he died, I don't think we had a resolution. We were speaking [to each other], but I had been off traveling in Mexico just before he died."

The constant and excessive drinking to numb the pain from his injuries deteriorated McKay's body and soul. He was taken to the hospital in early 1975. Stan Butchart sadly remembered, "I went over to the hospital to see him a couple of times. I bumped into Bill Winters, our doctor we had out there, and he said, when he went in to see Jack, Jack looked at him and said, 'I'm not going to make it this time, am I?' So he knew he was pretty bad."

Although it appears obvious in retrospect, McKay's problems started that fateful morning of 9 November 1962, upside down, trapped inside X-15 no. 2 on Mud Dry Lake, the weight of the aircraft pressing against his head and spine. But not everyone agreed. Paul Bikle succinctly stated, "What killed Jack was pretty obvious, and it had nothing to do with that accident. He had incipient diabetes, and we took him off flying [status]. He took some drinks and he died of cirrhosis of the liver." It was obvious when we spoke, Bikle was still dealing with the anger of losing someone whom he cared for deeply.

John remembers first hearing the news his father was in the hospital dying. He was half a world away, serving aboard the USS *Hancock* (CV-19), nearing Subic Bay in the Philippines. "A lot of the guys were just getting their sea legs and such, when I got a telegram from my sister, Joanne, that my dad was turning for the worse. Then I got another one that it was very bad. . . . I put in a chit to take emergency leave when we got to the Philippines. We were low on mechanics and, honestly, I was a damn good one, so they didn't want me to leave." John went to his commander, who "was a great guy, but I think he was pressured to deny [my leave request]. I remember just sitting out on the catwalk, thinking, 'This is bad shit.' So I went to the chaplain, and he looked at the situation, and said, 'I need more blood.' That's what he said. He goes, 'This just isn't bad enough.' I said, 'Thanks a lot, you little dick.' So I walked out of there. The next morning, my dad had about a week to live."

There wasn't much time, so John decided he had to do what was necessary to get home. "I [had] Joanne get hold of Forrest Petersen, who at that time was number three in the navy, as I understand it. I didn't want to do that, but I figured, if it worked, I only get one chance to see my dad before he dies." Suddenly the doors opened, and John got his leave home. "When I got back, I walked into his hospital room . . . and Dad was very happy to see me. Then he started losing it, talking about when he was a kid in Virginia, but he had a sense of clarity just for a minute when I walked in that room, and that was neat."

Soon after, his father passed away. John Barron "Jack" McKay, fifty-two years old, husband to Shirley, and father to three girls and five boys, died on 27 April 1975.

At his father's funeral, John was impressed that all the stops were pulled out to honor him. "It was absolutely outstanding. . . . I wore my dress whites. . . . They did a flyover with an F-104. That was the first airplane I ever crewed at NASA. It was [tail number] 812. . . . There was a gigantic wake that night at my Uncle Jim's. If you were a person who wanted to be seen with somebody in the world of aviation, that was the place to be that night."

The funeral was done well, but Sheri still has regrets her father's medical problems were not handled better while he was alive. "I think it definitely was brain damage that didn't show up for a while. When they did the autopsy, it showed quite a bit of brain damage. . . . I mean that part of it has to come out. I feel a responsibility at least to talk about it. . . . I know he told one of his friends, he wished he'd been killed in that accident."

John recalled what transpired after he returned to his ship in the Philippines. He said he was asked, "Why didn't you tell us who your dad was?" John was not happy with the question and what it inferred. "I said, 'Why'd I have to do that?' I think it was very innocent on their part, but it just goes to show that, if you know somebody, you can move things around a little bit. But that didn't have to happen. My response to these guys was, 'If my dad was a garbage man, it shouldn't have made a difference.' They looked at me and said, 'Yeah, I guess you're right.'"

Mark said that losing his father was a very confusing time. "I was in tenth grade, and at that time I couldn't concentrate on life, homework, girls, the whole nine yards. I bawled at his funeral, and it took me three years to get over it. My grades suffered, and I had to go to counseling. . . . To this day I still feel like I lost a part of him—and me—back then, and lost some direction, because I was quite young, and that's when you need your father." Now Mark understands more, saying, "He enjoyed and loved what he did. He wasn't the type to give up. He always went back to what he loved."

Opinions on McKay varied greatly. Flight planner Johnny Armstrong said, "What you saw was what you got with Jack. No put-on at all. He wasn't as precise a flier as some of the others, but Jack was what was referred to as a good stick-and-rudder man. He could bring an airplane back if it wasn't feeling well." Roger Barniki, who worked on the pressure suits worn by the rocket plane pilots, thought Jack was one of the best. "If I had to pick one who seemed to have a natural flight ability, it was Jack McKay. He was probably the most natural of the pilots I flew with. I don't know how better to describe it." x-15 operations engineer, John McTigue agreed, "Jack McKay would always bring the airplane home. He might not make the most elegant flight, but he landed probably the best of any pilot I ever flew with. He had the smoothest hand. Sometimes when we were in the [c-47] Gooney Bird, I couldn't tell when he touched down. I mean, he'd be coming across the lake a foot or two off the ground, then finally, he touched down and that was it. Jack was a natural pilot."

In his book, x-15 Diary, Richard Tregaskis quoted an anonymous engineer, saying about Jack McKay, "He's just a damn good Joe. He'll fly anything. He's the kind of test pilot that'll fly for you all day long. He doesn't like to criticize the airplane. If it'll fly, he'll fly it—and maybe enjoy it. Then he goes home, and maybe has a few on the way." McKay's son, John, laughed and said, "For the longest time there were six, seven, eight kids at home. And I'm not sure I'd want to come home to that." Having a family that big, he said, "should be against the law!"

When asked about the first thing that comes to her when thinking of her father, Sheri thought for a moment before she responded. "I can still feel where he kissed me on the neck." She took her left hand and rubbed the area lightly. "He would come home from work and sort of chase us around . . . and he'd grab me [and say] a silly little thing: 'I'm going to get you, chicken.' Then he'd kiss me on the back of my neck. I can still feel that."

Sheri then flashed forward to when she was a teenager: "Being hard-headed, sometimes we'd lock horns over whatever. I'd go stomping off to my room, and pretty soon I'd hear a knock on my door, and Dad would come in and say, 'Well, I know you're really angry right now, but I just want you to know one thing.' And I'd say, 'What's that, Dad?' And he'd say, 'You just remember, I love you.' It was so sweet."

Lost for a moment in thought, she looked up and said, "As an adult looking back, I would have to say he was an explorer and adventurer—the kind of person that goes from the known into the unknown, and that requires a lot of courage. I think people with a lot of courage are different."

6. Straight and Steady

You can't have that kind of a program and not support
it, because the last flight is just as serious as the first
flight. . . . Nothing changes. You just know a little bit
more about the airplane.

Robert A. Rushworth

X-15 flight 2-45-81 was set up to fail. No one felt that way when it was first
planned, and no one thought it possible up through the launch near Mud
Dry Lake on 1 July 1966. In retrospect, most of those involved with the
flight, including pilot Robert Aitken Rushworth, understood why the mission came to naught.

After the near disaster of Jack McKay's crash landing in aircraft no. 2 in
November 1962, the X-15 was rebuilt with one primary objective: speed.
Two external tanks were designed and fitted to the lengthened fuselage to
extend the burn time of the LR-99, and thus the top velocity, into the range
of Mach 7, and possibly beyond. Those tanks of liquid oxygen and anhydrous ammonia were full for the first time on flight 2-45, and this became
just part of the problem that occurred that summer morning after Rushworth dropped from the B-52 pylon.

Bob Rushworth spoke about the flight: "It was the first time with fuel in
the [external] tanks. . . . Everyone was concerned about the aircraft being
stable during release because we had to increase the launch speed for the
X-15." The weight of the external tanks with the fuel and oxidizer they carried equated to a 35 percent increase over an unmodified X-15. This meant
higher velocity was required to generate lift. "We couldn't go to 45,000
feet [to launch from the B-52], as we did normally to achieve that required
speed. With extra fuel, we had to launch a little bit lower than that."

Getting rid of those tanks also brought into play a new set of requirements with regard to flight planning. First was the possibility of the LR-99 not igniting immediately after release. This necessitated an abort and emergency landing at the launch lake with full tanks being dropped, then all X-15 internal fuel jettisoned prior to landing. Second, if the rocket engine started, then external sources were the first to be used, and once depleted the empty tanks were ejected as the flow switched to the supply inside the X-15 fuselage. Full or empty: those were the only options that were contemplated. What about in between? What if the tanks had to go when there was some amount of fuel and oxidizer still aboard?

Bob explained, "There was nothing in our flight plan that said you release the tanks half-full or half-empty. [We felt we] couldn't do that. . . . There was nothing in between." Harry Shapiro, a North American Aviation external tank engineer, related, "The drop tanks and their recovery system were kind of unique because you had to know if the tanks were full or empty. If full, you didn't blow them away from the aircraft, you let them fall off by gravity. If empty, you had to blow them away, so they wouldn't hit the aircraft."

The launch, at 11:02 a.m., went well. Attaining the extra B-52 speed proved to be no problem, and the LR-99 ignited into action as soon as Rushworth pushed the throttle to the 100 percent stop. Bob said, because of the addition of the tanks, "the pitch-up tendency was greater, and the stability of the airplane was less." His piloting skill overcame any inherent difficulties, and the flight proceeded normally, heading for 100,000 feet at Mach 6.

Inside the control room at NASA's Flight Research Center, engineers watched the incoming telemetry for anomalies. They were ready to jump if some abnormality crept in. About twenty seconds into flight, one of the engineers did just that, passing along to Jack McKay as NASA 1 that the fuel from the external ammonia tank was not flowing. Jack immediately radioed the information to Rushworth. The flight was less than halfway to the sixty-second depletion of the tanks. The conundrum of being half-empty or half-full was irrelevant, as jettison was only supposed to be an empty or full proposition. Doing so at a different time had unknown, and possibly fatal, aerodynamic consequences for the X-15. With no fuel flow indicated, there was no choice in the matter. Bob knew the predicament, replying to McKay, "Roger, understand," but then asked about the tanks. "What else to do for them?" McKay made it clear: "Shutdown. Tanks off, Bob." Rushworth had to shut down immediately, jettison the tanks, and hope for the best.

Aitken is an unusual middle name, unless part of your heritage happens to be Scottish, as was the case for Bob Rushworth. It signifies the name Adam on his father's side of the family. As for Rushworth itself, its origins can be traced back to the Norman conquest of England in 1066. The name evolved from a phrase to describe an enclosure of rushes, a type of hollow-stalked plant often used in weaving baskets or furniture.

Bob Rushworth joined the ancient family heritage on 9 October 1924 in Madison, Maine. His hometown had a rich history of its own, originally named Barnardstown, then changed to honor the name of James Madison Jr., the fourth president of the United States. Settled by the French, the area was cohabitated by the Native American tribe of Abenaquis; they fought together to stave off English encroachment of the area in the early 1700s. A brutal incursion by British troops on 23 August 1724 left more than two dozen warriors, women, and children dead, along with a revered Jesuit priest, Sebastien Rale, who had organized the Abenaquis against the onslaught. Afterward, the Abenaquis abandoned the area to settle in Quebec, Canada.

World War II was in full swing by the time Bob graduated Madison Memorial High School in June 1942. Still a few months under draft age, he decided to continue pursuing education rather than joining the military. He settled on the Hebron Academy, nestled about halfway between Lake Auburn and South Paris, fifty miles southwest of Madison. This was a preparatory school for college, but by May 1943 Bob and twenty-seven others left to join the war effort. This was partly by design from Rushworth and partly due to the unforeseen circumstance of being drafted, the only X-15 pilot to enter the military in this manner.

The previous month, he and a friend decided to enlist through an aviation cadet program. After finishing the tests, Bob was told he passed but his friend failed. They both returned to finish their last month at Hebron, and the first Monday after that happened, Bob received his draft notice. By June, even though he had already passed his aviation exams, Bob was an enlisted draftee in the U.S. Army.

Within a few weeks, the problem was sorted out, and Bob transferred into the flying program for which he had attempted to volunteer. Like so many pilots entering the war, he had his sights set on the heroic image of a fighter pilot, specifically hoping for the Lockheed P-38 Lightning, such as was flown by Joe Walker. Instead, Rushworth earned his wings in September 1944 and was assigned transport duty with the Douglas C-47 Skytrain, the military variant of the DC-3 airliner. His unit, the 12th Combat Cargo Squadron, was assigned to the India-Burma-China theater, often running supplies in support of British troops.

As the end of the war neared, his assignment changed to the Curtiss-Wright C-46 Commando, with the critical path over the Himalayas known as "The Hump." Starting at sea level from Brahmaputra, India, the Air Transport Command pilots made a hazardous 530-mile journey that tested the ceiling limits of the aircraft, topping out at more than 15,000 feet before dropping into south-central China, landing at Kunming Airfield on the Yungui Plateau at 6,200 feet above sea level. With the closure of the Burma Road by the Japanese in 1942, these cargo airplanes were the only lifeline for the Chinese army under the command of Gen. Chiang Kai-shek. Rushworth made his runs across the roof of the world unscathed; more than 600 airplanes and 1,000 men did not have the same luck and were lost over this route.

Bob separated from the military in January 1946, heading to the University of Maine for a degree in mechanical engineering in 1951. The school in Orono was so overcrowded with young men returning from the war that an off-site campus was opened at an old naval air base at Brunswick to accommodate the extra students. That lasted until his sophomore year, when Bob was able to transfer to the main campus of the university. Over that summer, he married his sweetheart, Joyce, who had recently graduated from nurse's training. Her job in that field helped secure a reasonably comfortable living for the young couple.

While still in school, Rushworth was reactivated at the beginning of the Korean War, transferring into the Air National Guard. He didn't have to go far, as his new assignment was with the 49th Fighter-Interceptor Squadron at the newly rebuilt Dow AFB, on the west side of the city of Bangor. Bob finally got his wish to officially transition from cargo planes to fighters, in this case, the Lockheed F-80C Shooting Star.

Not being directly involved in the war effort, Bob decided to see how far the military could further his education. He settled on the Air Force Institute of Technology at Wright-Patterson AFB, Ohio, achieving his second bachelor of science degree in 1954, this time in aeronautical engineering. Even with his education and flying experience, he complained later about slow promotions for unspecified "strange reasons." This may have come, in part, from his lack of overseas involvement, but then, not everyone was active in a theater of war. It was to take him a decade to move up from the rank of captain to major, but he later made up for that with a relatively quick promotion to lieutenant colonel while serving the air force on the x-15.

The path that led him to that coveted position opened when he applied for the U.S. Air Force Test Pilot School and was accepted into Class 56C. In an oral history project for the military, Rushworth was quoted as saying, "Probably the most important thing you get out of [the school] is being able to converse with engineers and designers on how airplanes fly, versus what they look like on the drawing board. . . . If one is good at that kind of a job, he can converse and make the engineer understand what he is doing while he is flying an airplane." After completing the course in January 1957, Bob was accepted at the Fighter Operations Branch at Edwards. For Rushworth it was the place to be, and in January 1958 he was selected for the x-15 program.

Paul Bikle ran rocket plane research at NASA. He admitted to not being enamored with the role of the Test Pilot School. "We didn't care whether they went to the school or not. It made no difference. The school, to me, has always been greatly overblown in the air force." Paul continued with his rebuke of the mystique, saying, "We had lots of good test pilots who never went through the school. The only thing I objected to was this aura of a man with excellence that they tend to throw on somebody because they graduated from there. I'd say, 'Hey guys, come over to Test Ops for the air force or NASA for a year [and you'll get] a lot more experience.'" One problem, Bikle noted, was too many graduates. "It gets like a lot of the schools, where you have more applicants for test pilot jobs than you can shake a stick at. . . . Only a small percentage of them really go into test flying anyway. There [aren't] that many jobs."

Bikle went on with a specific concern for Rushworth's selection on the x-15, which was raised by Bob White. "I can remember back when the air force first selected the pilots. First we had [Iven C.] Kincheloe and White. When Kinch was killed, I remember White came up to my office and said,

'You've got to talk to the general.' And I asked, 'What's the matter?' He [re-plied], 'Why, they're going to pick Bob Rushworth as the second pilot!' So I asked him, 'What's the matter with Rushworth?' [White said], 'He's just a goddamn cargo pilot!'" This was an especially interesting comment from White, since he, like Rushworth, had also flown the c-46. In White's case it was carrying troops during the Korean War, rather than Rushworth's car-go flights over the Himalayas in World War II.

There was even worse news to come as far as White was concerned. Bikle related, "White said, 'He's not motivated enough. We asked if he wanted to get on the [x-15] program, and he said he had to think about it for a week!'" Bikle said, "I've often thought about that when I moved over to NASA and they actually got to flying the airplane, since it turned out Rushworth was probably one of our most productive pilots."

The original plan called for the x-15 to have three pilots each from NASA and the air force, along with one from the navy. Adding the contractor pi-lot, Scott Crossfield, into the mix meant a total of eight. With the loss of Kincheloe in the F-104 accident, that left just two for the air force. The de-cision was made after the explosion of the no. 3 aircraft on the stand dur-ing the LR-99 engine run in June 1960 that no one else was to be added to the mix. NASA kept their three pilots, although Neil Armstrong was under-utilized for a long while after his first two missions.

Once the research flights were in full swing, there was no further quar-rel from White of Rushworth's qualifications. However, White continued his role as the primary military pilot, so Rushworth, like Armstrong, still had a long period of little to do with regard to the x-15.

Six pilots made first flights in 1960, starting with Joe Walker on 25 March and ending with Neil Armstrong on 30 November. Rushworth was in the number five position, sandwiched one week after Jack McKay and twenty-six days before Armstrong.

Rushworth's first was flight 1-16-29 on 4 November, with a scheduled 50,000 feet at Mach 2, using the eight-chambered LR-11 engines. There was nothing out of the expected during this familiarization mission through the local Edwards air space. He launched away from B-52 no. 008, under the command of Maj. Fitzhugh L. Fulton Jr. and Maj. Frank E. Cole, at 12:43 p.m., and landed on the Rogers lakebed a bit less than nine minutes later, at 12:52 p.m.

This was the beginning of an x-15 career that spanned thirty-four flights and fifty-one attempts, more than any other pilot who flew the sleek Inconel-X aircraft.

Less than two weeks later, Bob turned around and did it again. On mission 1-17 he completed a near-perfect second flight. The only discrepancy was the lower LR-11 shut down prematurely. Rushworth did a quick restart, finishing the profile pretty much as planned at Mach 1.90 and 54,750 feet. It was nearly eleven months until his third flight, 1-23, on 3 October 1961.

Through the intervening time, Bob continued his duties with the air force, evaluating several Century-series fighters, such as the McDonnell F-101 Voodoo, the Convair F-102 Delta Dagger, and the Mach 2 Convair F-106 Delta Dart.

Also during this time, the United States began competing with the Soviet Union for supremacy in outer space. In the infancy of the race for space, the most important question looming in the minds of the American public was: Who was going to put the first man into space? As Vice President Richard Nixon had pointed out during the unveiling ceremony for the first x-15, most assumed the rocket plane was going to upstage anything the Soviets could muster. But that didn't seem to be enough. Rushworth recalled, "Had Sputnik not come along when it did, causing [President John F.] Kennedy to put so much emphasis and money into the Mercury program, I'm sure we would have led Mercury by quite a bit more with the x-15. . . . The emphasis was now on the [lunar] effort, and that meant Mercury had to go fast."

By the time Alan Shepard was launched on his 5 May 1961 suborbital flight with *Freedom 7*, the x-15 had already performed well thirty-six times. When John Glenn made the first American orbital flight on 20 February 1962 in *Friendship 7*—only the third flight in the entire Mercury program—the rocket plane was poised for mission forty-nine. Bob put it in perspective, saying, "We had done everything with the system except the altitude work by the time [Mercury] flew, and, obviously, with all the equipment we had on board, they used about the same thing: same pressure suits modified a bit, same reaction control system modified to handle that vehicle, in other words, a lot of commonality."

As Rushworth mentioned, the pressure suit was a critical component of the entire x-15 system. Roger Barniki of NASA and Ralph Richardson from the

U.S. Air Force were instrumental in the success of this pressure suit. Before them, it was David M. Clark, a man with a vision of how to make women's girdles, who made it all possible.

Latex rubber was first manufactured into elastic thread in the 1930s, then woven into a fabric called Lastex. The innovation served as the foundation of the modern ladies' undergarment industry. It led Clark to perfect a method of using the revolutionary fabric to automatically machine knit single-piece stretch girdles. He started the David Clark Company in 1935 in Worcester, Massachusetts, to capitalize on his invention. In 1941, as the entire nation got behind the war effort, Clark started a line of military flying gear, such as anti-g suits and ear protection, because he was, as Scott Crossfield called it, "a compulsive gadgeteer."

Originally, Clark saw this almost as a hobby and poured a lot of his own money into the various products. His reputation in aviation grew quickly, and eventually the women's unmentionable business was supplanted by this new line of work.

With some humor, Crossfield wrote in his biography of why Clark was so successful: "The structural loads imposed on a girdle, as we all know, can be tremendous, and a machine that can build a good one automatically is an amazing engineering accomplishment. . . . With his ingenious machine Clark had all but cornered the important, expanding girdle market. Braving new frontiers, Clark moved into manufacture of brassieres, which, considering *those* structural loads, was even more awesome."

Clark first came to Crossfield's attention in 1951, in relation to his test piloting duties at high altitude that were on the horizon. He said of his first meeting, "David Clark, the owner and president of the company, turned out to be one of the most interesting men I have ever met in the aviation world. . . . [He] seemed to be always in high-speed motion."

Roger Barniki, NASA's life support specialist, spoke of Crossfield's role in developing a full pressure suit for the x-15. "The pressure suits were being designed by David Clark, [and] being overseen with great input from Scott Crossfield. A lot of things the early suits had were Scotty's designs. You see in pictures he has a face mask dam in the helmet, and the other pilots didn't. The early suits had a roll-up bladder where it didn't have a zipper. It was two pieces; the zippers were only in the link-net."

That link-net was a crucial development for what became the rocket plane's MC-2, and later A/P-22S, suits. Where previous suits just ballooned out under pressure, with mobility all but impossible, Clark's material allowed free movement. He described the material as akin to a Chinese finger puzzle. For example, when an arm bent inward, the link-net contracts in the crook of the arm, while expanding at the outside of the elbow.

As Roger mentioned, Crossfield's suit was different than the model eventually used by the rest of the x-15 pilots. Scott's was the MC-2 with internal rubber bladders, like a bicycle tire inner tube to hold in the pressure. The A/P-22S featured pressure-sealing zippers. Ralph Richardson, Barniki's counterpart in the air force, said, "That was a major advancement, when they got zippers you could depend on." Ralph said that, up until that point, zippers were fairly unreliable. "I had one on in an altitude chamber when the zipper let go, which is not a very good thing. . . . We were over 100,000 feet. Sergeant Crow, [who was in the chamber as a second person for safety,] saw the zipper start to go. Before I even knew it, he used the dump handle and got us back below 35,000 feet." His quick action saved Richardson's life. If Crow had dropped the altitude too far, Ralph smiled and explained, "then you start hunting for new eardrums." Difficulties of that sort were no stranger for anyone working with pressure—or lack of it. "When you're playing that game long enough, you're going to have ear trouble."

Richardson spoke of another area that caused problems: "Most of the tests [at the Aeromedical Laboratory at Wright-Patterson] were done in an environment where it was air-conditioned. However, when you brought it to the desert, and it gets to 120 [degrees Fahrenheit] on the ramp, we ran into problems in a hurry. The first tests I ran with it, I'd wear that suit for an hour, and I could lose from ten to fifteen pounds. . . . It was just pure fluid loss."

Describing the pressure suit layer by layer, Ralph said, "You started out wearing long underwear. Its primary purpose was the wicking ability. . . . When you perspired, if there was ventilation, you got cold. Without it, you'd really be in a mess. Next was the air-tight rubber layer. Then over the top you had this link-net. . . . It was very effective, but it had to be custom-made for each individual. It retained the [pilot's] shape to get maximum mobility. The tighter the fit, the better that mobility. Over the link-net was the outer garment, the aluminized cloth, and attached to it was the integrated parachute harness."

Roger Barniki transferred from NACA's Lewis Laboratory in Cleveland, Ohio, to the High Speed Flight Station at Edwards—in the "Great Desert," as he called it. "The managers didn't know what to do with me. . . . They put me on the B-29 for the X-1 drops. I wasn't even twenty-one, and they had me flying as part of the launch crew. One day, somebody woke up to that fact, called me in their office and said, 'You've got to go get your mother's permission.' And I said, 'You've got to be absolutely kidding! I've got a kid on the way and you're not going to have me ask my mother to be on this airplane.'"

As the X-15 program was coming into being, Roger was sent to North American Aviation's plant in Los Angeles to learn the life support systems. "After we got indoctrinated at North American, the program moved up [to Edwards]. . . . My job then was following the suits over at the altitude chamber. By this time it was run by a reserve [air force] major, Ralph Richardson. He and I were like that," Roger said, clasping his two hands together.

Barniki and Richardson remained not only colleagues but friends throughout their lives. Richardson spoke of the circumstances leading up to that first meeting: "I was a fighter pilot with a physiology background. . . . I flew the P-51 and the F-86 [in Korea]. After [that], I went to Japan, where we set up the first altitude chamber in the Far East. . . . When I came home, I was sent directly to Edwards. I got there in 1957." This was just in time to get in on the ground floor of the X-15.

The way the pressure suit was created was typical of so much on the X-15. Official channels were often circumvented, and thus costs tended to be much lower. Ralph explained how they worked the system, saying, "We ended up with a machine shop. In the air force that's illegal. . . . They have a machine shop and individual places can't have them. We had a milling machine and we had all kinds of lathes. When we first got the big lathe, the [inspector general] came around and wanted to know what it was. I was kind of stuck for an answer, then I told him [the lathe] was an 'Injector Nozzle Sizing Tube.' They came down and put that nomenclature on the thing! The milling machine became a 'Neck Ring Forming Tool.'" For Ralph, it was a fine example of government bureaucracy at work. "We had all this stuff on the books, and they had a heck of a time with it after I left to sign off the records."

Full pressure suits from the David Clark Company, with a history going back to the early gestation of the x-15, were later incorporated into the American space program and were in use all the way through the end of the Space Shuttle. It is safe to say that whatever course is taken in the future, the heritage of the x-15 suits will be involved.

On 4 October 1961 Rushworth returned to the x-15 for flight 1-23. He had been looking forward to getting back to work on the research program, so it was time to once again become familiar with the aircraft. One huge difference from the two flights that came before was that this was to be his first time with the LR-99. Bob said, "Each flight was as exciting as the first and second ones I did—much more exciting when we had the big engine, because that was a Cadillac system." I asked Bob if it was a bigger kick in the pants than with the LR-11S, and he replied, "Was it ever! Even at half-throttle, it was twice as much."

A second big change was that Bob was responsible for the first mission to fly without the large lower ventral fin in place. This design feature had been added originally because at high angles of attack, the upper rudder could prove ineffective, as it was blocked from the airstream by the nose-up orientation of the fuselage. Having a lower tail surface of nearly the same square footage as on top was supposed to give full directional control. Wind tunnel data as the program progressed hinted this might not be the case.

Rushworth discussed the process that led to the decision to try it without the lower ventral, saying, "What the tunnel showed was that we had better control over the airplane at high angles of attack with the lower ventral off. In other words, there was too much control from the excess tail, and it would cause us directional problems. That only became important if we lost our stability augmentation and you had to manually fly the airplane without [that system]. It would have been almost impossible. With the lower ventral off, it became possible to fly it without stability augmentation."

Dropping near Silver Dry Lake and flying a nominal mission to 2,830 miles an hour that morning proved the wind tunnel data correct. Bob commented after the flight, "The roll mode appeared to be less sensitive than on previous flights." This was understandable and expected, due to the smaller overall tail surface, and was right where the engineers wanted it to be.

Over the next seven months, Bob White was still busy knocking off speed

19. Bob Rushworth is helped from the cockpit at the end of mission 3-10 on 4 October 1962. During his ninth flight Bob attained Mach 5.17 and 112,200 feet. Courtesy of the Armstrong Flight Research Center.

and altitude records for the air force and appeared less than willing to give the spotlight to his backup pilot, relegating Rushworth to again wait in the wings. Once his latest hiatus was finished with mission 2-22 on 8 May 1962, Bob finally appeared to settle into routine x-15 assignments, completing seven flights in a little more than five months. This was a huge increase over his first three flights, which took nearly a year.

Not long after this series, White decided to leave the field, completing his time on the x-15 in December 1962. Rushworth moved up into the prime pilot slot and also became the sole pilot for the air force, until Joe Engle arrived a year later to provide some relief.

Medical data on the pilots became important, especially after the potential difficulties with high heart rates. Rushworth became the first to have his electrocardiogram successfully telemetered in real time during a flight, with mission 2-27 on 20 August 1962. "Everybody at that time was a guinea pig," Bob said. "That was one of the goals of the program, to determine what the psychological and physiological effects of flying at high speeds and high altitudes were on the individual. . . . I think it bothered me less than it did the other pilots, because I had accepted it. Walker and White, and probably a couple of the other guys, just didn't want to have anything to do with it."

This was only months away from White's final flight, and based on Rushworth's assessment of White's feelings, maybe this emphasis away from simply flying higher and faster may have hastened White's reasons for wanting to return to active military life. Flying airplanes for evaluation or combat, in his book, didn't mean being personally wired for data. Rushworth, on the other hand, was perfect for this role, as he understood the totality of the research program and how, in some cases, the pilot was just as important as the aircraft.

The pulse rate controversy was definitely one that stuck in Rushworth's mind and was one he helped explain. "I remember they were quite concerned with the initial pulse rate at launch and didn't understand why it went down suddenly right after launch, then peaked again when we hit another high, [such as] in our trajectory, then again just before landing." For Bob, the reasoning was clear. "You're sitting there thinking about what's coming next and it's bound to peak out just before launch. Then you get busy flying your profile and it goes back down. The next big point in your trajectory, [the heart rate is] going to come back up again. And when you get ready for landing, you've got a lot of gliding to do—and thus a lot of time to think. . . . You've got one shot at it, and that's it." Naturally enough, a person's heart is bound to push more blood as their body responds to the need to think, reason, and perform.

Four astronaut qualification flights occurred in 1963. All but one of those were accomplished by Joe Walker, as he pushed the overall altitude of the x-15 to its maximum height of 354,200 feet. The other qualification was for Bob Rushworth, flying mission 3-20 on 27 June.

By mid-morning that day, everyone was in position near Delamar Dry Lake, and Bob hit the launch release from the B-52 at 9:56 a.m. The LR-99 acceleration was on the money, running for a fraction past eighty seconds on the mission clock. At his debriefing, he said, "The deceleration at that point . . . [was] practically nothing. I could hardly feel that the engine shut down—it was just quieter." Bob talked more of the sensation: "After [the LR-

99] burned out, you just went from four gs eyeballs-in to zero gravity. We were in that kind of a trajectory. . . . From there on, it was very comfortable."

His terminology of "eyeballs-in" may sound strange, but it is in common usage among test pilots, especially those who regularly experience high g forces. The idea is that the force of the acceleration pushing the pilot back during boost becomes "eyeballs-in." The feeling is similar, but much more pronounced, as accelerating on a freeway in a high-performance sports car. On an altitude flight such as this for Rushworth, when the engine shut down, there was no drag on the aircraft as it continued to sail through its ballistic arc above the atmosphere. Acceleration simply ceased. For a speed flight lower in the atmosphere, at the moment the engine stops, there is a great deal of atmospheric drag, which produces a negative g force on the pilot, pushing him toward the cockpit windows and to what is termed "eyeballs-out."

The x-15 continued arcing upward toward a peak altitude of 285,000 feet, almost 4 miles above that necessary for Bob to be referred to as an astronaut by his military service. "As I approached the top," he said, "I could not get a good reading on the altimeter because it's off-center, and [the instrument is] recessed, so that I don't see any of the markings on the gauge. . . . I had to estimate as to what I had." Bob didn't want to take any chances of getting misaligned for his reentry. "I just let [it] go over the top straight." Then he glanced out for a short and precious moment to actually see where he was: 54 miles above the planet. Bob nonchalantly remarked, "Took a quick look out and saw all the clouds all over the ocean, [and] what the ground looks like from up there, [then] got back in the cockpit."

Through the structure of the x-15, Bob admitted he "could just barely hear the [ballistic control system] rockets fire. They were very faint to me." Then, all was quiet as he coasted for a time downward toward the wisps of atmosphere that were about to put great pressure and temperature against the Inconel-X structure. The x-15 was required by the laws of thermodynamics to expend exactly the same amount of energy reentering as it did pushing above the atmosphere in the first place.

"It was exactly the time Joe [Walker as NASA 1] called Mach 5," Bob said in his post-flight debriefing, "my pressure instruments all jumped up and started to work. . . . I had been observing a noise which sounded like wind whistling, and I was trying to determine whether it was inside the airplane

20. Bob Rushworth lands x-15 no. 3 at the end of mission 3-20 on 27 June 1963. He is coming back to the Rogers lakebed after attaining 285,000 feet and his air force astronaut wings. Above the x-15 is the f-104 landing chase, which demonstrates their similarities in shape and size and why it was used extensively to simulate x-15 landings. Courtesy of Defense Audio-Visual Agency.

and [its] systems, or whether it actually was outside. . . . Apparently, it had become so quiet and I didn't realize it, that I [heard] the effect of going under some [dynamic pressure] again."

As Bob talked about becoming the third x-15 pilot to earn his astronaut rating, he said, "Almost every flight was a special flight." Once he dipped back into the atmosphere, it was time to bring it down for a landing at the home base of Edwards. Rushworth slid the skids onto the lakebed, but getting the nose down was not as smooth. "It was a six g jolt that hit you when the nose gear came down. It was kind of a sudden smash onto the ground." That sudden drop was caused by the fulcrum effect of having the rear skids so far back on the rocket plane.

On this fourteenth flight for Rushworth, he became the third of eight x-15 drivers to become an astronaut at its controls. Less than a month later, on 25 July 1963, Bob was asked to attend a special ceremony at the Pentagon in Washington DC to officially receive recognition for the accomplishment. With his wife, Joyce, by his side, his astronaut wings were presented by U.S. Air Force chief of staff Gen. Curtis E. LeMay. Also present were the

21. Bob Rushworth receives his astronaut wings in a ceremony at the Pentagon. *Left to right*: Sen. Edmund Muskie, Rep. Clifford McIntire, Sen. Margaret Chase Smith, Maj. Robert Rushworth, Joyce Rushworth, and Gen. Curtis LeMay. Courtesy of the Margaret Chase Smith Library.

two senators from Maine, Edmund S. Muskie and Margaret Chase Smith. Rushworth was the first person born in Maine to fly into space, which greatly interested Senator Smith, who was a vocal supporter of space exploration and served on the Aeronautics and Space Sciences committee in the Senate. With Bob's fame as a hometown astronaut, Senator Smith thought this might have political potential. She asked Bob if he might consider a run for Congress from his district. Joe Engle recalled Bob's response: "I'm terribly honored, but during my professional career I've become so accustomed to basing my decisions on facts, I don't think I could hack the transition into politics."

With politics now firmly behind him, Rushworth was not yet half-finished as the workhorse pilot on the x-15 program. Three more flights passed for Bob before the end of 1963. These were primarily in support of experiments in aerodynamic research. The first of these was 1-38 on 18 July, where samples of ablative material were mounted to the aircraft in anticipation of the return of x-15 no. 2 after it was rebuilt following the rollover accident at

Mud Dry Lake. The ablative was planned to protect the skin if the extra burn time of the external tanks was able to move the speed into the Mach 7 range. The second was on 7 November with flight 3-23, to test theories concerning a razor-sharp leading edge on the upper vertical tail.

By 5 December Rushworth was ready for a high-speed run in aircraft no. 1. He powered up to 101,000 feet on mission 1-42 and pushed the x-15 to Mach 6.06, or 4,018 miles an hour. The Mach number is calculated using several factors, primarily based on atmospheric height, composition, and temperature. This was first discovered by the nineteenth-century Austrian physicist Ernst Mach. Based on the conditions present on that day and at his altitude, Rushworth edged out White's previous record speed of Mach 6.04, achieved on 9 November 1961. However, because of the vagaries of the Mach calculation, White still held the actual record in miles an hour, at 4,093. These two records were the fastest ever attained in the unmodified x-15 aircraft. Pete Knight was later to surpass the number twice when flying in the advanced x-15A-2.

On 28 January 1964 the x-15 reached a major milestone, mission number one hundred. This number had not been achieved by a single rocket plane type since the xs-1 on 1 December 1948. Rushworth had the distinction of being the pilot who was assigned this magic number. Flight 1-44 was officially to test stability and control of the x-15 using only the speed brakes on the upper vertical tail surface, evaluating this technique for later flights where the lower ventral was to be replaced by the mounting pylon for the supersonic combustion ramjet experiments.

By this point, which was to be the halfway mark for the entire research flight series, the x-15 had been carried aloft 142 times, making for approximately one flight in three coming home to Edwards still attached to the mothership. To commemorate the special flight, Maj. Gen. Irving L. "Twig" Branch, the commander of the U.S. Air Force Flight Test Center since July 1961, served as the b-52's copilot for the flight. Maj. Russell P. Bement was the general's commander on the mothership that day.

Achieving one hundred research flights had been the original goal for the entire x-15 program when it was first proposed. Rushworth spoke of why this was the case: "By the time they [expected to] get to the hundredth mission, and hopefully satisfied all the requirements of the program, all the airplanes would have been expended. They didn't expect three airplanes to go through any more than 100 missions. That was a reasonable forecast based on the past history of airplanes just wearing out from this kind of propellant . . . plus taking it apart and putting it back together continually, not to mention accidents."

Their reasoning was sound. Bob continued, "The no. 3 aircraft blew up, but since it was so early in the program, they decided they just had to rebuild it. In the process of doing that, we put in the automatic flight control, allowing us to go to higher altitudes. They used half the extra parts of the airplane that were ordered . . . to rebuild no. 3." When no. 2 crashed and needed to be rebuilt, the rest of the structural spares were utilized. Rushworth said, "We were out of spares, but we still had three good airplanes. . . . We got in just about twice as many missions [than originally planned] simply by good management of the assets that were available."

On a flight around this time, an interesting mystery occurred. "It was something that we just didn't understand," Bob said. "Halfway through the flight, on the glide home, I just let the airplane fly by itself, and it began to go through a porpoising [up-and-down motion]. We looked at the data afterward and couldn't figure it out. So, we put the airplane up again to repeat that, [which it did]." Listening closely the next time, Bob thought he figured out what was happening. "I could hear it, and it was the relief valve on the propellant tank popping [open] and was acting as a little uncontrolled ballistic control rocket. It wasn't serious or anything like that; it was just something we couldn't understand and had decided to fly it again."

The no. 2 X-15 came back into service, now designated as the X-15A-2, in June 1964, and it was Bob Rushworth who was given the job of checking out the rebuilt aircraft's systems. The first time the A-2 was airborne was on 15 June, but only as a scheduled captive flight. The flight designation was 2-C-53, and Bob ran through all procedures short of launching. At the point of reaching LR-99 igniter idle, he then switched over to jettisoning propellants as if descending into the landing pattern, then pulled the landing gear release to verify the cable worked correctly with the lengthened fuselage. Bob found the lanyard had actually been made too long. Once the mated X-15 landed, it was detached from B-52 no. 003, and one modification was to shorten the cable. Unfortunately, it was shortened too much, which caused problems over the course of several flights.

The next time the craft was airborne, hopes for a launch were dashed as a malfunctioning auxiliary power unit controller caused an overspeed to the point the APU was destroyed. Two days later Rushworth was back near Hidden Hills, this time for a successful mission 2-32-55. The A-2 went through its paces with nary a care. Bob hit Mach 4.59 at 83,300 feet, then headed down toward the Rogers lakebed. He noticed nothing unusual in the handling characteristics with all the changes that had been made to the airframe. With the X-15A-2 now part of the rocket plane inventory, the program was back up to its full complement of three aircraft.

One flight was not enough to wring out the systems thoroughly, and problems were definitely forthcoming. The landing gear seemed to be the most troublesome for a while, starting with mission 2-33 on 14 August. This was the first time the A-2 was taken above Mach 5. The higher speed meant a higher heat pulse into the aircraft. The structure actually expanded. The gear was supposed to be released close to the ground at low velocity, whereas, on this flight, the nose gear released into the hot airflow at Mach 4.2. Flight planner Johnny Armstrong explained what happened: "In the design, they didn't adequately allow for temperature differences that occurred, so, as the airplane stretched, the pull cable point didn't stretch, which allowed the gear up-lock to be released as a result of tension that was put on that cable."

One of Rushworth's test objectives was to set up an oscillation without the stability system turned on. Bob said, "I had just shut off all of the stability augmentation on the airplane and . . . as I recall, I pulsed it once and didn't get quite enough of a pulse to create an oscillation, so I did it again. . . . The first cycle went down and came up, [then] the nose gear slammed down, and it began to pitch quite badly because of the initial conditions. I reached down and turned the stability augmentation back on, and the airplane handled well."

How was it in the cockpit when the nose gear popped out at high Mach? "Noisy," he replied. "It was right under the cockpit and just about under my feet. When that thing slammed down, I just couldn't believe it! All of a sudden there was much more drag than I should have had. . . . I could tell it had to be the nose gear." Was Bob worried the gear might actually melt from the heat? "I was more concerned about whether I'd be able to land, or would have to jump out. I wasn't concerned the temperature was going to create any problems. . . . If it was down and the gear had locked, I was going to try to land."

The problem became one of having enough energy left to safely return to Edwards. When the gear dropped, Bob recalled, "I was probably 150 miles out. . . . I never did have to put out the speed brakes, [and] I got back to the base straight in at minimum flying speed." Joe Engle, in Chase 3, caught up with Bob as he approached the runway and helped talk him down safely. "Joe assessed that the tires were not in very good shape and could go on touchdown—which they did. I had to make a very tight turn to get back into the field, and in so doing Joe Engle couldn't stay with me." Bob laughed as he said, "We determined the x-15 had a lot better flying characteristics and lift at that condition than did the f-104. Joe was stalling out on the outside of the turn, which should have been a better position, yet he couldn't stay with me. I was a little more desperate than he was!" Joe went out wide and came back in to pick up Rushworth on the final approach. "After skidding and rolling for about a quarter mile, the tires began to shred and go in all directions. Finally, I ended up on the rims, [and] it was a normal, smooth ride from there on."

Bill Albrecht, operations engineer for the a-2, talked about the aftermath of the flight. "We put in a greater expansion tolerance in the cable system." The action of pulling the gear release moved the cable ten to twelve inches, similar to pulling an emergency hand brake on a car. This created a problem for one of the pilots. Bill said, "Jack McKay confided in me he was having trouble with that cable. He told me it was too short for his stocky arm geometry. . . . We invented a drum that the cable went around. . . . Effectively, it was a spiral pulley. For every inch he pulled at the handle, he'd get maybe an inch-and-a-half or so coming off the drum. . . . It worked so well we just left it in there." Flight planner Jack Kolf added, "You didn't have to worry about McKay's strength, just his reach."

After all the changes to the landing gear release system, another captive flight was needed. A chase plane was specifically assigned to photograph the deployment of the x-15 gear, but when the time came, cruising at altitude, the photographer in the backseat of the f-104 somehow missed catching the moment on film. The airplanes all returned to Edwards, reset the x-15 landing gear, loaded back up again, and took off for a second attempt. This time the photographic data were properly gathered. Captive missions

2-C-61 and 2-C-62 were both flown within a few hours on 15 February 1965. As soon as the film was examined, flight 2-36 was cleared to proceed on 17 February to test the modifications under high Mach conditions.

At 10:44 a.m. Rushworth launched, lit the LR-99, and headed up to 95,100 feet. Slowing down from his peak velocity of Mach 5.27, Bob heard a noise and experienced a sinking feeling that something was wrong. It happened at Mach 4.3, almost exactly the same speed as when the nose gear extended previously. He was still too high and too fast for anyone else to be close enough to see the outside of the X-15. He radioed to Engle in Chase 4, "A little bit strange here, Joe. Can you get in here as quick as possible?" Jack McKay at NASA 1 asked Rushworth, "What do you think came out?" Bob admitted, "I don't have any idea, [but] it felt like the gear again."

There was a hesitancy in Bob's voice as he tried to reason through what had occurred. He tested the A-2's response to his commands, making sure he had full control of the situation. "Something is real weird," he radioed.

When Engle was able to get in close enough for a look, he verified the nose gear had not deployed again, but one of the gear at the rear of the aircraft had come down. Bob said, "The right main skid came down at the same point, [while] doing the same thing. The skid lock broke off and gave me all this drag."

As Rushworth approached touchdown, he pulled the deployment lanyard, and Engle verified the rest of the gear came down as planned. Joe then called the final moments, reassuring Bob, "Two feet . . . one . . . beautiful, you're down." Was the gear going to handle the landing loads? Engle watched closely, then called, "It's okay, you're holding. Good show, Bob." The exasperation was evident in Rushworth's voice when he said over the network, "Boy, I'll tell you, I've had enough of this." McKay chimed his agreement, "You and me both, Bob. Very good flight."

Once he was safely on the ground, he opened the hatch, exited, and had a private moment with the X-15. He recalled at a conference years later, "When I got back home, I got out of the airplane . . . and symbolically gave it a boot. I didn't realize anyone was watching. Usually, there's no one out there at that point in time." Not long after, Bob returned to NASA to prepare for the post-flight activities. He continued his story, saying, "After I'd changed clothes and got back into NASA . . . I got a call from Mr. Bikle. . . . We hadn't even started the debriefing yet. He looked at me very seriously, and he said, 'What's this business about you kicking the side of the airplane?' Well, I knew right then it was about time I could start calling Mr. Bikle 'Paul.'"

Bob explained what happened and why he had kicked the x-15. Under the circumstances, Bikle understood Rushworth's frustration. After all that happened, the best thing that came out of it was a new friendship. Bob said, "From there on, I had that special relationship with Paul. It says one thing about the leadership on both sides, the air force and the NASA people who run those kinds of programs. We had a camaraderie that just wouldn't quit."

With Bob a bit fed up with the x-15a-2 at this point, John McKay took the next four flights. No more landing gear problems cropped up. Out of Rushworth's next six missions, he flew just one in the a-2, on 3 August, evaluating the ballistic control system. On mission 1-58 on 9 September 1965, Bob became the first and only x-15 pilot to achieve thirty flights in the research program.

By November Rushworth was reunited with the a-2, as the final element of the system was brought in to allow the x-15 to achieve its ultimate speed: the external tanks.

Testing of the tanks was a contractual obligation of North American Aviation, as part of the rebuilding of the no. 2 aircraft. Jim Robertson worked at NAA and designed much of the system. He remembered, "We couldn't afford to build additional tanks for the tests, so we developed these beams that had exactly the same weights . . . so they would behave exactly as the tanks would on the aircraft."

A large pit was built outside the main NASA building at the Flight Research Center. The pit was approximately ten feet deep and big enough to accommodate the twenty-two-foot-long simulated tanks. Robertson explained, "We filled that halfway up with redwood chips and big lumpy balls of fluffed-up canvas." The beams were mounted to the x-15, then the combination was rolled into position above the pit. "The aircraft was down on the skids and nose wheel, sitting on this platform, and was tied down with cables." Jim pointed out the physics they had to deal with. "Any time you try blowing any two things apart, it's an action and reaction principle, where the load is the same in both directions. With 30,000 pounds of ejection force on there, we were developing enough reaction with these tanks

188 | STRAIGHT AND STEADY

to lift the bird off the ground." In other words, the x-15 itself was to be very securely bound into position.

On 29 September 1965 the first test was ready. Jim spoke of cameras being used to record the event: "A photographer by the name of Don Peterson went up [to Edwards] with me [and] he set up all the high-speed cameras. We had this thing covered from a number of different angles." Each camera was aimed not only at the x-15 and tanks but also at a board with a measured grid. Running the film at high speed during the tests allowed Jim and other engineers to calculate the separation velocities. "From an instrumentation standpoint, we had what we called a velocity harp. It was a long sheet of Formica with a row of contacts on which you slid a wire. . . . With this, we knew exactly what the velocity was at the tank's [center of gravity] and at either end. We were measuring pitch as well as downward [motion]. . . . That's why we had a big rocket motor up at the nose."

A half-charge ejection was accomplished to see what happened if one of the two pyrotechnics failed in flight. Evaluating the data and film showed nothing unexpected, so a second test was set for 2 October with a full charge. The post-test report stated, "High-speed motion pictures showed good separation. During both tests, hydraulic and electrical power were supplied to the x-15. Subsequent inspection showed no damage to the x-15." The tanks were cleared for a test flight but were to be empty the first time out.

The x-15A-2 was mated with the flight set of external liquid oxygen and anhydrous ammonia tanks on Saturday, 23 October. Two days later the completed aircraft was towed to the Propulsion System Test Stand and made a successful ground engine run. Over the next week several problems cropped up, forcing cancellation of flight before the B-52 ever started its engines. Once all the issues were addressed, at 8:26 a.m. on Wednesday, 3 November, Maj. Charles C. Bock and Maj. Charles J. Doryland rotated the mothership off the runway and headed just thirty miles northeast to Cuddeback Dry Lake, with Bob Rushworth riding the wing pylon in the A-2. It was the start of mission 2-43.

After arriving near the launch lake, all systems were checked and ready for drop. Bob finished his checklist, verifying to Jack McKay at NASA 1: "I'm ready to go, Jack." McKay replied, "All set, let's go!" At 9:09 a.m. Rushworth dropped from the mothership. Jack confirmed, "Good light, Bob. . . . Real good heading and profile." Later, Bob said of the launch, "I got a fairly significant roll-off . . . that I didn't catch until I was over a good twenty degrees."

Even though the external oxygen and ammonia tanks outwardly appeared as mirror images of each other, they were actually of much different weights. Jim Robertson explained why: "Since [the x-15 didn't] require as much liquid oxygen as anhydrous ammonia, the whole forward end of the lox tank had big vessels inside that stored helium under pressure. This was used to drive the fluids out of the tanks and up into big, long feed tubes and into the internal system of the bird. That accounted for the differences in the empty weights." Later flights with full tanks had an even more marked difference. The oxygen tank weighed in at 7,494 pounds, whereas the ammonia tank was a mere 6,006 pounds. This asymmetry, whether empty or full, is what caused the excessive right-wing roll after drop.

Bob continued, "When I estimated I had gone supersonic, the airplane was flying without any buffet, and I made one roll input and got the impression that the roll stability was significantly less than I had expected from the simulator." It was the first time the lower ventral fin was installed in the past eighty-five missions, so it was thought he would have extra control in roll. The addition of the tanks somewhat negated the ventral's effects, so that proved to not be the case.

This flight was the only one of the program to launch from Cuddeback Dry Lake, but Rushworth didn't need that much air space to accomplish his task that morning. With no fuel in the external tanks, the primary purpose was to simply check the aerodynamics of the tanks and to verify their ability to safely separate from the x-15. Bob hopped up to 70,600 feet, and as he passed the designated speed of Mach 2.1, initiated the ejection charges to push the tanks away. It was a clean break, with the nose of the a-2 pitching down about five degrees. Seconds later Bob shut down the lr-99 and soon headed into the landing pattern.

For the tanks the ride down was not so smooth. They were supposed to yaw outboard, then pitch down and tumble, but instead they continued to fly forward before losing momentum. The observed trajectory was dangerous because the tanks could have recontacted the x-15 with disastrous consequences. The center of gravity had been miscalculated, and the drogue chutes were too small to stabilize the tanks properly. Both jostled about erratically. The forty-foot-diameter main parachute on the ammonia tank came out to slow descent onto the desert; however, the nose cap on the oxygen tank failed to separate because one of the two explosive bolts did not operate. That tank's parachute never released.

Both came down well within the bounds of the impact area predicted by the pre-flight calculations of Johnny Armstrong. The tanks were made for reuse, but the oxygen tank smashed into the ground with no parachute, making repair impossible.

When a helicopter later went uprange to airlift the tanks back to Edwards, the off-kilter center of gravity hampered the ability of the aircrew to remain stable in flight with each tank hanging from a cable. The experience and skill of the pilot got everyone home safely.

Considering the problems with the recovery system, the fixes were straightforward: installing a bigger drogue chute and moving the center of gravity farther to the rear of each tank. Initially, more empty tank missions were planned to work out the bugs, but none were flown.

With all the testing, there were always spare pieces lying around. That included spent rocket ejection charges for the external tanks. One day I had the opportunity to sit with a group of engineers and technicians from the X-15, and they told the story of a hapless machinist who came up against one of those charges when he decided he needed a new piece of furniture.

Bill Szuwalski started out, "There was a big bin at the back of the hangar, and we used to throw [the charges] out when we changed them." They looked like a simple pipe or tube and weren't marked for what they were. Of course, the charge was supposedly spent, so no one thought twice about having thrown it out. Bill went on, "Some guy thought it would look great as a lamp, [and since] everybody was going for lunch and left the hangar, he took one of the tubes and put it in a vise."

Ed Nice picked up the story: "Part of the pyro charge was still in it. [He] thought it was empty . . . and started cutting it. The hot blade hit the powder and set it off."

Now it was Bill Albrecht's turn: "It catapulted itself out of the vise and then was all over the hangar, spinning around, like when you let go of a balloon." Bill swirled his hands in a spiral pattern to emphasize his point. "There were airplanes in there fully fueled and everything, and it never hit any of them, just hit that [hangar] door." The charge impacted several metal I-beams more than thirty feet above the floor. The beams were burned and bent, and several panes of window were shattered. The machinist was uninjured.

Szuwalski said, "I came in from the hallway . . . and I had just opened the door when the thing blew up. . . . Some guys were running in; I was running the other way!"

Albrecht explained that in the aftermath, "we established a technique where no matter what it looked like when you looked at the nozzle, you weighed the rocket. If it weighed the right amount, it had propellant in it. If it weighed [a different amount], it might, so we treated it accordingly." Dave Stoddard added, "And they were also then marked 'inert,' because the air force had gotten pretty teed off about the incident." Ed Nice pointed out, "He never made his lamp!" In the end, there was one minor casualty. Albrecht said, "The worst thing that happened was when they went to repair the door. They had a scissor lift to put the workman up there to fix the window, and he fell off and broke his leg."

Jim Robertson said that since he was uprange at Mud Dry Lake so often in support of the external tank flights, some people decided to call him the "Mayor of Tonopah." He quickly added, "Which isn't a great thing." His wife pointed out there were more dogs than people in the town. Jim said you had to be wary. "At night, if you tried to go walking, you'd better carry a flashlight, because there were messes everywhere. People came into town in those old rickety farm trucks, and dogs would pile out the back end. Dogfights would take place right smack in the middle of town. All traffic stopped, and people would just wait until those dogs got through fighting. . . . To be the Mayor of Tonopah was nothing special. It was just an honorary name."

When it came time for the first test with the full tanks mounted to the A-2 on 1 July 1966, Robertson was right back in Tonopah getting ready for Rushworth to begin mission 2-45 out of nearby Mud.

Johnny Armstrong recalled, "One of the worst flights I've ever had the privilege of flight planning was flight 2-45." He went on to explain: "The external tanks had to feed into the internal tanks, then feed the engine. . . . The other concern we had was, if those tanks ever stopped feeding, or fed asymmetrically, well, we've really got a bag of worms. [The x-15] would just flat run out of control. . . . We had some 'el-cheapo' instrumentation that

22. The external fuel tanks on the X-15 were supposed to be jettisoned after fuel depletion. Unfortunately, on flight 2-45-81 on 1 July 1966, incorrect fuel flow readings forced an early ejection and an emergency landing at Mud Dry Lake. © Thommy Eriksson.

amounted to a couple pressure transducers . . . which indicated flow from those tanks. We set this thing up ourselves to be an unsuccessful flight right from the start, because the instrumentation wasn't any good. And that's exactly what happened to us."

The one unplanned contingency was to not have proper readings. The only thing seen in the control room was the telemetry that showed the ammonia was not flowing from the external tank. Jack McKay was NASA 1 for the mission, and he did not hesitate to immediately pass the bad news to Bob flying the A-2. Armstrong said, "Rushworth pickled those things off under the worst possible condition. We were concerned that if you did that at max-Q [maximum dynamic pressure], they might hit the airplane. . . . Very bad show to plan something that just flat didn't have a chance to succeed."

Bob talked about what happened as he jettisoned: "We had decided early in the program we weren't going to have that [tank ejection] problem." Nonetheless, that's what happened. "All of a sudden, the first thing is, we've got the problem [we didn't plan for]. What they told me over the radio was the oxygen tank was feeding and the ammonia wasn't, so I had to use the dual ejection charge. When I jettisoned the tanks, it sent the airplane into some very violent oscillations. Probably took about seven to ten cycles before the airplane began to fly level again." In his post-flight report, Bob said, "The sharpness of the tanks going off was quite impressive. I didn't really think that they were going to move the airplane as much as they did." Bob admitted, "Fortunately, the [stability augmentation system] stayed on. If that had shut itself off, we would have lost the airplane."

The extreme motions of the A-2 were possibly due to the right-side ammonia tank coming off just a moment earlier than the separation of the heavier left-hand oxygen tank. When all the data were evaluated, it was discovered the fuel was flowing exactly as it was designed to do. The "el-cheapo" transducers had sent erroneous information, which could have crippled or destroyed the X-15.

As the tanks dropped away, they experienced the same gyrations as on the previous mission under the drogue chutes, but the main parachutes both opened this time and properly slowed their descent. The dump valve at the rear of each tank worked, and excess propellant was vented overboard. At the moment the tanks hit the desert, the cords attaching the parachute to the tanks, called "risers," were supposed to automatically cut, but that didn't occur. High winds caught the parachutes and dragged both tanks "over large distances." They were later recovered basically intact, and were both able to be refurbished and reused, even after their battering trip across the ground.

Rushworth then made an emergency approach to Mud Dry Lake, but because of the circumstances, he had a lot of excess energy to dissipate. "I was not really at high enough altitude to jettison the full internal load of fuel [when] they told me to shut down. I really could have kept the engine going and flown the damn thing home! But we didn't know. What they saw was no feed on one tank, and that meant I was using internal fuel. . . . As it turned out, both tanks were feeding, and the internal fuel was full when I started to jettison."

Johnny Armstrong explained the solution to the bad pressure transducers. "After that, we got better instrumentation. They put some paddle switches in there that were just kind of little blades that sat in the flow stream. Whenever there was any flow at all, they deflected." He then said, "That was Rushworth's last flight. It was very anticlimactic." After being one of the few pilots to exceed Mach 6, this last time out, Bob flew to only 1,061 miles an hour, equating to Mach 1.70 at 44,800 feet. Had this originally been scheduled as his final flight? "Yes," Rushworth told me. "I was already packed and had just about checked out of the base when that flight finally came off. . . . I left on the sixth [of July], and that's when I went to the National War College."

That same day, the x-15A-2 had left Mud and was returning to Edwards, loaded on the back of the long, flat trailer. With other vehicles running guard in front and back on the highway, the convoy headed home. Some of the roads through both the Nevada and California deserts were tight and twisty, up and down hills. The convoy drove slowly to avoid difficulties. This was not always the case for other cars and trucks they met on the road, especially when angry motorists felt they had been stuck too long behind the strange-looking black rocket strapped to the flatbed.

Heedless of any danger, a couple driving a pickup truck with an over-cab camper decided they'd waited long enough, pulling out to go around when it appeared clear to pass. The problem was the road at that point was narrow, and the driver didn't notice the x-15 had a set of thin, strong, Inconel-X wings sticking out on either side. As the truck started to pass, the rear edge of the left x-15 wing contacted the front of the camper. The driver accelerated quickly to pass this "obstacle," and the wing acted like a can-opener, tearing the top of the camper right off.

The convoy stopped to check the damage. The heavy structure of the x-15 was unscathed. The driver later tried to sue for damages but was quickly denied any recompense when his disregard for the obvious hazard was brought to light.

During the course of flying the a-2, the highest altitude Rushworth achieved was 208,700 feet on 3 August 1965. Exactly one year later, and one month after Bob left the program, Pete Knight flew the aircraft to 249,000 feet. This was the highest attained for this modified vehicle. Rushworth said, "[The a-2] was a different-feeling airplane, and more difficult to control in reentry because the fuselage now was longer and we didn't have the authority from the horizontal [stabilizers] to hold the nose up when we got back into the atmosphere. So it was a different airplane to fly, and we just decided that it wasn't worth flying at that high altitude anymore, so we stopped it." Flight planner Jack Kolf said, "We never flew with [ballistic control] jets after we got the airplane back. . . . They took [the system] out. It was just one less thing you had to worry about."

From that point, the A-2 stayed with speed runs to about 100,000 feet. One of its primary duties was to fly with the scramjet test engine, which was supposed to come on line around 1965. Bob recalled, "That was going to be my program when the whole thing originally got started. I was going to do all the research flying and all the investigation of the engine itself. . . . I got terribly interested in scramjets and what we could do with them, but it just got delayed, delayed, and delayed. The only part of it that I got to do were the first flights dropping the external tanks."

The company selected to deliver the scramjet was the Marquardt Corporation. Their business in this area collapsed, necessitating a new supplier, Garrett AirResearch. Designs showed promise, but it was too late for Rushworth, so the X-15 testing of the engine was eventually taken up by Pete Knight. Bob remembered that even this plan never materialized. "We could run [the scramjet] between Mach 7 and 8, then start the engine and shut it down to see how much control we had over the fuel flow, what the shock waves were going to do, and things like that. It was a good, detailed research program, [but] because the [scramjet] engine didn't come along it went sour and was dead before the X-15 program was finished."

At the time Rushworth finally left the program, the war in Vietnam was in full swing. He said, "All the tactical outfits began going to Vietnam, so I opted to go to school first. . . . I graduated in 1967, then went to join the Tactical Air Command with the Vietnamese situation." Bob arrived in Vietnam in March 1968, flying as the assistant deputy for operations of the 12th Fighter Wing on the McDonnell F-4C Phantom II. Over the course of the next year, he completed 189 combat missions. Fifteen of those went into North Vietnam, as had missions flown by fellow X-15 pilot Bob White.

In his air force oral history interview, Rushworth said, "I took hits when I didn't even know it. On my last flight, I got hit in the wing by ground fire that created a problem, and had an emergency landing. Unfortunately, I was flying the commander's airplane—which had just been repainted. When 'Buckshot' [Col. Floyd White] looked at it, he said, 'Why don't you go to Hong Kong on leave, and then go home?'" Rushworth was hoping to finish two hundred sorties before returning stateside, but his commander felt there were too many pilots and not enough airplanes.

Bob made a comment that is very relevant to today's world, saying, "We military do a very bad job everywhere we go because we assume too much of the war effort, instead of letting them [the indigenous forces] take the challenge. I think every war we have been involved in that didn't come out right was because we . . . didn't get the people involved to support themselves. . . . If we can't . . . get them to fight their own battles on their terms . . . then it's ridiculous for us to go in and do the fighting for them. . . . When we try to give it back to them, they are going to lose; it's just that simple."

Assigned to Wright-Patterson AFB in Ohio, Bob was manager on the Maverick air-to-ground missile program. From there it was up to Systems Command as the inspector general for about seven months, then back to Edwards as the commander for a couple of years. "I got there in March 1974 and left in November 1976. The two times I was at Edwards were probably the busiest time periods of any activity that ever went on. . . . We had an awful lot of programs going through there. . . . When I was the commander, I had all the new generation airplanes."

Rushworth was very concerned about one type of testing when he was in charge. "One thing we in the air force just psychologically didn't like at that time were spin [recovery] tests, because every goddamn spin program we ever got involved with ended up in a crash. Shortly after I got there, the F-15 [Eagle] was just going into its spin program. . . . I walked in and told McDonnell that I was going to have a pilot monitor their testing and he had a go/no-go say. I said, 'You ain't going to fly until we get this cleared up.'" The contractor didn't like Rushworth's attitude. Bob continued, "They went back to their management and called the air force and said, 'What about this new guy out at the base that says he wants to control our flying activities?' And they were told, 'Well, he owns the base. I guess he can do that.' Very shortly after that, I ended up with five spin programs going on, not at the exact same time, but over the same period, which I didn't like at all."

Following his second assignment at Edwards, Rushworth stayed in the military just five more years before retiring as a major general in 1981. He was then the vice commander of the Aeronautical Systems Division at Wright-Patterson.

Eventually, he settled in Camarillo, a sleepy town close to the Pacific

Ocean, not too far away from the Mojave Desert where he had flown the x-15 for so many years. Of course, he didn't move there too quickly. "It took me about six months to get out here, to see if I could play golf in every part of the country that was available at the time."

A favorite hangout for members of the x-15 during its era, be it mechanics, engineers, managers, or pilots, was a place called Juanita's Bar in Rosamond. It was at the junction of the east-west road linking Edwards to Highway 14 and Sierra Highway, north of Lancaster and Palmdale. It seemed an appropriate location to hold Bob Rushworth's retirement party. Johnny Armstrong said there was a special memento presented to Bob that day. "There was a placard that Bill Albrecht made up to put in the cockpit of the A-2 that said, 'Do not extend the gear above Mach 4.2.'" They gave him a replica of the plaque, along with "that little paddle switch that had the two lights that showed flow to the external tanks."

In all my interviews with other pilots for this book, I never heard of a man more highly regarded by them all than Robert Rushworth. He took on assignments and made them the best. His longevity with the x-15 helped prove his devotion to the art of flight test and to the x-15 research program itself. Bill Dana said it best: "Rushworth was always my personal hero of the whole bunch. Before I was even flying, I knew Bob. I watched him through all thirty-four flights, [and] participated, at least in some way, on all of them." Bob's professionalism was always of the highest priority. Bill continued, "The guy was never out for the glory, and he never was out on the speaking circuit saying, 'Hey, look at me!' He just was here doing great work. I think that trait followed him through his whole career. Bob probably did more work and got less glory out of both the x-15, and out of his air force career, of anyone that's ever lived. He was a real bona fide good guy."

Bob reiterated this idea that flight test took precedence when we spoke at his home in Camarillo. "We only test fly aircraft probably one-tenth as much as a normal airplane. It's slow, and there's a lot of work between flights, a lot of preparation. It's more serious than just flying a stock airplane. Not that it wasn't just as dangerous flying that same one later on [when, and if, it became operational]. But when you initially get on the test phase, there are more people paying attention to 'it' as an airplane and 'it' as a system to be looked at and controlled."

Late in his career, Bob had an opportunity to fly the then new Rockwell B-1 Lancer bomber, built by the company that used to be known as North American Aviation. "I guess there were flights available [on the test program]. They were taking up senior people, but I didn't have any strong, urgent desire to get up in it. I would have had to ride in the back seat. Can't drive, don't want to go."

7. Skipping Out

The job was finding ways to get the data that you
wanted to get.

Neil A. Armstrong

To most of the world, Neil Armstrong was an enigma. As the first human being to set his space-suited boot on the surface of another world, he instantly became one of the most recognizable people in history. Yet he successfully avoided the spotlight, as he staunchly refused to discuss his personal life. There is something very positive to be said about someone who had the world at his feet but never took advantage of the situation. The same cannot always be said of people with whom he came in contact. A starry-eyed kid might have asked for his autograph in good faith, but so many others have made huge amounts of money selling Neil's signature that he had no choice but to turn down everyone. A bizarre incident from 2004 highlighted this problem: Armstrong's long-trusted barber sold his hair clippings for thousands of dollars. How would anyone have reacted to this type of behavior?

Paul Bikle was a friend of Neil's, not only through their working relationship on the x-15, but also with their shared passion for glider flight. Many years before Neil passed away, Paul talked about his post-spaceflight personality, saying, "He's been successful in divorcing himself [from public scrutiny]. . . . Even if it wasn't a characteristic of Neil, I think it was smart of him [to do], because the more you try to cash in on something like that, the more criticism you're going to draw. I think a hundred years from now, he's going to look better than he would have if he'd tried to cash in on it."

Neil had always been primarily interested in one thing: flying. To him, this was an important vocation, one that transformed the world. When someone with a genuine interest wanted to speak to him about this subject, he was usually happy to do so. Technology is precise, and there is no room for mistakes, only for new exploration of boundaries. He was most comfortable when immersed in the elements of aerospace engineering, equations, and design.

X-15 pilot Bill Dana explained his take on Armstrong around the same time I spoke with Bikle: "[Neil] is very reserved with people he doesn't know. When you get to know him, he's really warm and interesting. . . . I think he's a lot more like [Charles] Lindbergh than you'd possibly believe, both very bright men who did something amazing. I think they probably found the right person for [*Apollo* 11]. Neil is sharp and a dear friend. I just think he's a super guy, but until you get to know him, he's going to appear to be very aloof."

When I had the opportunity to sit with Armstrong and discuss the X-15, he was interested in my project, but reticent. His opinion of interviews was they are imprecise and prone to error in memory, and actually the worst form of historical documentation. He stated, "This is really one of the problems with the interview format." His point was certainly valid. Facts must always be checked and verified. A weakness in our age of instant global access to information is that one person may state something incorrectly, or even go so far as to manufacture information, then these "facts" are taken up by countless others as absolute, without anyone doing independent verification. This was something Neil avoided by pointing researchers to reports and other documents created at the time of the event.

Throughout Armstrong's career, he had a focus of purpose, and that precise focus, in some instances, is what got him in trouble when he flew rocket plane research flights on the X-15.

Although always listed as Neil Alden Armstrong's official birthplace, Wapakoneta, Ohio, is about seven miles southwest of the actual location. His birth occurred just a half hour past midnight, on 5 August 1930, in the living room of his grandparents' farm, to parents Stephen and Viola. Records show the family moved on average more than once a year before finally settling in Wapakoneta when Neil was fourteen. He attended four schools, then graduated from Blume High School in 1947.

Aviation was an interest that took root and stayed with him all his life. According to an interview Armstrong did for the NASA oral history project, he said, "I determined at an early age . . . that that was the field I wanted to go into, although my intention was to be . . . an aircraft designer. I later went into piloting, because I thought a good designer ought to know the operational aspects of an airplane." He recalled his first solo as a pilot was flying from a grass field in Wapakoneta at sixteen.

Heading west after high school graduation, Neil enrolled at Purdue University in West Lafayette, Indiana. He was provided a scholarship through a law passed the previous year, commonly called The Holloway Plan, which was a way to train new officers for the U.S. Navy following post–World War II attrition. The name derived from Adm. James L. Holloway Jr., who was the navy's chief of personnel at that time and was instrumental in the plan's implementation.

Armstrong took advantage of the program to pay for college, not necessarily through any intent to start a military career. Halfway through his sophomore year at Purdue, he was called to active duty. Neil was unhappy about leaving in the middle of his studies, as the program was supposed to allow two years' uninterrupted school before reporting for flight training. As is often the case, the military operates in their own time.

Neil left for Pensacola, Florida, to become a fighter pilot, completing his training soon after the Korean War began in June 1950. As an ensign he became proficient in the Grumman F9F-2 Panther, and his VF-51 fighter-bomber squadron was off the coast of Korea by mid-1951, aboard the aircraft carrier USS *Essex* (CV-9).

On 3 September 1951 Neil was flying at about 500 feet through a North Korean valley as wingman for lead pilot John Carpenter. The area was booby-trapped with wires strung across the valley to catch low-flying planes. Carpenter's missed the wires, but Neil's did not. The wires sliced through the Panther's right wing, taking about six to eight feet off the tip. Armstrong was high enough to recover control of the Panther and head south before he ejected, luckily back within friendly territory near Pohang Airport. Once reunited with his squadron aboard the *Essex* in the Sea of Japan, Neil immediately returned to duty, flying a total of seventy-eight combat missions before the end of the war, earning three Air Medals.

A year later, Neil completed the Holloway Plan's requisite three years of active duty, leaving the navy as a lieutenant (junior grade). He kept his commission in the reserves for another eight years. Armstrong returned to Purdue to complete his aeronautical engineering degree. Reports show his experience in the navy may have been one of the best things to happen with his academic career, as his grades improved significantly. For the first time, he became seriously focused on his goal. In January 1952 he graduated, deciding to take on the job as an experimental test pilot with NACA. He could have gone with an industry job rather than a government one, but the pay was not as important as the benefit he saw in being on the cutting edge of research.

Although Neil initially applied to NACA's High Speed Flight Station at Edwards, there was no slot available, so instead he was accepted at their Lewis Flight Propulsion Laboratory in Cleveland. Stan Butchart talked about first meeting Neil and how he was instrumental in Armstrong's eventual move to Edwards: "I met Neil on one of my trips back to Langley [Research Center, Virginia]. . . . I guess it was 1955. . . . Ed Goff, who was head of flight operations at Lewis, said, 'Hey I want you to meet this young pilot I just hired.' I turned around and this twenty-four-year old kid [Neil Armstrong] was standing there."

About an hour later, Ed took Stan aside and asked if he knew of any openings at Edwards. He went on to explain how he had hired Neil based on his outstanding credentials, but he really didn't have an opening at Lewis. Stan said, "I came home and told Joe Walker, [and] within a month or two Neil was out here. I think he showed up in about April or May. He and I became very close buddies because he found out I was an Eagle Scout, and he was, too. It kind of bonded us."

The test pilot position that opened up came about because Scott Crossfield decided to leave NACA around that same time. During our interview, Neil recalled, "I applied for a job at what was then the High Speed Flight Station. I transferred there from Cleveland. . . . The [x-15] project was already initiated, [but] contracts weren't let, so I was asked to transfer from Lewis after Scott Crossfield announced that he was going to leave in order to take a job [with] whoever won the x-15 contract." Neil went on to say he never considered he was taking Scott's position. "I was clearly the junior guy coming into the office, and [Scott] was one of the senior people."

While still at Purdue, Armstrong met Janet Shearon. They became engaged not long before he left for Edwards, but she did not immediately come with him on the move to the high desert of California. Stan Butchart

SKIPPING OUT | 203

recalled, "Neil was a bachelor then, and he used to come by in the evening. Miriam would have him over for dinner, and my oldest daughter was in high school. She was having trouble with math, and Neil was a crackerjack at that, so he'd sit down and help my daughter with her math every night."

Neil's tutoring skills for Stan's daughter, and their shared history with the Boy Scouts, set the stage for him to ask Stan some important advice. "I felt real proud when Neil came and told me he and Jan were thinking of getting married. He told me about their plans, but she lived back east, and he was worried about the move to California. I brought my wife out from Maryland after we were married, and Neil wanted to know how that worked. . . . I told him, 'If you are in love, it doesn't make any difference where you live, really. This is where your job is, and she'll come.'" Neil listened to Stan's wisdom, Jan moved west, and they were married 28 January 1956.

Once at Edwards, Armstrong got quickly acquainted with as many aircraft as he could in the inventory. He recalled at the time he arrived there were seventeen being tested. That seems like a small number, but these were almost all different and exotic airplanes used for experimental data gathering and flight test. Neil's first assignment was to learn and test techniques using a P-51 Mustang fighter.

Roger Barniki, who flew with many of the pilots as part of his duties as a life support specialist, talked of seeing Neil fly. "I remember Neil was going up one day. . . . If you like airplanes, you've got to love the sound of a P-51 engine. Neil taxied out and went to the edge of the lakebed and ran up the engine. . . . I watched and I watched and I watched as he sped down the runway. He must have gotten about three or four miles out. . . . Then, all of a sudden, he went just straight up! I said, 'Damn it, I've got to get a ride in that!'" It took Roger two years, but he finally got to take to the air in a Mustang. "It's a beautiful ride—the sound of the engine, the power. It's a sports car, it's a Ferrari. And Neil was a master with flying."

With several air-launched rocket planes, such as the X-1 series and the D-558, Phase 2, Neil also became proficient with piloting duties on mothership drop missions of the B-29 and P2B-1S bombers, participating in more than one hundred launches. One of the pilots with whom he shared this duty was Stan Butchart.

Two years after becoming a research pilot for NACA, Armstrong was given

23. In 1956 Neil Armstrong sits at the controls of the "Iron Cross" ballistic control system simulator. Courtesy of the Armstrong Flight Research Center.

his first opportunity at a rocket plane flight. The x-1b had been modified with a set of ballistic control system rockets in the fuselage and left wingtip. This was an extension of a program to test the idea of using small jets of steam to change the attitude of an aircraft that was too high in the atmosphere for aerodynamic controls. These small rockets were essential for the success of future x-15 flights.

The first tests of the system were performed using a cruciform rig, basically two iron I-beams in the form of an X, with a simulated cockpit at one end and rocket nozzles on the tip of the right "wing." The entire apparatus was mounted to a ball joint at its center, so it could move freely to the pulses of the jets. Once the idea and plumbing were worked out on this "Iron Cross," the system was transferred to the x-1b for flight testing. Initially, Jack McKay flew all the NACA tests, but then Armstrong took over responsibilities for the last four missions.

His inaugural x-1b flight on 15 August 1957 was basic familiarization and ended with the nose gear failing, as it had on McKay's first mission. On 27 November it was time to run a checkout with the control rockets under operational conditions. That did not go as planned, as Neil related, "I had intended to use it, but the day we were going to try it out, I was dropped too close to the lakebed to make the acceleration to whatever Mach number I was going to pull up from, to go through a parabolic arc. . . . I concluded I'd be too far away from the field at the conclusion of the run, and so I canceled it."

Armstrong flew twice more on 16 and 23 January 1958, but the entire series was done at relatively low altitude and speed, so there was little, if any, effect on the rocket plane from the steam jets. The x-1B never flew again, and the ballistic control system was transferred to a yf-104 for further flight testing. Once modified, the aircraft was redesignated as the jf-104A and eventually achieved altitudes of approximately 83,000 feet.

Between Neil's involvement with the x-1B and dropping rocket planes from a mothership, his position at naca naturally put him in line for ground-floor work on the x-15. "I was involved in [the] centrifuge training, simulator work, project reviews, and so forth." Both Joe Walker and Jack McKay were senior to Armstrong, so Neil did not get first shot at the program. He was brought in as the third naca x-15 pilot, which still gave him a fairly early slot in line.

Armstrong's first flight in the x-15 was 1-18-31. He became the seventh pilot to qualify and was the last of the initial group selected prior to rollout on 15 October 1958. Scott Crossfield, from North American Aviation, had been first to fly in June 1959, then two sets of three pilots from both the military and nasa were added to the roster. The prime nasa and U.S. Air Force pilots, Walker and White, made their debuts in March and April 1960. The remaining four flew within a two-month period, from September to November 1960: Petersen from the U.S. Navy, Rushworth from the air force, and McKay and Armstrong from nasa. Neil's was the last of the first flights for nearly three years. Only five more pilots joined the program from that point onward: two in 1963 (Engle and Thompson), two in 1965 (Knight and Dana), and one in 1966 (Adams).

Flight 1-18 was the twenty-ninth mission of the program. Launch occurred at 10:42 a.m. on 30 November 1960, near the town of Palmdale, heading roughly northeast. Neil noted in his matter-of-fact manner: "The launch was smooth, with no noticeable pitch or roll excursions." Still flying with the twin lr-11 engines, each of the eight chambers had to be individually ignited. The right-hand, or number three chamber, of the upper LR-11 failed to catch. Neil informed NASA 1, who was Jack McKay for this flight, "I can't get number three lit off." McKay told him, "Roger, go ahead and proceed with the original flight plan, Neil." Armstrong wrote after his flight, "Cycling the chamber switch proved fruitless, and was shut down."

The profile kept Neil at a level altitude throughout, peaking at 48,840 feet. His speed was a sedate Mach 1.75, half what the x-15 had already achieved under Walker and White. Armstrong tested out the controls, noting the rocket plane's responses. He radioed, "This is a nice flying airplane." During the rocket burn, as he passed north of Edwards, he made a turn to the west. Following shutdown he turned southeast to line up for his landing on the lakebed.

Joe Walker oversaw the landing in his chase plane, calling to remind Neil, "Okay, clear on the ventral." Armstrong jettisoned the lower ventral fin, which had explosive charges and a piston to push it safely away from the aircraft. "There she goes!" Neil called, then, "Flaps going down." The wing flaps took about eight to ten seconds to cycle fully downward. Walker radioed, "Atta boy," as the x-15 touched down. Armstrong slid across the hard-packed dirt on the two rear landing skids and the forward nose wheels, until friction brought him to a stop more than a mile later. Neil jokingly radioed back to Joe, "Thanks, Dad." At the control room McKay offered his congratulations: "Nice job, Neil." He knew exactly how Armstrong felt at that moment, since Jack had his first flight in the black rocket plane barely a month previously.

Soon after the flight came to an end, a modification was made to the aircraft systems. This marked the most significant change in the outward appearance of the vehicle since the installation of the LR-99 rocket engine. When first constructed, each x-15 was equipped with a six-foot nose boom. The rear half of the tapering needle was painted with wide, alternating black-and-white stripes. The front was bare metal with two instruments attached that looked like weather vanes. One was mounted horizontally, the other vertically, giving the pilot pitch and yaw information.

Armstrong's first flight marked the end of the use of this boom, to be replaced by a device called the "Q-ball" or, sometimes, the "ball nose." The system was developed by the Nortronics Division of the Northrop Corporation. Bob Hayos worked for Northrop as a test engineer on the program, explaining, "The Q-ball had four diametrically opposed ports, or holes, about a central port located in the dead center of the sphere. Each port had a pressure transducer behind it with the vertically oriented ports reading out pitch pressures, and the horizontal ports reading yaw. The central port read out total pressure (Q)." In other words, the Q-ball gave the x-15 its sense of direction into the airflow.

The Q-ball could withstand the types of temperatures expected as the x-15 expanded its flight envelope. The nose boom would have melted. With the new system, Hayos said, "The entire internal ball and its drive servos were cooled by liquid nitrogen when high aerodynamic heating occurred. The Q-ball and enclosing cone was machined from Inconel-X. . . . The Q-ball proved to be a most successful device and later was also used on the nose cone of the . . . Saturn rockets."

The heating that the Q-ball had to survive at hypersonic velocities was certainly unique, estimated at 2,500 degrees Fahrenheit. One way devised to prove its ability to withstand that heat was explained by Harry Shapiro from North American. "The way they tested that thing for temperatures was by using an F-100. They put it in the plume of the afterburner!" The Super Sabre was locked down to the tarmac with the ball nose mounted to a concrete slab behind the engine, then the engine was run at full afterburner, providing conditions worse than taking a blow torch to the device.

Research engineer Eldon Kordes explained how different it was to work on the x-15 program in the 1960s, pushing state-of-the-art systems like the Q-ball, the LR-99 rocket engine, and the idea of manned hypersonic flight. "It was an experimental airplane and a lot of things were done that you couldn't get done today. You couldn't get them through the system. . . . We were doing things where nobody had ever been before." The development of the Q-ball was deemed worthy enough to be featured on the cover of *Time* magazine on 27 October 1961. x-15 instrumentation engineer Glynn Smith affirmed its importance, saying, "The Q-ball was a big deal when it came aboard. That was something awesome. We had to go through and retrain on how to calibrate the [system]—which was about a four-hour procedure—with an inspector looking over your shoulder."

Considering the importance of the Q-ball, it only took a few days to remove the old nose boom on aircraft no. 1 and have the new system installed and checked out.

Nine days after Armstrong's first mission, he was ready to move forward with flight 1-19 on 9 December 1960. This time out Neil had two fellow x-15 pilots, Petersen and White, flying chase, along with Maj. Walter F. Daniel. The launch went smoothly, but the number three chamber on the bottom engine was slow to ignite. With a target speed of less than 1,200 miles an hour, the reticence of the single chamber made little difference. Neil pegged his altimeter on 50,000 feet, barely straying 100 feet up or down as he powered through the engine burn, straightening the sides of a large circle to return to the Rogers lakebed. Jack McKay, again serving as the controller, radioed, "Real nice flight, boy!" Neil recalled that having the first flight on the system "was happenstance. I was not involved in the development of the Q-ball at all. Just happened that [it] came to the flight stage at the same time I was flying."

As with several other pilots, once Armstrong's checkout flights were behind him, he went on to other projects at NASA—most notably as a consultant on the x-20 Dyna-Soar program. He didn't return to the x-15 for more than a year but instead stayed involved with the x-20, shuttling back and forth to Boeing in Seattle, until he left for the astronaut office in Houston.

Flight planner Bob Hoey worked not only on the x-15 but also on setting up launch profiles envisioned for the Dyna-Soar. One of his responsibilities was to create simulated runs on the Johnsville, Pennsylvania, centrifuge. This is the same equipment used to initially train x-15 pilots and, later, Mercury astronauts for various stages of their flights into space and subsequent reentry. Hoey said, "We decided the [U.S. Air Force] Flight Test Center would be lead on the flight planning and test operations, and NASA would support us. I participated in the Boeing simulation of the boost phase, then we went back to the centrifuge and did some runs." It took Bob about three weeks to work out the bugs. "When I got through with it, we got the pilots in there and flew the program. I actually got to ride the wheel once [simulating the launch on a] Titan III and the whole thing." Bob laughed as he remembered, "As I was strapped in, I asked, 'Aren't you going to put all these monitors on me and see what my heart is doing?' [The flight surgeon] looked at me and said, 'You're a flight test engineer, aren't you?' I said, 'Yeah.' He said, 'Well, you're expendable.'"

Hoey explained how the x-20 was originally envisioned: "It was an interesting program from a lot of standpoints. When we started out on the x-20, it was an extension of the x-15. It was going to go Mach 17, or something like that. We were going to launch it out of the Cape and land at Grand Bahama, then gradually build up [speed and altitude]. If everything went well, we might [have gone] orbital—once around." With all the political machinations, the program kept expanding to seemingly do all things. Hoey said, "By the time the program ended that orbital mission was the first flight. The risk was going way up, and the pay-off was going down." Instrumentation technician on the X-15, Donald Veatch, explained, "These were military decisions. Boeing didn't wish it." Then, he joked of how lofty the goals were becoming: "I think by the time they finished, it was decided they were going to do a moon landing!"

In the end, the Dyna-Soar went nowhere. Walt Williams said, "There is a point where you can fund something to a certain amount, and it just can't get anywhere without a little more push. It'll just spin its wheels and never make any more progress. The X-20 program suffered from that. It didn't get a big enough budget, and it just floundered. All the money that was poured into it ended up being wasted."

Roger Barniki remembered, early in his career, going with Armstrong to Johnsville for the boost simulations. "Neil, [Forrest] Petersen, and [Stan] Butchart, and a couple of guys from Langley, we all essentially lived with [one another] for a month or so. Neil took me under his wing. I was a relatively young kid, checking into Howard Johnson's, and I remember Neil trying to get the price down for me. He got the price down, then walked away from the counter. Then I asked him, 'Can you change a hundred dollar bill?' Neil said, 'God, can't you wait a minute?'" Barniki laughed at the memory of this and many other incidents while in Pennsylvania. "Neil taught me to eat fried clams with ketchup, so we're sitting at Howard Johnson's eating fried clams. We also went to a German restaurant called Old Mill. Then later I learned about Indian food and curry. . . . This guy, even though he's only about two years older than me, had such an education and was really smart, really nice, and he gave of himself. He wasn't snooty about anything. I owe him, and a few others, for my career."

During the year when Armstrong was on other projects, the X-15 leapt past Mach 4, 5, and 6, as well as surpassing 200,000 feet for the first time in a manned aircraft. These milestones all occurred in the span of eight months, following delivery of no. 2 to the government after North American and

Scott Crossfield completed the contractor phase of the program. By early December 1961 the aircraft was checked out for its first mission, which co-incided with Neil's return to the x-15 program.

Neil spoke about the reasoning behind his hiatus between his second and third flights, saying, "The standard approach that [NASA] used in those days with any flight project was to first get the research work started . . . then they took time out to check out other pilots and get them, not to say, cur-rent in the airplane, but at least have flown it and have an understanding of the operation." Once that was done, the pilot could start to work on the project team for another segment of the flight program. Neil continued, "I happened to be working on the MH-96 adaptive auto-pilot, which was in-stalled in the [no. 3] airplane, so I was spending the same amount of time on that effort."

The MH-96 adaptive flight control system was another innovation of x-15 research. While in the lower atmosphere, standard aerodynamic sur-faces such as flaps, rudders, and ailerons moved the aircraft about its three axes. Once above the point where these controls were no longer effective, the ballistic control jets changed aircraft attitude. The transition between these two conditions was not absolute. One type gradually lost effective-ness while the other gained. The MH-96 system, built by the Minneapo-lis-Honeywell Regulator Company, adapted to these dynamic flight con-ditions. The pilot used a single input to move the craft, while the adaptive system automatically provided either the surfaces or the BCS jets, or both in combination, as the most precise control. Neil was one of the prima-ry engineers integrating the MH-96 with the x-15, so it was appropriate he also became the first pilot on the program to use it on the inaugural flight of aircraft no. 3, with mission 3-1-2.

Ironically, it was the ball nose system that caused the first attempt on 19 December to be aborted, not the new MH-96. The next day the launch went forward, with Neil dropping away from B-52 no. 003 at 2:45 p.m. The launch area was officially out of Silver Dry Lake, but was actually closer to Silurian, farther to the north. This lakebed, however, was too small for use in the event of an emergency landing, thus the designation of Silver.

In his post-flight report Neil said, "Several discrepancies were encoun-tered during the airborne analyzer check of the MH-96 flight control sys-tem. Most reflected errors in the reaction controls and the auto pilot system, and were not of concern for this flight. . . . At launch, all three axes of the flight control system disengaged, and a severe right roll occurred, with accompanying yaw and pitch excursions. After the engine was lighted and pull-up initiated, one yaw pulse was performed . . . after which the flight control system was reengaged in each axis without transients."

Radio reception was so intermittent that Captain Allavie, in command of the B-52, had to relay between Neil and NASA 1. Because of this, Armstrong had to accomplish mission objectives using the on-board clock, with no reminders from the ground. After 105 seconds, he shut down the LR-99, achieving Mach 3.76. The flight path passed west of the army installation at Fort Irwin and directly overhead of Goldstone Dry Lake.

There were still many bugs to work out on the MH-96. Armstrong attempted tests throughout the more-than-six-minute flight, but several modes did not work properly. The angle of attack refused to hold position when engaged, and the aircraft continued to roll to the left when he set a constant bank angle. This problem may have been his own, as he noted the roll trim knob was apparently set incorrectly. When he tried a test for pitch attitude, he failed to get the system to stay where he set it.

As Neil approached Edwards, the radio situation reversed for a time, with NASA 1 relaying to the B-52 and chase aircraft. Then the radio deteriorated again, making communications difficult to follow as each of the aircraft and NASA 1 tried to link with the other. In the end, Bob Rushworth in the landing chase talked him down. "Gear down, looks good. Good show, Neil." Armstrong acknowledged the important role of the chase pilots in making safe X-15 missions possible, again using his paternal quip to Rushworth: "Thank you, Dad."

Perfect missions were not expected with the X-15. It was made to explore new territory; otherwise there was little point in spending the money and time, let alone risking the lives of the pilots. Operations engineer Meryl De-Geer spoke about this idea, saying, "It was an incredible airplane, really. I mean, when you consider what it was trying to do with what people knew at the time . . . [we] had trouble believing a lot of the data. They thought there were bad problems with it because it wasn't what they expected."

Armstrong shared his thoughts on the importance of the MH-96: "It was quite a good system . . . [and] was able to hold the airplane [attitude] quite tightly. . . . It was a rate-command system, so it was fundamentally a different type of control. . . . It's a little bit like comparing a truck that drives with a steering wheel, and a caterpillar tractor that drives with levers. It's difficult to compare, because they do different functions. Nonetheless, having said that, my recollection is that most people thought the Honeywell system was one of the significant things [on the x-15 program]."

Life support specialist Roger Barniki felt it was his job to know each pilot he suited up. When he had watched Armstrong's P-51 flight, Barniki noted something special. "Neil was one of those guys I never saw with uncertainty. . . . Joe Walker was that way, too. There's just an air about them when they fly that you can tell their confidence level. . . . You can't be intimate in the sense of working closely with somebody and not care about their feelings, and what they do and how they do it."

Roger explained that all x-15 pilots were exuberant in their missions. "I watched them get ready for a flight and come back, and just have the biggest damn grin on their faces. They went out and they were doing something they loved, and it really showed how much they enjoyed flying. That was their life. There wasn't one of them that didn't feel that way, or didn't give me that impression when they came back." The only time Roger didn't see this was "if something screwed up in the airplane that was their fault. Then they were kind of quiet. . . . They didn't hide things." This was equally important from Barniki's side. "The thing was, you had a trust with the pilot. You never lied and you didn't mislead, because their lives depended on everything you did, and their feedback was very significant, too."

Even before he became well known, Armstrong was always a private individual. One manifestation of this was his decision with Jan to buy an abandoned ranger station in the area of Juniper Hills, at the south end of the Antelope Valley. Stan Butchart related, "It was said they had no running water, and [they] didn't. There's an old gent who lived across the street from me, Jim Powers. [He] built houses. . . . Neil picked him up and took him up there to talk about what they could do for remodeling. It had a loft in it, and it was kind of rustic, and it was clear back up at the end of a hill."

During this time, Armstrong was part of a carpool that made the nearly fifty-mile trek to Edwards each weekday. Stan laughed, "Neil had some British car [a Hillman convertible], and the battery was always dead in the thing. At night he would drive up and park it backward to face downhill, so when he got in it in the morning, he could coast to get the thing start-ed. Pretty primitive."

By this time, the Armstrong family included two children, their son, Eric Alan (better known as "Rick"), and daughter, Karen Anne. Barely past two, Karen was diagnosed with an inoperable brain tumor, which tragically took her life on Neil and Jan's sixth wedding anniversary, 28 January 1962. Joe Walker's wife, Grace, sadly recalled, "They had a real hard time over their little daughter. That was a long, sad time. . . . Nothing could be done. [Neil] was trying to distance himself from it. He was trying to not believe it was happening, which left it all up to Jan, and that was wrong."

Neil's way of distancing himself was to immerse further into his X-15 du-ties, completing flight 3-2 eleven days before Karen passed away. This was his first mission above both Mach 5 and 100,000 feet. His often methodi-cal engineer-speak seemed even more stilted in the post-flight report: "In-asmuch as the flight path angle was still in doubt and the possibility of a higher-than-planned velocity existed, the airplane attitude was maintained at a value which would insure deviations from the trajectory to be in the direction of lower dynamic pressure."

There was a two-month hiatus after his daughter's death before Neil came back for his next flight, 3-3. He had three aborts in a row for various problems on 29, 30, and 31 March, then a successful launch on Thursday, 5 April.

That flight was a bit late in getting started, since there were problems with Armstrong's pressure suit and securing an EKG monitor for his heart. Neil explained after the flight: "The EKG went sour, the physiological recordings went sour, and Joe wasn't getting very much of anything up at NASA 1 and, furthermore, I was getting a burn due to one of the temperature thermis-tor pickups on my leg, so we disconnected that system and decided to go without it."

Once at altitude near Hidden Hills, the launch proceeded normally until the LR-99 failed to ignite on the first attempt. Neil talked of the thoughts racing through his mind as he dropped unpowered: "You don't have a very big band of altitude to get started. You have to start jettisoning [fuel] in or-der to get it all out by the time you land. . . . [I said to myself,] 'Did I have time to have another start?'" As Neil went down the checklist, he recalled in his report, "All I saw was the igniter pressure go to zero, and silence. I checked around the engine instruments and switches. . . . I couldn't find anything the matter and restarted successfully, although that sure seems like it takes a long time the second time for that engine to light up. . . . I'd guess I was approaching 35,000 feet by the time I got rounded at the bottom." The familiar push of the LR-99 sent the x-15 skyward. Neil touched 180,000 feet before reentering and landing successfully at Rogers, seven minutes and seventeen seconds after launch.

Between this date and 21 May, Neil's usual logical and exacting flying techniques faltered. He made mistakes on one x-15 flight and two associated missions, all of which added up to enough problems that Paul Bikle came close to taking Armstrong off flying status, and he seriously considered removing Neil completely from the program.

Rocketing for the first time above 200,000 feet after a launch near Mud Dry Lake, Armstrong's next-to-last x-15 mission was initially flown in a textbook manner. The view was spectacular from such a high altitude. He radioed, "Looking out, [I] can see an awful long ways."

It was 20 April 1962, and one of the test objectives of flight 3-4 was to maintain a constant g force during reentry using the MH-96 system. Neil became so engrossed in doing so he forgot to also watch his altitude. The x-15 skipped back upward, losing aerodynamic control as Armstrong continued south at a high rate of speed. NASA 1 called a warning: "We show you ballooning, not turning, Neil. . . . Hard left turn." Again, as with flight 3-1, radio reception was undependable. Captain Allavie relayed the message from the B-52: "Hard left turn, Neil."

In Armstrong's debriefing, he explained the sequence of events: "At about fifteen or sixteen degrees angle of attack and four g, I elected to leave the angle of attack in that mode, and I was hoping that I would see the g-limiting in action. We had seen g-limiting on the simulator operation at levels approximately four g . . . and it wasn't obvious that we were having any [effect], so I left it at this four g level for quite a long time, hoping that this . . . might show up. It did not, and apparently, this is where we got into the ballooning situation." He continued, "I was apparently at an altitude above that which I had expected to be, and which caused me to go sailing merrily by the field. As I saw Palmdale going by I was in a ninety degree bank angle, and essentially full deflection on the stabilizers. . . . We were

24. Paul Bikle took over as the director of the NASA High Speed Flight Station on 15 September 1959, a position previously held by Walt Williams. Courtesy of the Armstrong Flight Research Center.

having no heading change."

Paul Bikle talked of the incident from his perspective, saying, "NASA 1 radioed to warn Neil of his predicament. He was headed over the base at about 110,000 feet. He didn't have much [dynamic pressure] or aerodynamic controls available, and he ended up out over Los Angeles someplace. He managed to get turned around, and came out over LA [heading back north] at about 45,000 feet near Mt. Wilson. . . . He was off our plotting boards [because] we never planned on going more than thirty miles south of the base." It was now problematic whether Neil could bring the aircraft safely back to Edwards. "He [came north] by Palmdale at about 7,000 feet, and we thought he might try to slide it in on the runway there, but I guess he thought he could still make it back to Edwards."

Aboard the fast descending x-15, Neil agreed with Bikle's assessment, as he recounted that "the only other alternative . . . would have been Palmdale and I didn't want to get into their traffic pattern. Mirage Lake was about as far away as Rogers, and Rosamond wasn't much closer, so I decided to head for the south lake [at Edwards]. It looked like we were in good shape."

Neil's estimate of making the lakebed was not as good as he hoped. NASA 1 called, "What is your visual estimate of your location?" Neil admitted, "Looks like I'm . . . in pretty bad shape for the south lakebed." NASA 1 agreed, "Yes, we check that. Have you decided what your landing runway is yet?" Neil wasn't sure what he could accomplish, saying, "Let me get up here a little closer." Once he had more confidence, he radioed, "Okay, the landing will be on runway 35, south lake, and will be [a] straight-in approach."

Jettisoning the lower ventral could have helped Armstrong's energy situation. He said later, "The impression that I had as we approached the field was that the airplane wasn't making good the [lift over drag] that we had practiced in F-104 approaches around Edwards. I was flying at approximately 270 knots, and had I suspected that we would have been tight, I would have jettisoned the ventral, but it didn't even occur to me." One of the chase pilots radioed the idea as the x-15 closed on the lakebed. "You can punch it off any time you want to, Neil, for drag." Neil replied, "Oh, I should have done that before, shouldn't I." No time for admonishments, the chase pilot agreed. "Yep. Start your flaps down now. . . . Okay, you're well in, go ahead and put her down."

In the control room, engineers and technicians watched the flight unfold. They checked the plotting board and telemetry signals, looking for problems. If something went wrong, it was their job to give advice to NASA 1, so he could pass it along to the pilot. Operations engineer John McTigue recalled, "I kick myself so many times for not telling him to jettison the ventral. . . . Why the hell didn't I think of that? I was thinking of many other things—like how am I going to repair it!"

Neil recounted how badly the aircraft behaved as he approached for the slowest landing ever made in the x-15. "When I was in the flare, I found that I had to use large pitch motions on the stick, and it was a real sloppy, flaring, touchdown . . . probably 165 knots, something like that." He touched down, with barely any airspeed left, not far inside the perimeter of the safe haven of the dry lake surface. Bikle added, "The first skid marks were about 200 feet inside the edge, on the south end of the lake!"

As he slid to a stop, NASA 1 joked, "The posse will get there shortly." Commenting on how far south of the main base the aircraft came to rest, the chase pilot added to the humor by saying, "In about thirty minutes!"

The H-21 helicopter pilot reassured Armstrong, "We'll be there, Neil."

It had appeared at one point, Armstrong might not have enough energy to make it all the way back to the Rogers lakebed. Several people thought momentarily about going into Palmdale Airport. Vince Capasso, who shared engineering duties with John McTigue, said that if Armstrong had tried to land on the concrete runway it would have ended in disaster. "The skids were designed for the friction between the skids and the dirt on the lakebed— much higher than the friction on cement. Problem is, he wouldn't have gotten it stopped." Vince likened it to slipping across an ice rink, saying, "That would have been a bad scene to have landed on cement."

The following Tuesday, 24 April, Neil found himself axle-deep in mud at Smith Ranch Dry Lake in a T-33 Shooting Star, with Chuck Yeager in his back seat.

Smith Ranch was the launch lake farthest from Edwards used during the x-15 program. With a lot of heavy winter weather, it fell to Armstrong to fly up to see if it could support an emergency landing for a mission scheduled the next day with Bob White. "In this particular case," Paul Bikle recalled, "It had been wet, and we sent some guys up there to check it out. Jack McKay was the primary one who ended up in the argument. Jack did an overflight, came back, and said it looked dry enough to him." Some of the air force chase pilots disagreed. Bikle said, "The pilots were arguing about it, and Joe Vensel, who was director of operations, picked Neil, because he wasn't doing anything." Vensel was insistent, and he was the boss.

Bikle said, "The idea was, Neil was going to fly a T-33 and land on the lake—which he wasn't in favor of particularly, but that's what Vensel said to do. . . . If Neil felt it was too soft he could punch it and take off again before it slowed down." The air force pilots continued to say it was a bad idea. Bikle continued, "One of the air force guys that said it was too wet was Chuck Yeager, who said, 'You're gonna get your ass stuck up there so deep you can't get out! But I'll go along.' So Yeager went in the back seat, and they went up [to Smith Ranch] and touched down, rolling along for quite a ways. They slowed down, then suddenly, went in over the axles! Neil was saying afterward that he could hear Chuck hollering and laughing!"

John McTigue told of his part in the unfolding drama: "I was part of the support team that went uprange for the x-15 flight. We landed in a [C-47]

Gooney Bird, and you could land it a lot better than they could with the
T-33. We drove over to them, and Neil and Yeager were sitting on the wing.
I can't remember his exact words, but Yeager said, 'You're never stuck until
you've been in military [power] for fifteen minutes!'" His reference was to
Neil's attempt to use full engine thrust to push them back off the lakebed
and into the air. It failed, only digging themselves deeper into the soft sur-
face. McTigue said, "Yeager was sitting there laughing like mad, and poor
Neil was kind of a little down, but it's one of those things."

The X-15 was the hottest aircraft ever built for both speed and altitude.
With Yeager's reputation, it was easy to understand why he wanted to fly
the rocket plane himself. The last time he had flown with rockets was in the
X-IA on 12 December 1953. He went out of control after surpassing Mach
2.4, nearly losing the aircraft and his life. At the time Yeager flew with Armstrong and got stuck at Smith Ranch, Chuck was the commandant of the
air force's test pilot school at Edwards. He made it known he wanted an
X-15 in his stable so he could train his students to become astronauts. This
type of thinking is what got the U.S. Navy kicked off the X-15. The research
program was just that, not a trainer for pilots to make one or two hops into
the upper atmosphere and beyond.

On 25 April, one day after the Armstrong incident, Yeager secured a
seat aboard the B-52 that was scheduled to carry Bob White on mission
2-22. His assignment was officially as copilot of the mothership, but it was
part of his plan to insinuate himself into the program. Cloud cover at the
launch lake forced an abort of the day's activities, so Yeager never witnessed
a launch firsthand. Milt Thompson remembered, "Yeager was pressuring
Bikle for years to try and get on [the X-15]. Then he brought [aviatrix] Jac-
queline Cochran over and was trying to convince Bikle to let her fly. They
were down in the simulator and everything. . . . Bikle was a very pragmatic person, and he wasn't about to do anything stupid just to satisfy somebody's desire. He looked at it and asked Yeager, 'What are we going to get
out of it?' In Yeager's case, there really wasn't much he had to offer."

Because of this rejection, Chuck launched his own program to acquire a
rocket plane for the school. This idea culminated in mounting a rocket in
a pod above the tail of a modified F-104 Starfighter, redesignating the vehicle as the NF-104. Three of them were eventually modified. On 10 December 1963, in an attempt at an altitude record in the plane, Yeager lost

control and had to eject. The NF-104 destroyed itself on impact as Yeager floated down on his parachute, a bloody mess after being hit by the separating ejection seat. The incident was famously described in the book and movie *The Right Stuff.* That was the last time Gen. Charles Yeager flew a rocket plane.

Armstrong's streak of bad luck continued for another month past the 24 April incident at Smith Ranch.

On the morning of Monday, 21 May 1962, Bob White met with Paul Bikle, starting a long chain of events. White spoke about how the incident began: "Bikle asked me, 'Bob will you do me a favor and go up to [Delamar Dry Lake]?' I flew up and made some passes, came back, and told him, 'It looks fine.' I added, 'But don't have anybody go up there and practice landings until they paint the black stripes,' because it was just like trying to judge your depth perception over a calm sea—very difficult." Bikle explained the beginning of the convoluted events that transpired over the next several hours: "A series of days of bad weather and delays led us up to a point where we were pretty anxious to go."

At that particular time, Armstrong, Bill Dana, and Milt Thompson were rotating up to Boeing on the X-20, so they would always have a NASA pilot there to sit in on the meetings and put their stamp on the Dyna-Soar. Bill had come back and Neil was on his way up, stopping by the pilot's office first to pick up stuff before he went to Seattle. Paul exclaimed, "Joe [Vensel] nailed him again! He said, 'Neil, get in an F-104 and shoot a landing [at Delamar], and let us know if you think it's okay or not.' Neil said, 'I have to go to Seattle.'" Armstrong was really upset and didn't want to go to check a lakebed. Bikle said, "I can understand that, but Vensel said, 'I told you to go on up there!' So Neil goes down, and he's really huffy and charges off."

Armstrong grabbed an F-104 and headed to Delamar. According to Bikle, "Instead of making a little pass to look at everything, he starts at 40,000 feet and dumps all the drag like an X-15–simulated landing. Neil poured down over the lakebed, flared just at X-15 altitude, rips out across the lakebed with the gear up until it's time to slow down. . . . He disappeared off our radar screens when he got down low. We didn't see or hear from him again until about forty-five minutes later."

When next he heard what happened to Neil, Paul understood the mess this simple task was turning into. "We get a call from Nellis Air Force Base, [Nevada,] and the [officer of the day] said, 'We got a guy from your shop here, and he just tore the arrester gear out of the runway on approach!' Then Neil got on the phone, and he said, 'Oh, I had a little problem up there. It's okay for an X-15 landing. I'm going on up to Seattle.' Vensel asked, 'What about the F-104?' [Neil said,] 'Well, you'd better send a trailer up for it. It's on the ramp out here.'" Vensel couldn't believe what he was hearing.

"There was a whole circus of events after that," Bikle continued. "[Neil had] gone up there, and when he'd put the gear down he wasn't quite high enough. We went up there in a C-47 later in the day and you could see where the wheels first touched the ground. The tire tracks were two feet apart instead of ten, where they're supposed to be. [The F-104] was so close to the lakebed that's the furthest they could come down!"

As the obscuring cloud started to clear over the spiraling events, Bikle explained what happened from the moment Neil dove his F-104 downward to check out the Delamar lakebed:

Neil realized what he'd done and he firewalled the thing. He gained enough speed to get off the ground, but while the engine was picking up speed, the ship kept getting lower and lower, and you could see these wheel tracks come together. They got to within about a foot of each other, which meant [the gear] was almost full up. Then you could see them spread out again, to where they were [fully extended].

In the meantime, he'd dragged one tip tank off and dragged the ventral off the tail, which released the arrestor hook back there. [Of] course, he didn't know that. It also tore out one of the hydraulic systems and the radio antenna, which left him without any radio, only emergency hydraulics, and red lights flashing all over the place in the cockpit.

So, he just goes over to Nellis, and he can't call on the radio. He knew he had some problems, but he didn't know what, so he figured he'd make a flaps-up landing. . . . Neil came in at about 230 knots at the approach end of the runway at Nellis, and this damn hook's hanging out the back. He snagged that [arresting] chain across the runway at 220 knots—just tore the chain out of the runway! His airplane went down the runway dragging all this junk and blew the tires!

Nellis was tied up about three hours getting the runway cleared off. . . . Then Milt goes up in a two-place [F-104B] to bring Neil back down, and he overshot the runway, blew tires off, and tied the thing up again! Bill Dana went up, and when he got up there the aerodrome officer called and said, "Hey you guys, don't send any more airplanes up here! We'll load all your guys in one of ours and bring 'em down!" Disgraceful display all round.

Bikle was livid that one of his pilots could create such havoc. It was bad enough to have to deal with it all within the x-15 program office, but having to now include another base in all that had transpired was simply embarrassing for everyone involved. Paul reported that his first response was, "I grounded Neil and made him ride ten hours with Joe [Walker] before he could fly again. Oh, he was pissed off. What I really didn't like about that one was that he was just going to go on up to Seattle without assuming any responsibility or telling us what had happened."

Bikle bore some responsibility for the incident, since White had warned him ahead of time that the lakebed needed runway stripes marked on it so a pilot could gauge his height properly. Forty-four years later, White still recalled the incident vividly. "I went up the next day to Bikle, and I gave him hell. I said, 'Paul, am I a member of this team or not? Come on, I told you. Now it's your embarrassment, not mine.' It was unfortunate, you hate to see it happen, but it made me unhappy to think that he's questioning my judgment. And here I am, one of the prime guys in the program."

With one thing and another seeming to pile up, Armstrong was looking at what he should do next. There was one last x-15 flight for him to make, coming two months after the string of incidents. On 26 July he was ready for flight 1-32, and Neil was determined to make it a good one.

Talking about the x-15, Neil said, "The [research] pilot's role was very little flying. Most of the role had to do with project planning, flight planning, data acquisition techniques, and so on. . . . The job was finding ways to get the data that you wanted to get. . . . The project teams would prioritize the data needs, and figure out what flight trajectories and methods of getting as much of that as practical, [and] as efficiently, as possible in each flight. . . . We were a team working together."

Preparing the aircraft for flight took many people. The one with the final say-so was the operations engineer for each aircraft. John McTigue filled

25. With the X-15 hanging from the B-52 pylon behind them, Neil Armstrong talks with operations engineer John McTigue (*center*) and Elmor J. Adkins. Photo by Bob Boyd. Courtesy of the Dave Stoddard Collection.

that role, saying, "I was the one who had to make sure the systems were going to work. I signed each airplane off prior to flight [and] my signature said, 'Yes, the vehicle's ready to fly.'"

After Neil was secured inside the no. 1 cockpit that morning, a ninety-minute hold was called as the C-130, which was to support the mission, was delayed. Finally, Fitz Fulton started the engines on B-52 no. 003 and moved into position. At 10:34 a.m. Fulton pulled back on the mothership's yoke, lifting off on a forty-eight-minute ride up and around Mud Dry Lake for the launch from 45,000 feet. Once dropped, at 11:22 a.m., Neil lit off the LR-99 and pushed to the highest speed he attained on the X-15, Mach 5.74, just 11 miles an hour shy of 4,000 mph. It was a routine flight to 98,900 feet.

Having flown the first four missions in aircraft no. 3, using the MH-96 adaptive controls, Neil noted the difference in returning to the older system on no. 1: "I can say with truthfulness that this flight was like going back to the horse and buggy control system."

Armstrong safed the vehicle as soon as he came to a stop on Rogers Dry Lake. He recalled there were certain sequences to observe, and he prepared for his suit to be disconnected once the canopy was lifted. "My recollection is that by the time I landed, they were always rolling right up beside [me], and I was going through the procedures. Usually, they had their head in there, [and] they were helping me to get out."

He climbed over the sill and down a short ladder to the dirt, then into the suiting van for the trip back to the main base. During the ride, technicians helped him remove the pressure suit, so by the time they pulled up to NASA, he was relatively comfortable and ready to debrief. Neil said, "We'd usually have a debriefing session [and] go through the flight. . . . I think, the most important part was informal. That was getting together with engineers and flight planning guys and looking at what data we got that was okay. Was that what was wanted? What gaps there were, or what kinds of alterations needed to be done in order to get the specific kind of data they were looking for."

Back on the lakebed, aircraft no. 1 was made ready for the tow. The nose wheels were removed and replaced with a different set, and the rear end of the x-15 was jacked up to put the ground-handling dolly in place of the skids. After a post-flight going-over, the vehicle was later prepared for transport back to the North American Aviation plant in Los Angeles. The contractor was ready to thoroughly inspect and update all the systems. They had the airplane for the next nine months, before it was returned and readied again for flight in April 1963.

When he was asked during the NASA oral history project why he left the x-15 for a possible position in Apollo, Neil said, "It wasn't an easy decision. I was flying the x-15 and I had the understanding, or belief, that, if I continued, I would be the chief pilot of that project."

Paul Bikle disagreed, speaking candidly about Armstrong's performance at Edwards: "At the time when he applied to the astronaut program, I didn't recommend him. . . . I think Neil knew—even though we were good friends—that I didn't think much of some of the things he'd done on the x-15. He was smart enough not to use my name as one of his references." Bikle said he was glad of Neil's choice to do this, as it eliminated the need for him to say anything negative if he were called. However, it didn't do any good, as Bikle recounted, "Gilruth, the director at Houston, called me anyway, asking my opinion, and I said, 'Neil's a damn good pilot, and he's one of the best engineers I know. However, sometimes he's thinking about

engineering when he should be thinking about flying.' I said, 'If I were selecting guys for what you're doing, I wouldn't select Neil.' But they selected him anyway." Bikle thought Neil's decision to leave was the best for all
concerned. Paul was reluctant to be responsible for grounding a good test
pilot and a personal friend. Maybe the change to a new job was exactly what
Neil needed to regain his edge after all the difficulties of recent months.

On 17 September 1962 Neil A. Armstrong was announced as one of nine
pilots chosen to be an American astronaut candidate. In deference to the
Mercury program's "Original Seven," with luminaries such as Alan Shepard and John Glenn, Neil's group became known as "The Next Nine."

Training for the goal of landing on the moon prior to the end of the decade
was arduous. It was made even harder in that at the time of Armstrong's
admission into the cadre, the details on how to get to another world were
sketchy at best. Mercury had lofted two suborbital hops, similar to what
was being accomplished in California on the x-15. Two Mercury orbital
missions had also been flown from Cape Canaveral, Florida, at this point,
with Walter M. Schirra being launched on the third, *Sigma* 7, two weeks
after Neil's selection.

The two-man Gemini was being prepared but was not yet fully defined
on how it would fit in with the eventual Apollo lunar landings. Armstrong
and his eight colleagues began classes that were set up on everything from
orbital mechanics to radio communications. The temporary Manned Spacecraft Center in Houston consisted of several hundred people and some rented buildings, close to downtown. The permanent facility, which had the
look and feel of a college campus, was under construction ten miles south.

Electrical engineer John Painter was asked to set up the radio class. He
said his mandate was to "cover everything an astronaut needed to know
about how to communicate with the radio ground stations on Earth." His
classes began in October 1962, soon after the new group reported to NASA,
and as best he could recall, they were located in temporary quarters on Telephone Road at an old refrigerator factory called the Rich Building.

Painter said of his class of test pilots: "None of them wanted to talk much
in class, running the risk of looking foolish to their compatriots." Armstrong stood out to him for several reasons. "Neil had something else the
others didn't. He had x-15 experience. So, he had flown in the near reach-

es of space. That gave him a leg up, so to speak, with respect to the others. They showed a muted respect for him in class. If he spoke, which wasn't much, they listened." In addition, there was another factor John remembered specifically: "He tracked you. As I moved around the room, talking, Neil's eyes always tracked me. And, it's like he knew my destination before I did. If I shifted my gaze, when it came back, there was Neil with his penetrating eyes. I sometimes thought he could read my mind."

Classroom work, survival training, launch and orbital simulations, it was a long road to active duty astronaut status. Armstrong met the challenges, eventually being chosen as the commander of his first spaceflight, *Gemini* 8.

In the x-15, the highest altitude for Neil was 207,500 feet, 56,500 feet short of achieving his astronaut rating. At 11:41 a.m. Eastern Time, on Wednesday, 16 March 1966, the Titan II booster lifted off with Armstrong and his copilot, David R. Scott. Minutes later, they were both safely in space, orbiting the earth, America's two newest astronauts.

Primary objectives to be developed for the Gemini program were rendezvous and docking with another spacecraft and performing extra-vehicular activity (EVA), or more colloquially, a spacewalk. Both these goals were set for *Gemini* 8. The first was to find the Agena docking vehicle, which had been successfully launched on an Atlas rocket about an hour and a half prior to Armstrong and Scott.

The docking went smoothly, but soon everything turned sour. A control thruster, similar to those used on ballistic flight of the x-15, stuck open, sending the combined spacecraft into a violent spin. Thinking it was the Agena's fault and not *Gemini*, Armstrong undocked. The spinning grew worse, to the point where both pilots were nearing blackout conditions. Neil's logical and methodical engineering background helped him run through contingencies quickly. Once he realized it was their own craft causing problems, he shut down the primary thruster system, and engaged the reentry rockets to finally bring them under control. Using this system was a last resort, but it saved their lives, even though the rest of the mission had to be aborted, including Scott's planned EVA. *Gemini* 8 splashed down in the Pacific Ocean less than eleven hours after it had rocketed into orbit from the other side of the planet.

Gemini 8's emergency situation solidified Neil as being cool under adverse circumstances, eventually making him a prime candidate for a future landing on the moon. Even after this life-threatening turn of events on orbit, he took it all calmly as just being a part of his job.

By the time Apollo finally got off the ground, it was as yet unknown how many flights were needed to complete all the checkouts required of the Apollo Command and Service Module (CSM), along with the Lunar Module (LM). With the end of the decade looming, several calculated risks were taken to meet President Kennedy's goal. Chief of these was sending *Apollo* 8, only the second manned Apollo spacecraft, to orbit the moon in December 1968. This success paved the way for an Earth-orbital mission to wring out any bugs in the LM with *Apollo* 9, then a repeat of this test in lunar orbit on *Apollo* 10.

Armstrong, along with his crewmates, Edwin E. "Buzz" Aldrin and Michael Collins, had been assigned to *Apollo* 11 months previously. The schedule said they might have the first shot at landing, but no one thought everything would go so well with the preceding missions. Chances originally appeared small that Neil's crew would be first. As each milestone fell away, *Apollo* 11 looked more and more likely to make the first attempt. On 16 July 1969 their Saturn V launch vehicle took Armstrong, Aldrin, and Collins into orbit, then the third stage fired up and sent them on their way to the moon.

On Sunday, 20 July, at 4:17 p.m. Eastern Time, after a nail-biting landing sequence, punctuated by computer overload alarms, Armstrong brought the Lunar Module *Eagle* to a safe touchdown in the Mare Tranquillitatis region of the lunar surface. Six and a half hours later, at 10:56 p.m., Neil Armstrong climbed down the spindly ladder on the *Eagle*'s descent stage to step off onto the surface. Neil always said that simply walking around was not nearly as much an accomplishment as the technical challenge of landing. However, his fuzzy but iconic slow-scan television image, and the words he spoke as his boot touched the moon, are the ones for which Armstrong will always be remembered.

Engineering and flight remained close to Neil's heart. He prefers not to dwell on the emotional aspect of the accomplishment. The difficulties presented by winged flight out of the atmosphere in the X-15 are, in many ways, a more significant achievement than the brute force of launching a spacecraft into orbit on the top of a large, tubular stack of metal and fuel.

Armstrong refused to cash in on his fame, as some other celebrity astronauts have done. He did make a foray into product endorsement when he agreed to be the spokesman for a short period starting in 1979, for the then struggling Chrysler Corporation. His reasoning was sound, in that he thought their engineering accomplishments in the Apollo program were a good connection to their automotive branch. However, he was often criticized for this role when he first appeared in a television ad for the company on 21 January 1979, which aired during Super Bowl XIII. A later television role was a better fit, as he presented a 1991 series titled *First Flights*. The shows were each about different airplanes throughout history, ranging from early cloth and wood to the x-15 rocket plane. Neil introduced each segment and participated in some of the flying.

Paul Bikle recalled many special moments with Armstrong. They both shared a love of soaring in gliders. Bikle set the world altitude record on 25 February 1961, at 46,267 feet, flying his Schweizer sgs 1-23e. He had to use a World War II–era oxygen mask to accomplish the flight. In 1986 Bob Harris exceeded Bikle's record but had to use a full pressure suit to survive.

When I spoke with Paul, we sat in the living room of his tiny house. He lamented how the liability in the private aircraft industry at that time was preventing him from selling off his record-setting glider, which was then sitting in pieces in his backyard, unable to be seen or used by anyone. That beautiful white-and-red glider is now fully restored and hanging on display at Hugh's Vintage Aircraft Museum in Hollister, California, created and operated by Bikle's son. Unfortunately Paul never lived to see that happen, passing away on 19 January 1991.

Paul talked of a chance meeting with Neil after he had competed in a sailplane contest in August 1973: "I finished up and was staying an extra night. . . . About eight o'clock in the morning, the lady running the motel called up and said, 'There's some guy named Armstrong out here that wants to see you.'" Paul went outside and saw Neil and his wife, Jan. "[Neil] was, as usual, driving a real old wreck of a car. . . . They'd stopped at the Schweizer Aircraft factory [near Elmira, New York,] to see what they were doing, since Neil was interested in sailplanes. . . . They told him I'd stayed over after the contest, so Neil . . . came over to the motel. He and Jan and I sat out there on the curb [talking] for a couple of hours."

Conversation finished between the old friends. Paul noted, "Pretty soon he drove away with his exhaust pipe dragging on the street. . . . The lady [came] around, and she asked me, 'That the Armstrong that was on the moon?' I said, 'Yeah.' She said, 'You'd think he'd have a better automobile.' That was her only reaction." I told Bikle about my interview with Neil and the leaks in his office during the rainstorm. Paul said, "I gather, if Neil's office was leaking, his office wasn't any better than his car."

Paul also recalled that Armstrong was willing to share his transgressions during the x-15 program to help make a point to other pilots. "After he got out of being an astronaut, Neil was in charge of the aeronautics program at the OART [NASA's Office of Advanced Research and Technology]. . . . I remember being there in a meeting and was just amazed, because they were talking about some flying discipline problems at one of the other centers." This was certainly a subject of which Armstrong was familiar. "Neil was trying to point out to the center directors that they had to get more involved, and, if something needed to be done, they better do it, because they were responsible." What really surprised Bikle was when "Neil tells all these guys at the meeting, 'Well, when I was back working for Paul, he didn't like what I was doing. He grounded me, and I had to ride ten hours dual with the chief pilot!' I wouldn't have thought he would have told that story. . . . Although it was very much in character."

Bikle then shared his most poignant memory: "One night in Washington DC, we went out to dinner together. We came out and stood on the sidewalk, and there was a real bright moon. Well, he just stands there . . . looking up at the moon, and he said, 'You know, I can't believe I went there, even now,' which I thought was a real honest statement from Neil."

8. On a Roll

I didn't mean to desert a sinking ship, but I figured I
had better launch before the pieces started flying.

Joe H. Engle

"Okay, I'll see you in a few minutes," Joe Engle called on the radio as he prepared to hit the release switch. "Three . . . two . . . one . . . launch!"

As he dropped away from the B-52 to begin flight 1-39-63 in the early afternoon of 7 October 1963, instead of immediately moving the LR-99 thrust control out of idle, Engle stared at his angle of attack indicator. Not long before the first launch attempt, the indicator had stopped working. The abort had been called, but then it started working again, so the countdown was reinitiated. It had taken sixteen minutes to turn the mothership around to get back into launch position. At the moment the shackles released, the critical indicator went dead again. In his pilot report, Joe said, "It took me a few seconds of staring at the thing to realize it wasn't going to work. I remember reaching up and rapping it with my hand. . . . It seemed like a good idea at the time."

With no engine thrust, the X-15 continued to drop. "I took my hand off the throttle . . . and by then I'm sure I'd dropped pretty low." Finally figuring he had other things to do than worry about an inoperative dial, he pushed the throttle forward. "I was slow coming up . . . so I [did] a delayed engine light as we had done in the simulator." A chase plane radioed, "Good light." This was quickly confirmed by NASA I: "Looks good, Joe. Everything looks real good."

The flight plan for his first mission was to keep the thrust at 50 percent. However, to make up for the small bit of lost time, Joe brought the engine to 75 percent thrust. With gs building quickly, Engle experienced the impression he was heading upward much steeper than planned, a feeling exacerbated by the higher level of thrust. This was one reason that on first flights the amount of power was kept low. "I was impressed with the steepness of the climb. I was locked . . . on the instruments, but when I looked out, it felt like I was going up real steep. I don't know whether it was the way the sun was coming in the window. . . . Everything else was checking out, so . . . I figured that was just because you don't generally climb at twenty degrees."

Engle finally dropped the throttle to the planned 50 percent. At about eighty seconds into the flight, he pushed over to level out at 77,800 feet. The LR-99 rocket stayed steady for another forty seconds before Engle shut it down at Mach 4.21. He was a quarter of a Mach number faster and 4,000 feet higher than planned. All things considered for a first flight, especially after the delayed engine light, he brought it in very close to target, highlighting the precision that followed him throughout his time on the X-15.

During the extended run at near maximum velocity, the X-15 was becoming hot. "As the airplane would heat up, and where dynamic pressure was getting very high, [it] was getting to be very, very sensitive." With his Kansas country charm, Engle equated this sensitivity to "milking a nervous mouse." He went on, "This nearly quarter-inch steel on the side would oil-can, just like it does when you'd set a can out on your driveway in the sun. It would go 'ka-wham,'" he explained, slapping his hands together loudly. "Bob Rushworth warned me about it ahead of the flight, because he knew that I was going to go into that flight regime, [but] even knowing it was going to happen . . . it really got your attention."

The airframe cooled as it slowed and descended, the popping noises that had jarred Engle "right down by the old knee bone" stopped, and he began to prepare for the final segment of his mission. This flight envelope, 14 miles high and nearly 3,000 miles an hour, was a new one for Engle, as it was for all first-time X-15 pilots later in the program. With the view from that height so different, he said, "It looked like . . . I was pretty high and hot at altitude. I was rolling it up one wing and then the other in an attempt to feel it out. It was apparent . . . I needed to get the nose down to get a steeper glide angle into higher density atmosphere."

Joe had a lot of confidence in the X-15, even though this was his first time aloft. "It handled so nicely and so cleanly, I figured it was better to do a roll and let the nose dish out and end up in the attitude I wanted, rather than push over to a real low angle of attack. I wasn't sure what the handling qualities would be [in that case], so I didn't think a thing about rolling it on over." As Joe said years later, "There's nothing more natural for a fighter pilot than to do a roll." He was happy with the way it turned out, as it accomplished exactly what he intended—to set up the energy for his landing approach.

Rolling the x-15 had been accomplished safely by several program pilots since the first powered flight by Scott Crossfield four years earlier. The difference in this case was that Engle was the first pilot to make an inaugural flight in nearly three years, and rolling the aircraft was not something that had been done recently.

Pete Knight later made a comment concerning Engle's roll maneuver that made it sound like Joe was unaware of some aerodynamic aspects of the x-15. Pete said, "He didn't have the slightest idea about the inertial coupling . . . [and] just thought it would be nice if he could roll the airplane." This is a dangerous phenomenon that causes an aircraft to tumble out of control in several axes at once. It had nearly cost Chuck Yeager his life in 1953 when flying the x-1A, and three years later did claim the life of Mel Apt in the x-2. What Pete neglected to mention in his criticism of Engle was the x-15's large wedge tail above and below the fuselage, which had been designed from the beginning to prevent this problem. In fact, these large tail surfaces actually provided too much control authority under some circumstances, necessitating the removal of the jettisonable portion of the lower ventral for a significant portion of the flight program. Joe explained that in test pilot school, "We were taught inertial coupling right and left."

By the time of Engle's first flight, rolling the x-15 was not standard practice, but no one thought to mention this to Joe before he set out. "The thing was," he said, "I didn't think a thing about it. Nobody said it was not a maneuver to do in the x-15." Engle's unseen barrel roll went without a hitch, as he knew it would, losing the altitude he wanted to set up for an optimum landing. In the bowels of the x-15, flight recorders dutifully charted the maneuver while Joe concentrated on getting the airplane onto the lakebed.

From above, the state of Kansas appears as a patchwork quilt. Nearly from edge to edge there are plots of green and brown farmlands, broken only by rural roads, lakes, and lazy convoluted rivers, with an occasional town and a scattering of cities. Almost exactly at the midpoint of the northern state line lies the geographic center of the contiguous United States, near a crossroad called Lebanon. Approximately a hundred miles south and east of this spot is Dickinson County. The county seat is Abilene, most famous as the birthplace in 1890 of the architect of the World War II D-Day invasion of Europe, who later became the thirty-fourth president, Dwight D. Eisenhower. Another ten miles east is Chapman, a community built primarily by immigrants from Ireland and their descendants, first settled in 1868. The town is located in a fertile valley along the Smokey Hill River. At one time this area had been home to numerous Native American tribes, including Pawnees, Delawares, and Potawatomis. On 26 August 1932 Joe Henry Engle arrived on the scene, born here to a local farming family.

Joe attended Chapman Elementary School. As he grew, he knew farming was not in his blood. Instead, he preferred to think of becoming a pilot. These daydreams were possibly due to the romantic stories coming out of China from a group formed by Claire L. Chennault, formerly a captain in the U.S. Army Air Corps. At the behest of Madame Chiang Kai-shek, Chennault went to the Far East to use air power to aid the Chinese in their war against the Japanese. He formed the American Volunteer Group, better known as the Flying Tigers, and was awarded the new rank of colonel while in their employ.

In the corn and wheat fields of Kansas, these exploits of battles in the skies in exotic places gave young Joe a goal that would transform his life. He built model airplanes and joined a group called the Junior Flying Tigers of Chapman. It was also a time when the world was plunging headlong into war. Later that same year, when Joe was a few months past his ninth birthday, America joined the fight after the Japanese attack on Pearl Harbor in Hawaii. By the time he was old enough to attend Dickinson County High School, World War II was finished. However, Joe's passion for flying stuck with him. In addition, he excelled at basketball and other athletics and was even reported to have sung in his high school glee club. Like many others who went on to fly the x-15, he was an avid outdoorsman who loved to hunt, fish, and hike.

When it came time to leave the Chapman area for college, Engle originally thought of going to Kansas State University in Manhattan, not far down the road from his hometown. KSU was the alma mater of his father, and with his sisters also choosing this university it had become something of a family tradition. However, Joe changed his mind after finding KSU did not have an aeronautical engineering degree program. Instead, he headed farther east to Lawrence, home of the University of Kansas. While attending, Joe joined the Reserve Officer Training Corps, better know as the ROTC. There he met and fell in love with Mary Catherine Lawrence from the Kansas City suburb Mission Hills.

Joe earned his degree in 1955, going on to receive his pilot's wings in the air force in 1958. His first assignment was with the F-100 Super Sabre in the 474th Fighter Day Squadron, part of the 9th Air Force and the Tactical Air Command. Soon after he arrived, the squadron was redesignated as the 309th Tactical Fighter Squadron at George AFB in California. A few years later, Joe decided he wanted to move into test piloting, so he applied for the Experimental Test Pilot School as a member of Class 61C.

One of the most influential air force pilots at Edwards during the early 1960s was Chuck Yeager. Engle had flown with Yeager when they were both stationed at George, where they flew with and against each other in mock dogfights. Engle's skills continued to impress Yeager, the test pilot school's commandant. They also shared a passion for hunting and other similar activities, forming a friendship that survived a lifetime. Yeager was quoted as saying Engle was "one of the sharpest we had in the program."

Once Joe graduated in 1962, he went directly to the Fighter Test Group at Edwards. This was fifteen months prior to his first X-15 flight, and Joe Engle had already unofficially become part of the program by flying missions in the important role as chase pilot. He flew five times in this capacity before taking the X-15 up himself, twice for Joe Walker and Robert Rushworth, and once for Jack McKay. Engle once said, "Chase was a lot of fun. I think the most fun."

About this time the air force earmarked Engle for the X-15, but to give him additional experience before coming aboard, sent him to the new Aerospace Research Pilot School (ARPS), where he graduated in Class III. This was the second phase of the original Test Pilot School, created to incorporate studies and flight experience designed to lead the elite of air force pilots directly into positions on America's space program.

Engle explained the process that led him to the X-15, saying, "I didn't really decide. I was . . . given the opportunity by a couple of the more influential people, including Col. Tom Collins, who was head of flight test at Edwards. . . . It happened as a result of Bob White's normal rotational re-assignment coming up, creating an opportunity for another air force pilot to be included in the cadre. . . . When Bob White moved out, [Bob Rush-worth] moved up into prime pilot. Then there was an opportunity to get someone to come in as back-up pilot." This was the perfect opportunity as far as Joe was concerned. "My desires at the time were to go fly that air-plane. . . . I could get a tour in there, and then still have an opportunity to fly, maybe, in the Apollo program. They were also talking about a winged orbital reentry vehicle." This would become the Space Shuttle.

Engle finished his course of study and flight training at ARPS in June 1963, then almost immediately was announced as a new X-15 pilot. He didn't have to move far since the air force pilots assigned to the program worked out of Fighter Test Operations, about a mile and a half from the NASA Flight Research Center.

Soon Engle was making his first X-15 flight, near the end of which he exuberantly rolled the aircraft, quickly lowering his altitude. The X-15 in-struments held all this information about his maneuver, and their secrets would not be divulged until the engineers got hold of these data, decipher-ing their meaning.

Joe was really enjoying this flight, as he stated in his post-landing report: "I think it handles nicer than the F-104, particularly in roll and pitch. In the pattern, the X-15 seems to be more maneuverable. . . . It's just a nicer feeling airplane." He felt comfortable and confident the closer he got to the touchdown point. "I couldn't tell too well when the skids hit," he said, "but if you'd happen to be relaxing at the time that nose [hits, it] sure draws your attention when it raps down. It would . . . wake you up." Although his first mission was completed, there were ramifications still to come for the rookie pilot.

By the time of the initial debriefing, all the flight data had yet to be de-coded. This was a responsibility of NASA. Glynn Smith was an instrumen-tation technician, and he said, even though he worked on the team that flew the fastest rocket plane, his job was rather sedentary. "Instrumenta-tion, you didn't go anywhere. . . . It was our job to take out all the drums with the data and airspeed, all that stuff, and get it developed. We devel-oped our own film. . . . Betty Love was upstairs, and the film that we processed we gave to her in rolls. They took that and figured out all those wiggly traces." Betty was part of a pool of women known as "computers." They manually did the type of data reduction that today would be relegated to an electronic computer. Her group read the oscillograph film, then converted it into meaningful data, plotting it for the engineers to interpret. Betty explained how she did her job:

When I started, there [weren't] any computers. So the film was read over a light box with a six-inch scale. You read from a trace, which was a reference line to the . . . film that had been recorded during your research flight. You converted the inches of deflection into engineering units. You didn't always know the reason you were doing it, because the computer section was different from the engineers. But you got busy and you did it. . . . It went to the engineers, who wanted it yesterday.

Roxanah Yancey was head of the computing section when I was first taken there. . . . I didn't have a mathematics degree like they wanted, but I had a degree in science, so she said she'd give me a chance. When she took me into the room it was all ladies. . . . She gave me a desk with a Friden calculator [a mechanical calculator with more than one hundred keys to enter data], a data sheet, a six-inch scale . . . , and a lightbox all my very own. Then she said, "You're part of the group." I said, "Thanks." I looked around and I asked her, "Well, aren't there any fellas here?" She said, "No, they don't have the patience."

Engle put the first X-15 mission behind him, going about his business, with the primary task being to prepare for his next research flight. In the course of reading these data, engineers found the roll information, dismissing it at first. Joe laughed as he related how they seemed reluctant to bring it to his attention. "When a couple of the engineers had queried me about it right after the flight, in kind of a lighthearted manner, they thought they had a malfunction of the oscillograph." What they had seen was the roll trace go off one side and disappear, then come back down on the other side. There were two explanations: either the equipment was acting strangely or else the aircraft had done a 360-degree roll. Since the engineers had already checked the recorder and found it appeared to be working fine, that only left the second option. "They were kind of halfway kidding, and I didn't think anything about it. I guess it was about a week later I began to realize and be informed that I needed to understand this was not a maneuver we

236 | ON A ROLL

had better be doing."

At an anniversary event honoring the x-15, Engle shared further details of what transpired: "Mr. Bikle, you know, was really neat. . . . You can't believe how great Paul Bikle was. He was a pilot, thank God. And Bob Rushworth said, 'Come on, we've gotta go down and see Mr. Bikle. . . .' We walked in and Paul was sitting there with . . . cigar ashes down his shirt. He was looking at something, and he just made me sweat for a while, and finally looked up and said, 'Engle, did you roll it?'" Joe answered immediately, saying, "Yes, sir, I did." Bikle wanted to know why Engle had done the maneuver. "I explained to him that I really thought I was going to overshoot, and I wanted to get the nose down, and just did a barrel roll, and let it dish out. He thought for a while, and he said, 'Hmm, well I'd [have] done the same damn thing. . . . Don't do it again, because everybody'll want to do it.' And that was the last I heard of it."

Joe volunteered during our conversation: "It's a beautiful rolling airplane, but, by the same token . . . I can understand why that's not something they would want the research program to be doing. . . . As it turns out, I think that's probably the reason there wasn't any reprimand other than a reminder that's not what's supposed to be done in the x-15. . . . Nobody had thought to say anything." When asked if they started briefing all pilots to not roll the airplane, he laughingly said, "Didn't have to!" He went on to finish the thought, saying, "I was a lot younger than the other pilots who were involved in flying the x-15. I think my attitude . . . was to keep my mouth shut and learn from them all that I could, because every pilot has their own characteristics and personality—like airplanes do—and you can really learn something from everybody."

Planning missions was the heart of the x-15 program. Each was meticulously timed, with the purpose laid out and no question as to what must be accomplished once the aircraft dropped from the shackles. With only a few minutes from launch to landing, there was no time for improvisation; either the goals were met or a new mission had to be planned and flown. Repeating a flight meant that other objectives would have to be postponed or canceled.

"One of the key jobs in the x-15 was flight planning," Paul Bikle related. "The air force always had about two or three engineers working with the NASA contingent on the x-15. There was a fellow by the name of Bob Hoey, who was one of the sharpest engineers on the program, and a guy named Johnny Armstrong. I think they did a lot more than they were given credit for."

Johnny Armstrong also praised Hoey. "The guy that first started flight planning was Bob Hoey. He, and a fellow named Dick Day, used to get in the [c-47] Gooney Bird . . . and fly down to [Los Angeles International Airport] where the simulator was at North American Aviation. The simulator was the key thing for flight planning. That was the tool." This simulator, often referred to as the "Iron Bird," was later moved to the NASA Flight Research Center. Hoey's and Day's travel time was then greatly reduced, although their workload increased dramatically as they immersed themselves in charting x-15 missions into unknown territory. Johnny joined the program as it matured, staying through to the end.

"When I got in the program," Armstrong noted, "there was a fellow by the name of Warren Wilson and myself. It was shortly after that when we needed somebody else, and we hired a guy by the name of John Manke. . . . Somewhere along the way, Jack [Kolf] came in. I stayed and lasted out through most of the program as a civilian with the air force. . . . It was a test-team concept [with] the only difference being I was paid by the air force and they were paid by NASA."

Johnny was raised in Alabama and joined the air force to see the world. He ended up at Edwards AFB. "I came out here in October of 1956 as a second lieutenant, spent four years as a flight test engineer working programs such as performance, and stability and control testing. . . . After that, I did a lot of F-104 work, [including] an F-104 zoom program that got an altitude record [on] December 14, 1959, [at] 103,395.5 feet."

This was a time when the Soviet Union and the United States were trying to one-up each other as quickly as they could. On 14 July Vladimir S. Ilyushin made a zoom flight of his own in a Sukhoi T-43-1, topping out at 94,661 feet and grabbing the international record altitude for a turbojet-powered aircraft.

The Soviet feat was considered unacceptable by the American military, so both the U.S. Navy and Air Force began crash programs to upstage the record as soon as possible. The navy chose a YF4H-1, and on 6 December Cdr. Lawrence E. Flint achieved an altitude of 98,561 feet. Close on their heels, the air force group made ready with their attempt, using an F-104C piloted by Capt. Joe B. Jordan and planned by Johnny Armstrong.

To make this flight possible, Armstrong utilized the maximum aerodynamics of the fighter and squeezed every ounce of thrust from the J79 engine. Many modifications were done to the airframe and engine, including changing out the entire tail assembly for a larger one from an F-104B, as well as increasing the afterburner fuel flow rate and maximum RPM.

Armstrong was painstaking in his work, as was Jordan in his piloting skills. Jordan took off, accelerated to Mach 2.36, then pulled up into a nearly fifty-degree climb. At the top of his zoom Jordan surpassed Ilyushin's official mark by 8,734.5 feet, a nearly 10 percent improvement in record altitude in just five months. The air force had thumbed their noses at the Russians, not to mention winning a small piece of interservice rivalry against the navy. The record was also significant because it was the first time in aviation history a vehicle had surpassed the 100,000-foot altitude mark after taking off from a runway under its own power. On 3 October 1960, President Dwight D. Eisenhower presented the Harmon Trophy to Captain Jordan for his record flight. According to Johnny, "Eisenhower said . . . 'Why couldn't he have gone the other half-foot?'"

After a brief move to NASA's Marshall Spaceflight Center in Huntsville, Alabama, Armstrong was persuaded to return to Edwards to join the X-15 team. "I got back here in early 1962," Johnny recalled.

Armstrong found the X-15 being run differently from the earlier rocket plane programs he had seen on his first stint at Edwards. The military ran the test series, then, when they finished with the vehicle—and if it happened to survive intact—they would turn it over to the National Advisory Committee for Aeronautics. The X-15 altered the way things were done from the start. The U.S. Air Force, NACA (later NASA), and the U.S. Navy all worked together in a way that was unprecedented up to that time. According to Engle, "The thing that you just cannot imagine is the awesome compatibility and working together of the entire X-15 team. [Everyone] blended together so beautifully and so perfectly. . . . It was an entire team of professionals, all focused on the next flight."

Overall mission planning, and specific flight planning, was one way this happened. Air force flight planner Bob Hoey had this to say: "We decided probably the best way for us to integrate into the program—as a joint program—was to get involved in the mission planning, because during that time period you've got to touch base with the technologies involved, all the systems: heating, stability, everything. All these experts have got something to contribute, and, as a mission planner, you have to take [everything] into account."

Johnny Armstrong further elaborated that there were three basic types of flights: altitude, speed, or heating. "Those were unique and different, and you just worked from those basic types to build your flight plan. You then took those ingredients down to the simulator . . . and worked with it on this really antiquated equipment—this old analog equipment—instead of all the digital equipment we have nowadays." Of course, in the standard of the day, this equipment was considered state of the art.

For speed and altitude, the flight planner plugged in all the necessary parameters to the simulator, and "in a couple of hours have in-hand what you thought the flight was going to be." But heating flights were more complex, as Johnny remembered: "Those were the pat-your-head and rub-your-tummy type of thing. They were more precise. You had to go up, find an altitude, roll into a bank, and hold an angle of attack within very fine limits to get the conditions you wanted. Those took significantly longer."

Engle spoke of planning a flight profile from his perspective: "We, as pilots, would be brought in . . . mainly to see if there were any aspects . . . that were either very demanding or that we'd want to build up to incrementally." Since Joe came later in the program, his input was different than what would have been the case with earlier pilots. "My involvement in each mission was to learn what the profile was going to be and what were the test objectives of the flight. Then I had to decide what piloting techniques would be needed to get the data . . . then start working on that profile in the x-15 simulator, all the while, keeping refreshed on possible malfunctions that could come up."

As time passed, each aircraft branched further away from its counterparts. By the time Engle was selected for the program, there were not three relatively interchangeable x-15s, but instead three distinct aircraft with their own quirks and personalities. The no. 3 aircraft had started this trend with the inclusion of the MH-96 adaptive flight control system. NASA flight planner Jack Kolf said, "We eventually evolved into an area where we specialized in one airplane. In the early part of the [flight] series, we took various types of programs and worked with them regardless of what airplane or design.

All three airplanes were different in little ways, so later you would special-ize in one [x-15] and learn its systems and work with that one a bit more."

The most drastic of these changes was realized after the rollover acci-dent in November 1962 and subsequent direction with the rebuilt no. 2. In many ways, it was seen as an entirely new aircraft, and the flight planning reflected that, immediately intriguing Johnny Armstrong. "I wanted to work on the x-15A-2," he said, "particularly when it came back from the modifi-cation, primarily because the air force was interested in the scramjet tests . . . and things like that—an awful lot of interesting experiments on board."

The new role of the A-2 was an area where Joe Engle was originally to be-come more involved. Robert Rushworth was the air force project pilot for the modified x-15, and once he planned to leave the program, Engle was to be his replacement. This did not work out the way it was intended.

Armstrong's role in planning each flight was also coordinated with an op-erations engineer, whose job it was to make sure the aircraft could physically do what the plan dictated. "He certainly would give you a different view," Armstrong said. "In my way of thinking, there were several main players, the flight planner, the ops engineer, and the pilot. They kind of came [in] a little group. . . . [The operations engineer] technically had charge of the airplane [and] directed what needed to be done."

Some of the men who accomplished this aspect of the x-15 included Bill Albrecht, Herm Dorn, and Vince Capasso. Meryl DeGeer also moved over briefly from the lifting body program. "When we lost the one lifting body [M2-F2], I got assigned to x-15 no. 1 as an ops engineer. I had three flights and three aborts. That was my time on the program. Then headquarters decided to rebuild the lifting body that had crashed, and I chose to go back to that."

Following his first familiarization flight in October, Engle set to work on his next mission, 1-41, set for 14 November. The only deviation from the plan was, for the first seven seconds under rocket power he pushed the throttle to 100 percent thrust instead of the planned 75 percent. Joe said he prided himself on flying the profile as close as possible, specifically emulating the un-compromising work of air force x-15 pilots Bob White and Bob Rushworth.

Two months later, on 8 January 1964, as Engle was preparing to be launched from the B-52, the mothership started developing problems of its own. Joe seemed so engrossed during the last few seconds, preparing for the drop, that when he felt the vibration from a B-52 engine running roughly, he thought it was some bad turbulence, so he kept going with the checklist. Joe didn't pick up on a chase plane telling him about a bad B-52 engine, so he radioed back, "Roger. Three . . . two . . . one . . . GO!"

At the same moment he fell from the pylon, the B-52 engine shut down. After the flight, Engle made a remark about "deserting a sinking ship" and how he thought it was a good idea to take the X-15 away from any problems with the B-52. His reaction was similar to the incident in 1956, when the P2B-1S mothership for the D-558-2 Skyrocket had a runaway propeller, and Stan Butchart had to drop Jack McKay even though Jack had wanted to abort the launch. In the case of X-15 flight 1-43, the B-52 luckily suffered no serious damage. Engle completed his first flight faster than Mach 5, going over the top of his arc at 139,900 feet. Seeing the inky blackness from that height, he remarked, "Sure a dark sky up here."

Prior to launch on this mission there was some leakage of steam from the ballistic control system, leaving frost over the tiny X-15 windows. This may have contributed to a problem with glare that Engle experienced as he was going uphill. Almost immediately after getting squared away on his flight path, he radioed back to NASA 1, "Going to have to do something about that sun." The window frost acted in the same way a dirty car windshield can be annoying when the sun hits at the right angle by diffracting the light.

X-15 pilots on previous flights had experienced some troubles when the sun would come directly in the windows, but they did not seem as bad as that experienced by Engle. At first, no one made the connection with the frosted window panes, instead, placing all blame on the reflective layers on the spacesuit helmet visor. Engle told his debriefers, "It was just a little difficult with all these reflections going around within the face plate. . . . It's kind of like looking through a one-way mirror, it impairs your vision a little bit, but not to the point where you can't fly the airplane." In the end, it was most likely a combination of both the frost and helmet reflections that clouded Engle's vision.

After the flight, Joe also suggested that an additional source of sun glare and reflection might be the silver outer spacesuit layer, and maybe something to cover it up might help. True to form, the crew preparing him for his next flight in April decided to present Engle with their solution. "They had a black bib made up to cover the suit," Engle said. "It had 'The Black

Knight of Rosamond' put on it." When asked if he actually used it during a flight, Joe recalled, "I've still got it at home, but I don't think I ever wore it."

In April and May Engle performed two exemplary missions in aircraft no. 1, both of which were to further expand his experience at higher altitudes. On the first of these flights the difficulties centered on new ballistic control system rocket motors, a balky inertial navigation system, and an overheated auxiliary power unit.

Prior to flight 1-46, a navigation system from Sperry Rand was installed in the aircraft. Perry Row and Ron Waite, both project engineers, said in their operations report, "Reliability of the Sperry Inertial System continues to be a serious problem, as demonstrated by the preflight activities."

This device was made up of two major components, a stabilizer and a computer. While still attached to the B-52, the system received location updates. Once the X-15 launched, the Sperry device took over. The stabilizer was primarily a gyroscope that sensed any changes in acceleration or direction. These data were then sent to the computer, which used the information to calculate the exact position of the X-15 as it moved along its flight path. Each time a computer and stabilizer were installed, one or the other component failed, requiring removal. Then the process started over. By the time they were through, five pieces of equipment were changed out, totaling three different pairs of stabilizers and computers, before one set worked well enough to be deemed ready. Their diligence paid off, and the unit performed acceptably during flight.

As for the ballistic control system rockets, the old motors had been removed at the end of their useful life and replaced with new ones which were lighter and supposedly better. However, this did not turn out to be the case, as the new motors tended to leak. These had to be replaced by much heavier machine-forged motors. Due to their small size, the weight penalty was modest, but these motors were critical to maintain the attitude of the X-15 during its time outside the atmosphere on a ballistic trajectory. On Engle's flight with the bad BCS motors, the leakage did not affect his reentry.

A third problem occurred, this time with one of the auxiliary power units, located in the equipment bay behind the cockpit. Sometime during the flight, one of the two APUs began to overheat. A small fire broke out, although it was apparently of short duration. If the fire had continued, it could have destroyed the APU, spreading to the surrounding equipment and necessitating an emergency landing. Because it extinguished itself quickly, the damage was contained to the APU itself. The reason for the fire was likely leakage of the volatile hydrogen peroxide used to fuel the units. Numerous troubleshooting runs after the aircraft was safely back in the hangar refused to expose the leak source.

An additional test on this flight was a set of reference lines placed on the inner pane of the windshield. These lines provided a backup horizon reference for the pilot, which could be useful while climbing to altitude. Engle had mixed feelings about them. "The lines on the canopy, I think, would be good to verify that you haven't over-rotated grossly. . . . As far as flying the profile, I don't think you could tell accurately enough. It requires too much time, and too much attention, to look out and verify that you have it right on the horizon." Lines of this type were eventually used on the windows of the lunar module to aid the astronauts in their moon landings.

The next flight, 1-48, was Engle's fifth mission in the X-15 and competed for resources with an all-out nuclear "attack." The May 1964 joint military exercise was given the name "Desert Strike." Before it was through, it involved tens of thousands of army and air force personnel in the largest mobilization of forces since 1945.

The scenario was that the mythical governments of Nezona and Calonia had a disagreement over water rights in the Colorado River area that bordered the two "countries." Political tensions escalated to the point of launching a mock nuclear exchange. This was done to test the responsiveness of the various military units engaged by the U.S. Strike Command. Thousands of soldiers were marched through desert terrain at temperatures reported as high as 105 degrees Fahrenheit. Air force fighters and bombers attacked various bases, and special oil-drum devices were created, releasing mushroom clouds to simulate atomic blasts, adding to the realism.

Time magazine lambasted the administration of President Johnson for even allowing the exercise to occur. In their 5 June 1964 edition, they summed up the costs in dollars, manpower, and equipment. "Although considered a war 'game,' Desert Strike ran up costs that smacked of the real thing. The two-week exercise consumed some $60 million . . . involved more than 100,000 men, 780 aircraft, 7,000 wheeled vehicles, [and] 1,000 tanks. All were deployed over some 13 million acres of [the] California, Nevada, and

Arizona landscape. Air force units operated out of twenty-five airfields from Texas to Oregon." The worst part, *Time* concluded, was not only the cost to the government but in the lives lost. "The exercise caused or contributed to the deaths of thirty-three men, including six in aircraft crashes, five by drowning, five in truck accidents, and two sleeping soldiers who were run over by a tank."

With the resources being consumed and redirected throughout all this, it is a wonder any x-15 operations could continue. In addition, regardless of the conflict between Nezona and Calonia, the Armed Forces Day at Edwards went forward as planned on Saturday, 16 May. The no. 1 aircraft was used as a static display, directly delaying Engle's flight from the 14th to the 19th. Joe joked about this allocation of resources: "It sounded like we were going to have to recruit some aircraft from Desert Strike to get some chase airplanes."

Once the public retreated from Edwards and the chase planes were rounded up, the flight went off with no problems to speak of, giving Engle his highest altitude to date of 195,800 feet. This would remain his personal record until the following May on flight 3-42.

Each mission instilled new confidence in Joe's skills as a research pilot as he honed his technique. A point of pride was to maintain the x-15 in as level an attitude as possible as it dropped away from the B-52 pylon. Due to the position between the fuselage and inboard engine pod of the bomber, and the resultant slipstream, the x-15 would naturally roll off to the right. In his post-flight comments, Engle noted his performance in the matter: "I was concentrating more on roll than on previous flights. . . . You have to hold a little bit of left aileron in to keep from getting a roll-off at launch, and then you have to get it out as soon as you drop away from the B-52. It's just a matter of anticipating how much you have to have and anticipating when to take it out. If you wait till you see yourself start to roll, then you get a roll. . . . There's no real problem, just a matter of refining the launch [and] getting it as smooth as possible."

It was around this time the modified x-15 no. 2, also known as the A-2, was being readied for its first flight since the rollover accident. The checkout flights were originally the responsibility of Robert Rushworth for the air force and Jack McKay for NASA. Since Rushworth had more experience than

Engle on the rocket plane program (twenty flights as of June 1964 versus Engle's five), he would do his part of the work to wring out any bugs with the new systems, while Engle was gaining enough stick time to eventually take over the slot. This plan was supposed to allow Rushworth to leave to pursue his military career outside the x-15 program. With this in mind, Joe was nearly always assigned to fly chase each time Rushworth flew the A-2.

On the second such mission, on 14 August 1964, the first problems with the landing gear on the modified x-15 started to crop up. The heat build-up at Mach 5.23 seeped into the structure. As it expanded, the brackets holding the landing gear cable stretched the cable beyond what it could handle, releasing the nose gear into the slipstream at Mach 4.2. Rushworth knew he had problems but didn't know how serious until he got low and slow enough for Joe, in Chase 3, to catch up with him. Engle explained, "When something like that occurred, you weren't really close enough to the x-15 to know in any detail what happened. When that gear came out, our initial concern was where to head to pick him up. We didn't know whether he had enough energy to get back to Edwards or if he might have to go into Cuddeback. So, the first concern was whether to set up the energy in the chase plane to head out and pick him up en route." In this instance, Rushworth's piloting skill got him all the way back to Rogers Dry Lake, which is where Engle rendezvoused to help bring Bob home safely.

Joe radioed to Rushworth as he pulled in close enough for a good look, "Bob, your tires look pretty scorched, I imagine they will probably go on landing." Rushworth decided to stay with the aircraft, bringing the x-15 in for a perfect touchdown. Within three hundred feet of nose gear contact, the tires disintegrated. Engle recalled, "The tires were burned off, so it rolled out on the rims. That was a little concern, too, whether they would dig into the lakebed." The gear mechanism itself stayed intact, and Rushworth was able to bring the rocket plane to a safe stop—after a very short rollout of 5,630 feet.

The x-15A-2 had several such difficulties with the nose gear and air scoop door, as well as the rear skids, before everything was ironed out and the research program could again move forward. Joe Engle witnessed each of these incidents from his perspective as a chase pilot, but never flew the A-2 himself before leaving the program in late 1965. Rushworth's plans to find a replacement air force project pilot on the advanced x-15 continued to be

26. Moments after launch on mission 3-35 at 1:16 p.m. on 28 September 1964, Joe Engle acceler-
ates away from B-52 no. 003. On this flight he achieved Mach 5.59 at 97,000 feet. Courtesy of the
Armstrong Flight Research Center.

frustrated.

Engle's eventual decision to move on from the X-15 program and not take
over the A-2 as planned was still in the future. For now, the idea remained
for him to gain further rocket plane experience. Since actually flying the X-15
was limited, he spent a great deal of time simulating missions, both on the
ground and in the air. When asked how often he used these methods, Joe
said, "At first, an awful lot. . . . I would go down on the weekends and sit
in the simulator to get more familiar with the cockpit. We had . . . an awful
lot of F-104 flights to the various dry lakebeds." Using that Mach 2 needle-
nosed fighter was about as close as a pilot could get to simulating the real
deal, at least for the last minutes of flight and landing. Some pilots recalled
doing hundreds of these special X-15 simulated flights using the F-104.

For five out of his next six actual X-15 flights, Engle moved from the no.
1 to the no. 3 aircraft. This vehicle had the special MH-96 adaptive control
system built by Minneapolis-Honeywell, which was brought to fruition on
the program by Neil Armstrong. True to its problematic nature, the first time
Engle used the system on flight 3-30, he experienced several malfunctions.

One recurring theme, which came back on this flight, was sun glare. Engle was asked specifically if he felt launches should not occur under some circumstances, to which he answered, "Certainly not a recommendation not to launch. We have a worse problem going at any time of day out of Mud Lake—that's right into the sun." Joe had not launched from Mud that day, but the one time previously, on flight 1-43, definitely left an impression. With a late morning to midday launch time as standard, heading nearly straight south as was required at Mud meant going into the sun, no matter the season. Engle never launched out of Mud again, instead, almost exclusively flying from a point farther east near Delamar Dry Lake.

Before the no. 3 aircraft was released for flight again, it had another important job: becoming a movie prop for a training film on x-15 rescue techniques. The life support supervisor for NASA was Roger Barniki. "One of my jobs was putting together a film and a book on training everybody for pilot rescue. I worked with a gentleman named Tom Fiscan. He looked a hundred years old, like Mr. Magoo, and had worked for one of the big aircraft companies. The two of us put together a treatment of a script, but I didn't know anything about pictures, and he said he did. He had worked for Howard Hughes's company for a long time and did movies. He put down the script and put together the storyboard. We took a lot of footage [on 13 and 14 July 1964]."

Once the basic procedures were shot, from that point forward the project appeared as a comedy of errors. Roger continued, "It needed to be narrated and cut and put together. . . . We're setting up to shoot these narration scenes and someone says, 'Okay, who's directing this?' And this guy says, 'Well, I am.' So we're sitting there, and he's sitting there, and everybody's looking at each other. I asked if somebody was going to say 'Action,' or whatever. We spent a half day being idiots." Barniki left the production, returning to see it once it was supposedly completed. "The one thing is this is a technical film. It can't be nice as far as aesthetics, [so] they show this film, and I say, 'Gee, that's a great film—except that it's technically, horribly incorrect.' I spent the next couple of weeks recutting it, then showed it to the NASA director, Paul Bikle. We also had the training book [to go with the film]. The guy did a beautiful job on the handbook." Barniki had to recut the film himself, and this time it worked much better, as Roger related: "The film turned out [so] the firefighters, mechanics, and other people could understand and see what was going on. That worked out and was used for the rest of the program." While all this was transpiring, after the filming was completed, the no. 3 aircraft was returned to flight status.

On 29 July Engle was back in the x-15 cockpit. As with so many of Joe's flights, it went just as it had been planned and simulated. Engle was very proud of the fact the word "nominal" was attached to most of his research missions. "Not every flight was a record flight, and obviously there was an awful lot of very good, solid engineering data that was retrieved from the x-15. I recognized that and tried as hard as I could to fly as accurate a profile as I was given. . . . It was satisfying to slide out on the lakebed after a flight, especially one where nothing had gone wrong, and you really knew all the data you brought back was what the engineers were looking to get from that particular flight."

For one flight on 10 December, Engle returned to the no. 1 aircraft. His primary goal was testing methods of guidance control planned for use on the x-20 Dyna-Soar. Also, both outboard wing panels were coated with a substance called thermopaint. This was done to study the effects of heat and shockwave impingement due to the installation of pods on each wingtip. The special paint reacted only if certain temperatures were exceeded, in this case 1,100 degrees Fahrenheit. The outer half of the left wing was painted a dark green, while the same area on the right was a light green, giving the look of a car that had its fenders replaced but had not yet been painted to match the rest of the vehicle.

Each x-15 was outfitted with numerous cameras, including one behind the right shoulder of the pilot to record the instrument panel readings during flight. Amazingly, this camera was not set to record at a constant speed, so finding a correlation between a specific frame of film and a time from launch was all but impossible. In addition, the aperture setting was often left too wide, and the film was so overexposed as to be unusable. The engineers were finally frustrated enough that, after flight 1-51, they insisted both of these problems be fixed before future flights. This decision was critical in tracing the problems encountered several years later, on the only fatal flight in the x-15 program.

On 2 February 1965 Joe made his first flight of the year and the fortieth for the no. 3 aircraft. It marked his highest speed in an x-15, Mach 5.71, or

3,885 miles an hour. His subsequent flight on 23 April, mission 3-41, was made at a slightly lower speed, but also at an altitude of less than 100,000 feet. This was the altitude range where the A-2 was expected to make its speed runs in the future, and various samples of ablative materials were being tested on sections of the fuselage and flight controls. The late April mission was originally scheduled for mid-March, but the weather had other ideas, delaying the mission seven times before it cooperated. When Engle dropped from the pylon just after 9:44 a.m., he noted, "It was a real smooth launch. In fact, I was so surprised by it, I had to snap out of it to get the throttle on."

Between flights 3-40 and 3-41 Joe added docent to his list of skills, as a special VIP came to visit Edwards. Vice President Hubert H. Humphrey came to inspect the base on 2 April. Joe spoke of the experience, saying, "I was the new guy on the program, and Bob [Rushworth] said, 'You're going to go down to the hangar and stand by the x-15. . . . Just stand there, you won't have to say anything.' God bless his soul, the old liberal soul, [Humphrey] was a neat guy. You couldn't help but like the guy. He was so friendly and had that smile on his face all the time. He came up and he said, 'Well, what have we here?' It scared the hell out of me because I didn't know how to talk to a guy like that." Joe recovered quickly, replying to Humphrey, "It's the x-15, sir." The vice president thought a moment, then said, "Oh yes, yes. And how many squadrons of these do we have?" Engle wasn't sure what to say, but his training quickly kicked in when he replied, "Not very many, sir." Joe explained, "I didn't want to tell him we didn't have any squadrons. [Humphrey] said, 'Well, we need more of these. I'll fix that when I get back to Washington,' and I said, 'Well, thank you very much, sir.' I didn't even know how many we needed at the time."

All of Engle's missions since the previous July had been at relatively low altitude, with only one exceeding 100,000 feet. His next series of flights once again pushed Joe's experience higher. With flight 3-42 on 28 May, Joe jumped nearly 14,000 feet above his previous record, going over the top at 209,600 feet. A new instrument had been installed that was supposed to better predict the maximum altitude for a flight based on real-time conditions. As the LR-99 engine approached scheduled burnout, the instrument showed he was going to be too low. Assuming it was correct, Joe ran the engine long enough to boost his speed by an extra 100 feet per second and

27. The LR-99 rocket engine is put on display, and Joe Engle is assigned to show it off for the visiting dignitary, Vice President Hubert H. Humphrey. Courtesy of the U.S. Air Force.

brought the angle of attack up slightly. These actions should have brought him back on his ideal profile. Instead, the predictor was off, not his flying. Following what it told Joe to do caused him to overshoot by 10,000 feet.

On 16 June Engle pushed X-15 no. 3 to 244,700 feet. When Joe started reentry, a small pitch oscillation began, possibly due to his stick inputs. He quickly realized what was happening and took his hand off the controls.

As the x-15 system automatically started to damp the motions, Joe slowly started putting in new commands. The ride smoothed out and went without further incident—until he came in on final approach for landing.

Somehow, an f-4 Phantom II was on the ground in the way of where Engle was to do his post-landing slideout. Jack McKay at NASA 1 called up, "Joe, got a little message for you. An f-4c is sitting right in our runway 18. You'll have to clear him, over." Engle replied, "Oh, I got him, just off to the left." It's unknown how and why the fighter jet was in this position, but Engle had no way to go around in the unpowered x-15. His best option was to steer the rocket plane out of the way as best he could once he touched down. Joe later said, "As soon as I touched down I put forward stick in and then started all the way over with right aileron to start the airplane veering to the right. . . . It seemed to work real good."

Less than two weeks later, on Tuesday, 29 June, Joe Engle attained his maximum altitude in the x-15 program. Attached to the aircraft for flight 3-44 were experiments that measured boundary layer noise at hypersonic velocity, along with a horizon scanner and a radiometer. The most important reason for the flight, however, was to evaluate piloted reentry techniques from space using a winged vehicle.

Lt. Col. Fitzhugh Fulton and Col. Harry Andonian took Engle aloft using B-52 no. 008. The x-15 launch occurred at 10:21 a.m. near Delamar. A slight misalignment of the LR-99 engine moved the aircraft left of the intended course as it boosted upward. Engle instinctively corrected the problem. He said, "Not a big or a violent correction, but I eased in a little bit with the rudder pedal. I saw the needle start back toward the middle, and then I started paying attention to the shutdown conditions for velocity and altitude and didn't notice it again."

The radio communication as Joe's trajectory peaked told the rest of the story. NASA 1 informed Engle, "Profile is beautiful, Joe. Track is real good. Okay, we have 260,000 [feet] and you can start your roll maneuvers." The ballistic arc continued upward, topping out at 280,600 feet. NASA 1 radioed, "Engine master off and congratulations, Joe!" Engle had become the fourth x-15 pilot to surpass 50 miles in altitude, thus earning his astronaut wings. Joe replied, "Thank you!" NASA 1 confirmed, "You just did it!"

While at zero g outside the atmosphere, the x-15 was now a spacecraft. The ballistic control system jets were required to stabilize the vehicle and to

28. At high altitude, such as with the three astronaut qualification flights of Joe Engle, the x-15's aerodynamic controls were useless while outside the atmosphere. A set of small rockets in the nose and wingtips stabilized the vehicle to provide correct orientation for experiments and to set up for a safe reentry. © Thommy Eriksson.

put it into proper position for data gathering. Minutes later they were also used to align properly for a safe reentry. Even though there was no air to carry the sound, I asked Joe if the jets were audible inside the cockpit when they fired. "Yeah, the nose jets were. Especially the nose-down firing jets. . . . It may have been more of a translation of the vibration sent through the structure, and that's what was heard. I'm sure that's what it was." He went on to explain how the sensor in the nose of the spacecraft was affected: "I do remember it was very apparent that the nose jets were firing, because they were in close proximity to the Q-ball, which sensed that and tried to point toward the balanced dynamic pressure. . . . When the forward jets fired at high altitude, with practically no air molecules around to disturb it, the Q-ball would chase the plume of the jets." When I asked what the jets looked like when in operation, he said, "As I recall, it was just a white flash."

Coming downhill on reentry, Joe hit his peak speed of Mach 4.94. At that altitude, he had the same experience as other x-15 pilots who flew a similar profile, telling his debriefers, "I got the impression that we were going to have to really put the binders on to get stopped. It looked like [Edwards] was right down between my legs." Years later Joe recalled, "You got up there so high, and, being ballistic, you were going to come right back down. . . . Even coming into the pattern . . . you just weren't used to that perspective. Looking down at where I was going to land, it gave me the impression I had a lot more energy."

Joe's touchdown and rollout occurred just over ten minutes after his release from the B-52. A chase pilot called to him as Engle stopped in a cloud of dust. "Well done, ol' buddy!"

A few weeks prior to Engle's first astronaut qualification flight, James McDivitt and Ed White had completed the mission of *Gemini* 4. White had become the first American to perform an extra-vehicular activity, or spacewalk. When they de-orbited and splashed down in the choppy waters of the Atlantic Ocean, there were officially 26 naval vessels, 134 aircraft, and more than 10,000 personnel from the Department of Defense involved with the recovery operation. The X-15, with its pilot-controlled reentry and lakebed runway, used a few people and vehicles from NASA and the air force. Sitting in the X-15 on the lakebed, Engle radioed a slight dig at the much larger operations for Gemini: "Okay, you can call the carryall [van]. I can get out of the capsule here." To which NASA I replied, "Rog, we don't want you to get seasick after that!"

At two months shy of his thirty-third birthday, Engle became the youngest American to fly into space. Gherman Titov still holds the record for the youngest for any country, as he was about a month shy of his twenty-sixth birthday when he orbited in *Vostok* 2 on 6 August 1961. Joe Engle's age record for the United States stood until Sally Ride flew aboard the Space Shuttle *Challenger* on 18 June 1983, beating Engle by almost ten months.

Joe Engle, however, still holds a very special record in the chronicles of spaceflight. Within less than four months, he went on to fly twice more into space with the X-15. No one has ever made three spaceflights in so little time. The closest astronaut to accomplish this feat was NASA X-15 pilot Joe Walker, who made three qualification flights in slightly more than seven months, January through August 1963.

Engle shared his experience on how it felt to be at such high altitude, saying, "It was really spectacular! In the X-15 you had two very small windows, and you really were aware of how dark space was up there and how very bright the Earth was. That was a neat, spectacular thing to see." According to Engle, "The only drawback was that in the x-15 you just didn't have but a few precious seconds to see and absorb that, and log it in your brain so you could remember. And even those few seconds generally were very busy . . . going to or returning from whatever attitude the experiment called for. It was one of those things where you had to glance out and just capture it in one quick snapshot in your mind."

Contrary to any test pilot joking about what they tended to call the "spam-in-the-can" programs of Mercury, Gemini, and Apollo, there was still the innate understanding that if they were chosen by NASA to move to the astronaut office in Houston, they were "deserving occupants of the top of the pyramid," as Tom Wolfe wrote in *The Right Stuff*. Joe Engle was already thinking along those lines and was close to making a decision to leave the x-15, as Neil Armstrong had done three years previously. In the meantime, there were still missions to be flown at Edwards.

On 4 August Joe entered the cockpit and was sealed in for flight 3-46, but it was aborted prior to the B-52 starting its taxi for takeoff. The next day Joe and the entire x-15 team were again frustrated with another mechanical malfunction, grounding the flight just as the B-52 prepared to start its engines. It was not until 10 August when the x-15 was successfully dropped from the mothership. Less than a second after launch, even before Joe had pushed the rocket engine throttle forward, the yaw damper dropped offline, doing so a total of twenty-one times throughout the mission.

As Engle continually reset the troublesome damper, he made one of the few mistakes in his time flying the x-15. The toggle switch that had Joe's attention was located at the bottom center of the instrument panel, below the horizon indicator. About an inch to the right of this switch was another, which changed the adaptive damper controller to a mode called css, or control stick steering. Approximately thirty seconds into the flight, Joe was reaching to reset the yaw damper and instead hit this second switch, activating the css mode without realizing he had done so.

NASA engineer Joseph LaPierre explained what happened in his report: "With the control stick steering mode engaged, normal acceleration and pitch rate are fed back as the primary loop signals. . . . The system is designed to provide adequate damping and an essentially constant normal acceleration response." LaPierre criticized Engle for not noticing what he had done. "It is difficult to understand why no pilot comments were made during the flight, and very few made during the post-flight briefing. . . . The pilot . . . did not follow established ground rules by not going to the alternate [flight] profile. The double malfunction, of yaw damper and CSS on, placed the pilot and the airplane in an undesirable situation."

Despite the difficulties, and the less than glowing post-flight report from LaPierre, Engle achieved his second astronaut qualification only six weeks after his first. His ballistic arc peaked at 271,000 feet, almost directly over the California-Nevada border, at more than Mach 5. By the time Joe came to a stop on the Rogers lakebed, he had traveled 284.1 miles from his B-52 launch point in less than ten minutes. This was his farthest flight in the program, and his last in X-15 no. 3.

By this time Engle had made it clear he was looking to move out, necessitating his replacement. Milt Thompson and Joe Engle had started on the X-15 program together, and they would end up leaving at nearly the same time. NASA replaced Thompson with Bill Dana, and the air force brought in Pete Knight to finally relieve Rushworth. They were announced as new project pilots on 26 July, two weeks prior to flight 3-46. Joe moved on with preparations for his final mission in the X-15.

A special technical conference was slated to commemorate the one hundredth time the no. 1 aircraft was carried aloft, hopefully to launch. This was the only one of the three airframes that accomplished this numerical feat. It was to be the sixty-first actual launch, making the full flight number 1-61-100. Engle was given the assignment, with a launch date scheduled for 8 October, as the conference was underway. The event reportedly drew more than seven hundred scientists and engineers and was supposed to be the ultimate flight demo for all attendees. As the mission moved forward, a ballistic control system upward-thrusting rocket motor leaked excessively, and the flight had to be aborted prior to launch, instead receiving a designation as 1-A-100.

The conference, and Engle's flight, also coincided with one of the most bizarre incidents associated with the X-15: reports of a supposed mass incursion of UFOs over Edwards on the nights of 7 and 8 October, no doubt so aliens could check out the amazing rocket plane in flight for themselves. One report gave the "facts" of what happened, created by a "researcher" whom I call Mr. Smith so as not to add any credibility to his report. "It seems that twelve luminous UFOs came right down low . . . over a secure military runway. These craft were all sighted visually by air force personnel and by several types of radar. . . . The air force scrambled several jet fighters after them. . . . During the event the possible use of nuclear weapons even became an issue."

In response to a Freedom of Information Act request, Smith was given copies of several hours of audio tape, supposedly recorded from the Edwards air traffic control and other sources during the reported time of the sightings. The tapes apparently did not satisfy Smith, so he reedited them himself. In his words: "When I realized what had been done, it presented a challenge which made me determined to find out what was hidden within that chaotic mass of sound. . . . After many months and countless hours of laborious research, cataloging, and editing the snips and pieces of the audio tape, I was able to organize the sound so the conversations could be understood. I had successfully restored the tapes to their original and correct sequencing."

To do all this with no correlation from any other source was laughable. There was no reason to believe anything was out of order, except the undeniable fact that Smith felt them to be. He concluded, "The resulting reconstructed tapes are now ironclad, documented proof of the existence of extraterrestrial UFO visitation to this planet. I have sent copies of my finished product directly to a number of major government agencies. . . . Not a single official agency has tried to debunk or discredit the event, or my presentation of the tapes." They probably wouldn't know where to begin.

Surprisingly, Smith did not make the connection that all the UFO activity was simply the arrival of the conference attendees—probably from Area 51.

Unaware of the alien visitation, Engle came back to try again for what was now flight 1-61-101, several days after the conclusion of the conference. On 13 October the B-52 never left the runway due to uncooperative weather. The next morning, all was ready, but a slight delay was made because of an XB-70 mission that took off at 11:00 a.m. Less than an hour later, the B-52 was finally climbing for altitude, and at 12:46 p.m. Engle hit the launch switch, then rocketed upward for his final X-15 mission and third astronaut qualification flight. He attained an altitude of 266,500 feet.

As with what seemed to be the majority of Joe's flights, he and the sun just did not get along. The post-flight report stated, "The lateness of the launch and sun angle produced serious reflections on the instrument panel from the pilot's silver suit, and required the pilot to look directly into the sun." Engle confirmed, "Actually, the hardest thing during the flight was seeing the instruments during the climb portion."

As Joe approached the lakebed for the final time, Pete Knight in Chase 4 verified the x-15 was in proper order. At 12:55 p.m. the rear skids of the rocket plane touched down on Rogers Dry Lake, and Pete said, "Very good. . . . Beauty!" After close to a minute of deceleration, mission 1-61 was concluded.

There was one last duty for Engle to perform before he left Edwards to pursue his career as a NASA astronaut. Three weeks after his flight, he accompanied Bob Rushworth as the landing chase on flight 2-43, featuring the first use of the external tanks on the x-15A-2.

Previously, Joe had been heavily involved with preparations for the first tank flight. Jim Robertson, who was a North American engineer on the external tank program said, "I worked several times with Joe Engle. . . . All these fellows would do so from time to time, but Joe seemed to be down the most. He was the guy I spent, oddly enough, the most time briefing on the tank system and how it worked."

As the first flight for the A-2 neared, Engle faced a commitment. "Before the airplane got ready to fly, I had to make a decision whether to stay in there and complete my tour [on the x-15] or go on into Apollo." During Rushworth's flight, Engle was able to watch the operation of the full configuration of what was originally supposed to be his special project. Instead, Joe never took to the air in the A-2, making his choice to try for the moon.

Five months later, on 4 April 1966, Engle was officially announced as part of Astronaut Group 5. Nine of the "Original 19," as they called themselves, would go to the moon during the Apollo program. Three landed as lunar module pilots, four others orbited as command module pilots, and two swung around the moon on the aborted *Apollo* 13 mission. Of the rest, one resigned and another was killed in an automobile accident, while the others eventually flew into space on Skylab, the Apollo-Soyuz Test Project, and the Space Shuttle. Joe Engle came close to adding a fourth moonwalker to the list.

When Joe arrived at the Manned Spacecraft Center in Houston to report to NASA, the two-man Gemini program was in full swing. Six orbital missions had flown during its first year of operation. During that time American experience in space had literally skyrocketed. Ed White opened the hatch on *Gemini* 4 and floated alongside for twenty minutes, the first fuel cells were used for electrical power, two manned spacecraft rendezvoused, and short duration missions gave way to extended stays on orbit of nearly fourteen days. On the downside, the first real space emergency occurred just weeks before Engle's arrival. The *Gemini* 8 mission had come close to killing Neil Armstrong and Dave Scott when their spacecraft went out of control with a stuck attitude thruster not long after performing the world's first orbital docking. By November four additional Gemini missions were flown, and the program was officially completed. Apollo waited in the wings to take men to the moon, with the first Earth-orbit test flight scheduled for February 1967.

Engle entered NASA too late to participate in Gemini, but being part of a moon-landing mission was definitely in the cards. Nine months after Joe was selected as part of this elite group, the *Apollo* 1 spacecraft caught fire on the launch pad, killing the crew. The program ground to a halt.

Once Apollo was moving again, Engle was in a unique position with his three-time X-15 astronaut status. He was the only person to ever join the NASA program rated as a flown astronaut prior to his selection. As he completed training and moved into flight operations, he was initially assigned to the support crew on *Apollo* 10. This flight was the final dress rehearsal for the first lunar landing, descending to 47,400 feet above the surface. The crew was commanded by Thomas Stafford, with Gene Cernan as lunar module pilot, and John Young as command module pilot. Both Young and Cernan eventually flew missions to the moon of their own with *Apollo* 16 and 17.

For Engle, being part of the support crew brought him into the official crew assignment rotation and also to the attention of Cernan. After he returned from *Apollo* 10, Cernan chose Joe to be the lunar module pilot with Gene's backup crew for Al Shepard's *Apollo* 14 mission. Ron Evans completed the ensemble as the command module pilot. Being backup was supposed

29. Joe Engle was slated as lunar module pilot on *Apollo* 17. This crew served as the backup crew for Alan Shepard's *Apollo* 14 mission. Engle was bumped from *Apollo* 17. *Left to right*: Joe Engle, Gene Cernan, and Ron Evans. Courtesy of NASA.

to lead to becoming the prime crew for a landing three missions later.

While preparing for the flight of *Apollo* 14, a strenuous schedule was imposed. Cernan, Evans, and Engle, as the backup crew, were able to let off a little steam by designing a very special mission patch as a humorous dig against the prime crew. They used this patch to highlight the fact that, besides Shepard's fifteen-minute suborbital flight aboard *Freedom* 7 in 1961,

Al's crew, including Stu Roosa and Ed Mitchell, were all making their rookie flights into space, whereas Cernan was already a seasoned veteran of a lunar voyage.

The alternate mission emblem used two endearing characters of cartoon fame: the Roadrunner and Wile E. Coyote. The coyote was shown with a gray beard, representing Shepard as the "old man" of the space program, who was swooping up from Earth toward the moon. Cernan's crew, represented by the Roadrunner holding a "1st Team" banner, had beaten the coyote to the moon, planting the American flag. The words "Beep Beep" across the top of the patch antagonize the poor coyote. Gene Cernan explained in his book, *The Last Man on the Moon*, "All the way to the moon and back, even on the lunar surface, whenever the [*Apollo* 14] crew opened a box, bag, or locker, out would float a 'First Team' mission patch. . . . Perhaps the most repeated phrase on the private radio loop during the flight of *Apollo* 14 was Shepard's annoyance when still another patch would suddenly appear. 'Tell Cernan,' he growled, 'Beep-beep, his ass.'" Engle was always known for his sense of humor, and was delighted in having a hand in the interplanetary prank.

Regardless of their comic turn with the patch, Cernan's crew did an exemplary job, which then earned the prime crew slot for *Apollo* 17. The only problem in their way became the lack of political will for future lunar landings.

After the near disaster of *Apollo* 13, with the crew barely making it back to Earth following an oxygen tank explosion that crippled their spacecraft, many people started to worry that a fatal accident was an inevitability. President Nixon undercut the program, and Congress had already been slashing NASA's budget since before *Apollo* 11 made the first landing. The roster of flights originally extended to at least *Apollo* 20, but that was canceled on 4 January 1970, leaving *Apollo* 19 as the final mission. Both the 18 and 19 missions did not make it through the rest of the year. They both got the ax on 2 September.

Apollo 18 was to be the first actually carrying a career geologist to the moon, Harrison "Jack" Schmitt. Since the reasoning behind going to the moon—besides beating the pants off the Soviets—was ostensibly to study lunar geology, Schmitt needed a ride, and *Apollo* 17 was the last one off the launch pad. Cernan's crew was in jeopardy. To satisfy scientists, the possibility existed that the *Apollo* 18 crew, commanded by Dick Gordon, would bump Cernan's. Instead, the idea was proposed to switch out only the lunar module pilot, which would give Schmitt a spot on the surface with Cernan but would leave Joe Engle without a flight. In an interview conducted by Eric M. Jones for the *Apollo Lunar Surface Journal*, Gene Cernan tried to clarify the situation:

I think Joe Engle knew the handwriting was on the wall. Because I wanted to keep my crew together, I had done some fighting [with chief of the astronaut office Donald K. "Deke" Slayton] for Joe to stay on the mission, and had a lot of sympathy and support from people [to do so]. But only to the point where I was asked [by Deke], "Are you telling me you will only fly with Joe? If you're telling me that, then it makes my decision a lot easier." And I told Deke, "No, I can't tell you that. I've fought too hard for the opportunity to have my own crew and to be in the left seat."

[Schmitt] trained hard, he studied hard, and he worked hard. Jack was not an "adequate" lunar module pilot, he jumped in with both feet and was an outstanding lunar module pilot. Now, contrast that with Joe Engle. Joe was born with a stick and rudder in his hand. Joe Engle is probably the finest stick and rudder aviator that I've ever flown with in my entire life. He could make an airplane do things that I don't have guts enough to make it do. So Joe was an outstanding aviator. In contrast, Joe was only an adequate lunar module pilot.

On 13 August 1971 the official announcement was made: the crew of *Apollo* 17 would consist of mission commander Gene Cernan, command module pilot Ron Evans, and lunar module pilot Jack Schmitt. At thirty-three minutes past midnight, Eastern Time, on 7 December 1972, *Apollo* 17 made a spectacular nighttime liftoff from Kennedy Space Center, Florida. Four days later, Cernan and Schmitt powered their lunar module, *Challenger*, into the Taurus Littrow valley of Mare Serenitatis, while Evans orbited overhead in the command module *America*.

Engle talked about his feelings on losing his chance of going to the moon: "Oh, I was disappointed. I obviously wanted to go and had trained in the backup crew to cycle into that spot. By the same token, the Apollo program was primarily a geologic study of the moon's surface. Jack Schmitt had a doctorate degree in geology and was experienced in that field from the technical standpoint. It was a logical decision to make. When . . . Schmitt's flight was canceled, the desire to move him up a flight in order to get him on the surface of the moon was understandable, but disappointing from my standpoint."

Being pragmatic about the situation also helped Engle see another opportunity. Joe explained the situation, saying, "For me to have been considered for the early Space Shuttle flights because of my background and experience in testing the x-15, for instance, made sense to me, too. It's just having the nose on the other side of the fence. You know you've got to look at the reasons for those things and put your personal desires secondary to what's good for the program."

Evidenced by his repeated success over sixteen flights of the x-15, Engle's strength definitely was with winged space vehicles rather than the ungainly contraption of a lunar module balancing its tail on rocket exhaust. He was in a prime position to be instrumental in the development and flight test of the Space Shuttle system. "It was right after Apollo was finished," Joe said, "[when] I had an opportunity to make my input to the guys that selected the crews as to whether I wanted to be on Skylab, or start working the Space Shuttle in the very early essential design phases. I think because of the rationale where I had convinced myself it was the right thing to put Jack on *Apollo* 17, I now felt where I could contribute the most would be in the design and development of the Shuttle."

Not only was Joe invaluable in these early phases, but his expertise was put to excellent use over the course of atmospheric test flights with the prototype Space Shuttle *Enterprise*. The missions were officially called the Approach and Landing Tests, or ALT. NASA modified a 747 cargo plane to carry the *Enterprise* atop its fuselage. The idea was to fly the shuttle off the 747 and land on the runway at Edwards to test the low-speed handling characteristics of the giant winged space plane. Joe said, "That was a fun thing, when we flew the ALT tests on the Space Shuttle, and flew off the top of the 747. Fitz Fulton was flying the 747, and he had dropped me [eleven out of my sixteen flights] in the x-15. When we got to the ALT program, [my copilot] Dick Truly and I made a big thing out of the first time I had Fitz under me in the 747, that it was [eleven] to one now, and counting. I was going to drop him as many times as he dropped me." There were only five ALT missions flown, two of which were commanded by Engle, which meant he still owed Fitz nine more drops to even the score.

Two missions into orbit were commanded by Engle: *Columbia* on STS-2 and *Discovery* on STS-511. His first flight marked the only time when the vehicle was flown manually from de-orbit burn to landing. Flying reentry three times with the x-15 made that achievement possible.

In 2003, at a conference celebrating the centennial of the first powered human flight by the Wright brothers, Joe spoke of how his age reflected in his career. "I really feel lucky to have gotten in on both programs. [Television personality] Dave Hartmann told me, 'You were damn near too young to fly the x-15 and damn near too old to fly the shuttle,' but it happened anyway. [Chuck] Yeager was not quite so sensitive. He said, 'Yeah, you just crawled under the covers in time for the x-15, and [you] got into the shuttle before they kicked you out of bed. . . .' I'll tell you, everything under those covers was absolutely great, and I wouldn't trade it for any other career in the world." Joe Engle put it very succinctly, saying, "The x-15 was the greatest airplane I've ever strapped my butt into."

9. Inconel Meets Celluloid

It's happening so fast. The whole flight depends on
how well you perform in the first eighty-two seconds,
because that's how long the engine burns. So the success
of the mission is established right there, and you're just
trying like hell to keep up.

Milton O. Thompson

According to the x-15 publicity machine, the purpose of the program was to take American astronauts 100 miles into outer space. The fact the x-15 was not capable of such a journey mattered little.

North American Aviation built the three x-15s, and they knew better than anyone what the altitude goal of the program was: 250,000 feet (just past 47 miles). Yet a September 1959 special report from *Skywriter*, the publication of NAA's Los Angeles Division, helped keep the propaganda afloat.

The occasion marked for this edition was the first powered flight on 17 September by Scott Crossfield. "x-15 Ready for Man's Conquest of Space," read the bold headline, with the subhead of "Powered Flight Signals Start of Probes beyond Atmosphere." Scott's mission achieved Mach 2.11 and 52,341 feet, a good start for a research craft that was to eventually traverse the skies of the desert southwest higher than 67 miles, exceeding 4,500 miles an hour. By the second paragraph of the lead article, the publicist was in full swing: "Ultimate destination of the x-15 is to an altitude approaching 100 miles and a maximum speed in excess of 3,600 mph." Interesting how the altitude is overstated, while the velocity is too low.

Atop the page is a composite photograph showing the pilot's-eye view from inside the x-15 cockpit. The upper part of the instrument panel is in the foreground, while through the rectangular windows is the curvature of the earth, snow-capped mountains and sparse clouds far below on the right and a crescent moon with twinkling stars looming ahead.

This edition was published two years after the shock of the Soviet Sputnik, and all Americans needed to believe in the infallibility of the United States. Pie in the sky was exactly what everyone wanted on our menu, and the x-15 was supposedly poised to deliver on that promise.

Hollywood was not far behind North American's public relations writers. If it was good for the country, it was perfect for the silver screen. First proposed by screenwriter Tony Lazzarino in October 1958 as *Exit*, his movie was originally slated to follow the x-2. Since that rocket plane program had already concluded and no x-2 aircraft survived, U.S. Air Force deputy chief of the Pictorial Branch, Maj. Stockton B. Shaw, suggested a switch to the brand new x-15. Lazzarino jumped at the opportunity, rewriting his script into *Time of Departure*. It was later cut down to a much simpler *x-15*.

Both the air force and NASA decided on full cooperation with the motion picture company, especially since it originally had Bob Hope associated as producer, a huge celebrity name. It was then taken over by Frank Sinatra's Essex Productions to bring the movie to fruition with a total budget of $422,800. Government cooperation extended to providing two technical advisors, Capt. Jay Hanks from the air force and Milton Orville Thompson from NASA. Even though it was still very early in the x-15 program, the U.S. Navy was not invited to participate.

The tagline for the movie, when it had its gala premiere in Washington DC, was this amazing, yet erroneous, proclamation: "Actually Filmed in Space!"

Crookston, Minnesota, was the town where Milt Thompson was born on 4 May 1926, but it apparently had so few memories for him that he failed to even mention the location in his 1992 biography *At the Edge of Space*. The community is near the northwest border of the state and appears to share more commonality across the line with nearby Grand Forks, North Dakota. It is a beautiful area with a sparse population, cut by a meandering river that continually snakes back on itself as it traverses westward from Lower Red Lake, through a series of benign rapids, onward to the Canada-bound Red River.

Instead of embracing Crookston, Milt talked of his hometown as Pontiac, Michigan, where the family moved when he was young. It was from this local airport that he had his first taste of flying in an open-cockpit biplane, around the time he became a teenager. Unlike most other pilots who flew the x-15, Milt admitted to never being consumed by any passion for flight. In his book, he said hunting, fishing, and ice skating were more to his liking, then added, "Flying was not everything, but I did get turned on by it."

At the time he was still a young boy, Milt had trouble connecting with the world of aviation. In his book, he wrote, "I built model airplanes by the carload and had them hanging all over my room. I could not get them to fly properly because no one had ever told me how to ballast the model. . . . I became so disgusted that I would build them and then take them up to my second floor bedroom, set them on fire, toss them out the window, and watch them go down in flames." Understanding the dynamics of flight came much later for Milt, in college.

Approached by U.S. Army aviation recruiters in 1943 while a senior in high school, Milt got momentarily excited about the prospects of joining World War II as a fighter pilot. The reality was Thompson would graduate in June as a seventeen-year-old, and when the army found out, they were no longer interested. The U.S. Navy held no such reservation, so he decided to join the naval aviation cadet training program, one that also provided a year of accelerated college before entering the military.

Milt arrived at Milligan College, outside Elizabethton, Tennessee, and admitted to being shocked by the racial segregation and also by the nature of many of the people in the foothills of the Great Smokey Mountains. "I will never forget those hillpeople walking down from the hills in single file with the man in front and then the woman and all the kids following her. The man was usually carrying a rifle, and the woman was carrying a baby and chewing tobacco." The college had a heritage as a Christian liberal arts school, but during the war the entire facility had been voluntarily turned over to the navy cadet program.

The war continued unabated, with future x-15 colleagues Joe Walker and Bob White in Europe, Forrest Petersen and Jack McKay in the Pacific, and Bob Rushworth in China. Milt finished at Milligan and entered the navy to begin military training. Coinciding with Milt's September 1945 completion of his six-month basic course at Chapel Hill, North Carolina, the war drew to a conclusion, and everyone was offered the opportunity to leave the military. Because of the slow process of mustering out, Thompson, who had originally thought of taking the government up on its offer, instead decided to stick with it and go for his pilot's wings.

Thompson was of the age that happened to leave him between wars. A majority of the twelve x-15 pilots served somewhere in a war zone, be it World War II, Korea, or Vietnam. In NASA's official press release, which is now oft repeated in multiple sources, Milt "served as a naval aviator in World War II with duty in China and Japan." Considering how he was still in training when the war concluded, why has this belief persisted? Was it a feeling within NASA of wanting all their premiere pilots to be regarded as war heroes? Or maybe it was some overzealous NASA public relations hack who simply took it upon himself to create a myth. Maybe this unknown writer was a colleague of whoever later wrote the story about the excessive height performance of the x-15.

By the time Thompson entered advanced training at Pensacola, Florida, he was flying a Consolidated PBY Catalina, a seaplane often remembered for the ocean rescue of downed pilots. Milt talked of his enjoyment of the airplane but also about how slow it was: "Flying is just the boring interval between takeoff and landing." That idea was certainly highlighted when he performed his first trainings off the deck of the light aircraft carrier, the USS *Saipan* (CLV-48).

By the time Thompson went on active deployment in mid-1946, his assignment put him in the Grumman F8F Bearcat. This airplane really got his attention, certainly more so than the lazy Catalina with a cruising speed of just over 100 miles an hour. He stated flat out this fighter was, "The most impressive aircraft I have ever flown."

Milt took a round-the-world jaunt, courtesy of the U.S. Navy, meeting his future wife, Therese, on a stopover in San Diego. They dated for a year, finally marrying in June 1949. Five months later, in November, Milt took his final leave from the navy, heading then to the University of Washington in Seattle to pursue his engineering degree. By the time he graduated in 1953, they already had a family that included three children. This proved a perfect incentive to move directly into a job at nearby Boeing, first as a structural test engineer, then as a flight test engineer on the new B-52 Stratofortress bomber. It was certainly steadier work and more lucrative than when he had taken breaks between quarters at college flying pest-control spraying missions over the forests of Oregon and dusting crops south of the California border near Mexicali.

It only took a few years before Milt decided to try for a job at the NACA High Speed Flight Station at Edwards. He cites his inspiration as hearing of the exploits of Scott Crossfield, who had served with Milt in Seattle in their naval reserve fighter squadron. Thompson, however, also admitted that Scott did not help in his job hunt, never answering any of his letters of inquiry. Milt wrote, "It was several years later that I found out Scott had a very low opinion of me. In fact, he considered me to be a dumb idiot." Later in life they apparently became friends of a sort.

De E. Beeler, from NACA, visited Boeing to obtain B-52 test data, and Thompson arranged a conversation, which ended with Beeler offering Milt a job at Edwards. It was the opportunity he had been waiting for, so Milt packed up the family and headed south to the Mojave Desert. His first day on the job was 19 March 1956. Less than two years later, Milt moved into a research pilot slot, one of five at NACA at that time.

As 1958 drew to a close, NACA transitioned to NASA, and the x-15 had rolled out to public adulation. As the x-15 began its flight research program in 1959, the next step in rocket plane evolution was already being explored and designed: the x-20 Dyna-Soar. Joe Walker and Jack McKay were the primary NASA pilots on the x-15, leaving Neil Armstrong, Milt Thompson, and Bill Dana to be assigned the early stages of the x-20 in a role classified as pilot-consultants.

Milt talked about the Dyna-Soar, saying, "That started out to be just another research airplane. It was never supposed to go into orbit. It was to be launched on the Titan I out of Cape Kennedy. The first flight would have been a flight up to about Mach 10, and we would have landed on one of the islands down near Bermuda. The maximum speed was going to be on the order of Mach 18, and that [suborbital] flight would have landed in Brazil." Early glide tests of the dynamic-soaring vehicle were to be launched from the same B-52 pylon then utilized by the x-15. It is also possible some sort of rocket engine pack might have been mounted to accelerate the x-20 into supersonic range, extending the aerodynamic knowledge before committing it to a Titan I launch and such high Mach numbers.

Each passing day seemed to add some new capability to the Dyna-Soar, bolstered by the impetus of the upcoming Project Mercury manned space-flights. Milt said, "During that period of time, the Mercury program was charging along, and they were getting all kinds of publicity, so the air force felt, 'Well, maybe we should go into orbit, too, with the x-20.' There was a lot of politics in the air force, because they were trying to sell the Titan III booster." That rocket could have indeed put the x-20 into orbit, so what better way to sell the booster than to say it was a critical item on the path to beat the Soviets in outer space?

As the scope grew, pilots such as Armstrong and Dana foresaw the disastrous turn the program had taken, deciding to leave before the all-but-inevitable cancellation. A joint program built on the foundation of the good relations between NASA and the air force on the x-15 was disintegrating. At the optimistic time of the official announcement of pilot selection, there was but one civilian name—Milt Thompson—and the rest were from the U.S. Air Force: Albert H. Crews Jr., Henry C. Gordon, Russell L. Rogers, James W. Wood, and William J. "Pete" Knight.

"There was a lot of money invested in that program," Milt pointed out. "And there was still a lot to go. I think by the time it was canceled, there was something like $700 million invested in the x-20 and about $500 million to go. If you figure that period of time of the 1960s, that was a hell of a lot of money." That investment had purchased a lot, including laying the framework of the initial x-20. Milt said, "They were building pieces of hardware. In fact, we used the inertial platform in the computer for the x-15. They had already built [auxiliary power units] and tested them, and they were actually constructing the first vehicle."

Politics and public perception also played a large part in the x-20 end game. The war in Vietnam was taking away research money, and the general mood in the country was that, with plenty of competition with the Russians, why create more with our own military having a space tug-of-war with NASA?

One of the worst reasons was one Milt spoke about: "By the time they canceled it, everybody was pretty confident they could do it. It was said, 'Well, if you're confident we can do it, then why [actually] build it?'" As has been found repeatedly, especially with regard to research programs involving spaceflight, the hubris of thinking all the details have been worked out does not necessarily mean programs can be safely and routinely accomplished. The x-15 proved this on almost every flight, many of which would have never returned intact if a man had not been in the cockpit completely in control of the aircraft and the situation.

In the end, the scope of the x-20 had been broadened so much there was no mission left to perform. Milt said, "Even though we did go into orbit, it was strictly a research vehicle, [with] no military mission. . . . Right after that, they made the MOL [Manned Orbiting Laboratory] program, think-ing they did have a military [objective], which was more of an observation kind of mission." Like the x-20, MOL floundered and eventually died. Milt never understood how short-sighted the military service was with regard to putting humans on orbit. "It's unbelievable to me they haven't really ever come up with a manned military mission."

With the demise of the x-20, all six astronaut candidates had to find oth-er work. The air force had assignments for all their pilots, including Knight, who eventually went to the x-15. Thompson had a secure test pilot slot at the NASA Flight Research Center, but it was a bittersweet moment to see the Dyna-Soar falter and fade away. He still maintained his humor about it all, saying, "I was one of the world's first unemployed astronauts!"

Almost everyone at Edwards who worked for NASA in the early 1960s had some involvement, if only peripherally, with the x-15 research program. This was certainly true for Thompson, who routinely flew the "Iron Bird" simulator and other tasks. However, since he was not directly involved as a project pilot, Milt had more time available, in some respects, when not hopping back and forth to Seattle for consulting on the Dyna-Soar. Because of this, he was selected for an extra duty during 1961, as the NASA techni-cal advisor on the motion picture x-15.

He jumped at the opportunity to work with a Hollywood production telling the x-15 story. When I asked him about those months with the likes of action star Charles Bronson in an early starring role and the beautiful Mary Tyler Moore in her first starring role in a movie, Milt leaned back in his government-issue gray office chair, put his hands behind his head, laughed, and said. "It was a lot of fun!"

By the time Thompson was asked for his expertise, Tony Lazzarino's script had been through several iterations, including changing his collab-orator from Max Dioguardi to James Warner Bellah. When the script was

first presented to NASA in late 1959, with an ambitious orbital flight of the
X-15 as its fictional centerpiece, the agency had been less than enthusiastic.
Byron A. Morgan, NASA's motion picture production officer in Washing-
ton DC, stated flatly, "A motion picture depicting a false mission for this
research airplane would seriously injure the X-15 informational program.
As a consequence, we will not cooperate on the production of this script."

Lazzarino reined in his story to a point, bringing it more in line with the
reality of the X-15, but he still insisted on exaggerating some points, such as
the ultimate altitude expected of the rocket plane. A memo from Clotaire
Wood, a special assistant to NASA's deputy administrator Hugh L. Dryden,
tried to inject some sanity on this point when he wrote, "Please don't use
100 miles. The X-15 will reach altitudes of fifty miles or more. While [the
LR-99 rocket engine] has the power to boost it up to about 100 miles, the
recovery from such a flight path would not be within the flight limits." From
a technical standpoint, Wood was absolutely correct, but from the view of
a Hollywood screenwriter, the magic triple-digit altitude was just too sexy
to resist.

Both government agencies, NASA and the U.S. Air Force, eventually gave
a green light to proceed. Milt explained his involvement, saying, "They requested permission to come to Edwards to film the project. It took a lot
of support from the air force and NASA, too. It was questioned how much
NASA ought to devote to that, but they finally decided to go ahead to cooperate fully." Like other movies with a military theme, this saved the production company a lot of money. "We and the air force made people and
equipment available. As one of the technical advisors, I was primarily trying to make sure what they did as far as the flying part of it was accurate.
We obviously didn't have any impact on the story, because it was kind of
a love story stuck with a bunch of soap opera kind of things. But, as far as
the flight part of it, I think it was reasonable."

Milt recalled being involved for approximately three months overall. "I
think they were here on-site probably a couple months, and after that we
went down and spent one month in Hollywood during the cutting process." Did he have any input into how to make the film better? "Oh, yeah,
a lot of stuff! It was completely wrong [on many things]. . . . They really
didn't understand airplanes at all. The guy that wrote the script had apparently never been around an airplane."

30. X-15 "pilot" Matt Powell (played by actor David McLean) gets pointers on how to fly the X-15 from the real research pilot and the film's technical advisor, Milt Thompson, leaning in on the right. Courtesy of the Armstrong Flight Research Center.

The director's spot went to Richard Donner, a newcomer with only a few television credits to his name, such as *Wanted: Dead or Alive* and *Route 66*. Besides Bronson, starring roles went to David McLean, famous for his Marlboro Man commercials, and James Gregory, a perennial character actor who had recently been seen in the pilot for Rod Serling's *The Twilight Zone*.

For the production of *X-15*, North American Aviation also wanted a piece of the pie, so they supplied a full-scale X-15 mockup for use by crew and actors. Milt said, "It was a joint program, so they had to share the wealth. . . . The mockup was seen hanging on the B-52 in a lot of those shots. It was pretty detailed, as far as it went."

Capt. Jay Hanks filled the technical advisor's slot from the air force. "He made sure the uniforms were on properly," Milt laughed. The value added to the movie because of Milt and Jay was never calculated. The same was true for the other government hardware and facilities loaned out and the spectacular flight footage provided, which would have been impossible to otherwise obtain. State-of-the-art special effects filled in where no camera could follow, but these were obvious model shots, never engaging the audience except in small doses.

Filming began in early 1961 at Edwards. Roger Barniki shared his experience: "Charles Bronson . . . was out there doing a lot of weight lifting [between scenes]. The guy . . . who I would call the Joe Walker character [David McLean], he fizzled out, never saw him in anything after that." Roger wasn't just watching; he also appeared in the movie, as did many other real rocket plane crew members. "We were running around for a couple of days with pancake makeup. You probably won't recognize me . . . because I've got dark hair, and I'm thin!" He talked about one scene that stood out, saying that as the life support supervisor, "it was my job to go up and put the pilot in [the cockpit]. We were at the bottom of the stairs, [and] normally I'm the first one up, so I'd go on up, no problem. Once the camera's rolling, next thing I know, I have one hell of a pain in my side, and guess who's in front of me?" It was James Gregory, playing Thomas A. Deparma, based on real-life X-15 project manager Paul Bikle. "He was something! You learn that you're not the [movie] crew, you're just some jerk that works on the airplane."

Florence Barnett, wife of X-15 crew chief Larry Barnett, said, "I remember the story that Larry told me about playing cards with Bronson under the wing of the B-52." At the time of our interview, her husband was having memory problems, so Florence passed on what she could, relating to Larry: "You told me Bronson was a really nice guy." Her husband agreed, "He was, yeah." Florence went on, "[Bronson] spent a lot of time with you out on the flightline, mingling with the guys. He was very down-to-earth, unlike Lee Majors, who was a real prima donna when he was out there for the *Six Million Dollar Man*. You didn't care for him much, but you said really good words about Bronson."

Billy Furr, from the X-15 rocket shop, said, "When they were filming, I had to go down and shoot some liquid nitrogen vapors all over the ground. I was on the opposite side of the airplane." Billy was simulating the fueling process, producing the characteristic fog that often enveloped workers getting the X-15 ready for flight. He smiled and said that he enjoyed providing cheap special effects.

Harry Shapiro recalled the production team coming to the North American plant in Los Angeles to shoot various scenes, one of which immortalized him in the movie. "They took a couple of shots of the [NAA] engineering department. . . . There was one little shot where I'm walking from one spot to the next and putting something in a drawer, looking in the drawer, then going on." He also helped stage various conditions on an X-15 instrument panel. "We had to take some indicators into the lab and simulate changing altitude. . . . We brought these up to a high altitude, then quickly took the pressure off it and made this needle go around real fast."

With big fanfare, *X-15* premiered in Washington DC on 21 November 1961. The theater lights dimmed, the audience hushed, then a bright, blue sky appeared on the screen. Soon the scene shifted and started to fly through a bank of white, fluffy clouds. Jimmy Stewart, one of the best-loved American film stars, and himself an experienced pilot for the U.S. Air Force, began the narration: "On December 17th, 1903, man made the first successfully controlled flight in an aircraft, thus breaking a barrier that had existed for millions of years. Today, man, with his intelligence and reason, has suddenly come to the crossroads." The scene dissolved to a grove of trees in the early evening, then tilted upward, dramatically racing through the branches to focus on a spectacular Milky Way field of stars and nebulae.

As the camera moved upward, Stewart continued, "Some believe that the guided missile and electronically controlled space vehicles are the ultimate answers to spaceflight. The recent orbital and suborbital achievements have been spectacular and extremely important. However, man will never be satisfied in the undignified position of sitting in a nose cone acting as a biological specimen."

From that point, the audience was taken to the great expanse of Rogers Dry Lake. In the far distance, through rippling waves of heat, the B-52 made a long takeoff roll with the X-15 under its wing. "And now the X-15 is ready," Stewart said, "manned by a pilot who will make all the decisions for accurate control in flight and reentry and recovery. X-15 is the key to an operational procedure that will be directly reflected in the spacecraft and the spaceflights of the future." As the mothership came closer and flew overhead, the powerful engines drowned out all other sound. An F-100 and F-104 took off to catch up with the B-52. Another scene shift and the audience saw the view from the bubble window on the right side fuselage of the bomber, the X-15 just feet away under the wing. A beautiful and ethereal symphonic score from Nathan Scott crescendoed; the title, list of actors, and production credits all passed by as the x-15 prepared for launch.

Through the course of the movie, we were introduced to three fictional x-15 pilots: Matt Powell from NASA, Lt. Col. Lee Brandon from the air force, and his backup, Maj. Ernie Wilde. Played by David McLean, Charles Bronson, and Ralph Taeger, they are broad representations of the real-life Joe Walker, Robert White, and Robert Rushworth. We also met their wives and girlfriends and had deep conversations about the psychological effects on the pilots of testing a hypersonic research rocket plane.

With various machinations, the story wove its way through mission aborts, the explosion of x-15 no. 3 during the LR-99 engine run, and eventually to a perfect Mach 5 mission with banner newspaper headlines. Pithy dialogue was used throughout, such as Deparma explaining to Col. Craig Brewster, played by red-headed veteran actor Kenneth Tobey, how they must watch the pilots for signs of stress. He said, "We've got to know the score every minute, and if a breakdown should show up in any one of them, we've got to be able to spot it. And if it doesn't show, we've got to be sure the reason it doesn't is because it never happened." During a press conference, real-life reporter Lee Giroux asked, "Sound minds and a sound body?" To which Deparma answered, "Oh, something more than that, I would say, the complete scientific dedication to the job at hand."

During the penultimate action sequence, Matt Powell experienced an in-flight emergency with the engine. Flying in the F-100 chase plane, Brandon stayed with Powell to see the rocket plane safely onto the lakebed. Brandon's plane was damaged by flying debris from the x-15. In protecting Powell, Lee stayed too long with the crippled Super Sabre. The aircraft fell onto the lakebed in a horrific crash, the only fatality of the movie. The production used the film footage of a real F-100 accident that took the life of its pilot.

After Lee's death, recriminations were soon replaced with a new resolve to get the x-15 into space. Preparations for the final flight took place in a briefing room with a blackboard listing the various personnel involved. Fictional character names were interspersed with real ones, including Jack McKay and Bill Dana. The technical advisors were given the nod, with Captain Hanks as Chase 3 and Thompson as Chase 4. Screenwriter Tony Lazzarino promoted himself to C-130 pilot.

Powell successfully launched and lit the rocket, heading his x-15 upward to a spot in space that the real aircraft could never safely match. Tom De-

31. Maj. Anthony Rinaldi (played by Brad Dexter) assists Lt. Col. Lee Brandon (played by Charles Bronson) from the x-15 cockpit following a successful mission into space. Courtesy of the Armstrong Flight Research Center.

parma radioed, "NASA 1 to all stations. Matt Powell has successfully made his exit and is in his coasting pass. He's over 100 miles up!" Cheers and back slapping broke out in the control room over his announcement. As the x-15 dropped back into the atmosphere, Jimmy Stewart closed the film, narrating, "Fact has overtaken fiction. The pathfinders have marked and illuminated a trail in space for the others who soon will follow. This is the beginning of the natural extension of man's capabilities."

The house lights came up at the end of the film, and, for the most part, audiences were underwhelmed.

NASA operations engineer Bill Albrecht explained the typical reaction from those involved with the real thing. "A gang of us were invited to see the movie, and [the producers] gave us dinner beforehand and everything. Then we went to the first showing. We sat there, and we were laughing. The producer was behind us, and he asked, 'Why are these guys laughing?'" Bill's answer: "It was because of Hollywood." Project engineer Meryl

DeGeer agreed, "There are several scenes that are kind of funny." Research engineer Eldon Kordes said the reason was clear: "They didn't want any facts."

Not all the x-15 guys were completely critical of the movie. Wade Martin from air force quality control gave Donner's work a passing grade: "Oh you know, it was glamorized of course. . . . What the heck, I didn't think it was that bad." Harrison Storms, the man at North American who created the x-15, also attended the premiere. He gave a capsule critique, saying, "It showed how hard it was to do things. The documents of the flights were well done."

After all his work to try to bring clarity and accuracy to the production, what was Milt's final review? He thought a moment, then said, "Well, I didn't mind it too much, because I like the flying scenes . . . but the overall story was hokey. . . . If they had stuck to it and made it a flying story, period, without all the other stuff, it could have been very good. Some of the shots were just fantastic."

For those who worked on the movie, it did little to springboard their Hollywood careers. Richard Donner returned to the small screen for the next fifteen years, with various shows like *Have Gun—Will Travel*, *The Man from U.N.C.L.E.*, and *Gilligan's Island*, before finally getting back to directing movies with *The Omen*. Donner then became a major motion picture talent with 1978's *Superman* and the *Lethal Weapon* series.

From his time as the Marlboro Man, David McLean was given a free and unlimited supply of cigarettes from his employer, Philip Morris. He was diagnosed with cancer in 1964 but continued to work as an actor for the next seventeen years. During that time David was instrumental in the fight to ban television advertising that glamorized smoking. That fight was successful, taking effect 2 January 1971. Amazingly, McLean survived with cancer for thirty-one years, succumbing in 1995.

Of other notables, this was Frank Sinatra's third movie produced out of seven in his career. Instead, he stayed primarily with acting and singing and garnered much better reviews. Ralph Taeger had a mildly successful acting career in television. James Gregory was possibly the most successful x-15 actor outside Charles Bronson, working continuously on the big and small screen until his retirement in the mid-1980s. Of Nathan Scott's sixty film and television composing credits, x-15 stood out as one of the most dramatic and moving. Other projects for him included *Wagon Train, Drag-*

net, and more than a decade on *Lassie*. Lazzarino's cowriter, James Warner Bellah, went directly from outer space to the cowboy West, penning the major Hollywood release *The Man Who Shot Liberty Valance*, starring John Wayne and *x*-15 narrator James Stewart. Lazzarino himself had no further connection with any Hollywood-created motion picture production. Milt returned to his regular job flying airplanes for NASA, leaving his film career behind, at least for a time.

On the horizon was a completely new type of flying vehicle, one without any wings at all. All the lift was derived from the airfoil shape of the aircraft body. The term "lifting body" was completely descriptive of these contraptions, which looked more like flying bathtubs than aerospace research craft. There were usually a few appendages tacked on, such as a vertical tail or two—sometimes even three.

"The lifting body—I grew up with that right from the start," Milt recalled. "So, I just felt much more comfortable with them. . . . I sat in the wind tunnel with it while we were testing. By the time we got ready to fly, there was no question whether it was going to do everything it was supposed to do."

The concept had been introduced as early as 1921 from designer Vincent J. Brunelli. The idea caught hold in the innovative environment fostered by Paul Bikle at the NASA Flight Research Center in the early 1960s. Milt said, "When the Dyna-Soar was canceled, there were a number of us here that were still interested in a spacecraft which could fly back and forth, so one of the guys got interested in the work that had been done on those lifting bodies . . . but nobody felt that confident in them. So one of our guys . . . built a couple models and flew them around, and we finally convinced our director to let us go ahead and build a [full-scale] wooden one."

The person interested in the idea to whom Milt referred was lead engineer, R. Dale Reed. He started working on the project on his own time, bringing in others who became excited by the prospects. Milt convinced Bikle to go ahead with construction of a prototype, which Paul authorized without ever notifying anyone at NASA outside the Flight Research Center.

Gus Briegleb fabricated the plywood outer body of what became the M2-FI at his local glider factory. The interior was a truss work frame of aluminum, with landing gear from a Cessna 150. A large bubble canopy broke through the flat plane of the top of the fuselage, along with two vertical tails at the edge of the body. Small horizontal surfaces provided pitch and roll control. The entire budget was a mere $30,000. Glynn Smith, an instrumentation technician on the x-15 who also worked with Thompson on the first lifting body, said, "Milt was the main gun on the M2-FI. We were more personally involved with it [than the x-15] because there were only ten or twelve of us in the whole operation. That was instrumentation, mechanics, pilots, and everyone else."

As the vehicle came together, the team had to make some decisions. The entire M2-FI was transported north to the Ames Research Center in Mountain View, California. The lifting body was mounted into one of its large wind tunnels, started up with Milt, Dale, or Ed Brown inside at the controls, and the aerodynamic response was noted at simulated airspeeds up to 135 miles an hour.

These tests gave enough confidence in the M2's airworthiness to proceed with the next step: towing it across the Rogers lakebed. This was accomplished with a 1,000-foot-long rope connected to the rear of a 1963 Pontiac Catalina. Bill Straub hopped up the power in the Pontiac enough that the muscle car was capable of 110 miles an hour, even with this ungainly, white tub leashed behind like a reluctant dog. Numerous tow tests, starting in April 1963, eventually got the lifting body about 20 feet off the ground in the car's rooster tail of dust. On 16 August Milt and the M2-FI graduated to using a NASA C-47 for a tow aloft to 12,000 feet. Descending at 3,600 feet per minute once the rope was released, Thompson had precious little time before he flared for a perfect touchdown, at the end of the proof-of-concept flight for the lifting body.

"We mostly flew that pretty successfully," Milt commented. "It was just to demonstrate that these odd-ball configurations could withstand the heating [and] could also be maneuvered and landed just like a conventional airplane. What we intended to prove didn't require really high speeds."

In June, two months prior to this first lifting body flight, Thompson was officially announced as the next x-15 pilot. Joe Walker was relinquishing his spot, so Milt filled the vacancy. He had flown four times previously as a chase pilot for the x-15, first on 17 July 1962, for Bob White's flight 3-7. Now it was time for Milt to get into the cockpit himself. However, Thompson found he was in the unusual situation of now having two major programs with which to contend. This conflict was to eventually come to a head with a confrontation between himself and his boss, chief NASA test pilot Joe Walker.

On 29 October 1963 Milt Thompson was finally ready for his first x-15 mission, 1-40-64. Lt. Col. Fitz Fulton and Col. Gay E. Jones flew B-52 no. 008 to the Hidden Hills area, where Thompson launched at 12:42 p.m.

Milt compared flying the x-15 to riding a bull, which came partly from his reported shock after being released from the mothership's shackles. When we talked of this first flight, Milt said, "I got into the program a little later, so I didn't get the benefit of some of the early pilot training. By the time I got in, they felt it was routine. You can just put a guy in it, and you can make a flight, and no problem." He went on to explain the differences between the "Iron Bird" and the real thing, saying, "We did our training on a big simulator down here in the hangar. It had all the actuators for the control systems like the real airplane, but you'd sit in the cockpit, and you'd smoke a cigarette and drink a cup of coffee, and you'd fly the flight. And you'd do this for hours and hours, so when I got up in the [actual] airplane and got ready for launch, the first thing that happened was a tremendous jolt as you drop off the hooks."

Thompson was the ninth pilot to experience this phenomenon in the x-15. Milt said the sensation of lighting the LR-99 was another thing that captured his undivided attention. "You're heading back like this [he demonstrated, by leaning way back in his chair], then looking at the instrument panel in a manner in which you've never seen it before! So it was really a big shock. Of course, we had enough training and practice that we were able to do it. I guess I'm famous for the comment that it was the first airplane I was glad when I shut the engine off!"

Several years later I had the opportunity to ask Bob White how he felt about Thompson's comments. Bob, as one of those whom Milt had accused of not warning of the launch experience, defended his position. "You're not fired off, you're just dropped. . . . Why should he be so surprised? They've watched airplanes all the time, from the x-1 dropping out of the bomb bay to the x-2." For Bob, it was just not a big deal, and he thought it was nothing so earth-shattering it had to be shared. "So they drop you and you start flying. I can't imagine Milt's feelings, because there was a countdown. It's 'three, two, one,' and as soon as you drop off [the B-52] you start flying. That's all. It's instinctive."

Whatever Milt's feelings about his first launch, he performed well during the flight. On mission 1-40, his objective was Mach 4 at 74,000 feet. He exceeded the plan by one-tenth of a Mach number and 400 feet in altitude. He said later, "I think the X-15 was a pretty impressive airplane, no doubt about that."

After burnout, and during the glide back to Edwards, Milt noted the X-15 was not staying where he wanted it. In his post-flight report, he explained the problem was his own. "I also noticed a tendency for the airplane to roll off and finally decided I was standing on a rudder, so I got off that, and things felt pretty good." When a debriefer asked if he thought he was on the planned flight path, Milt said, "Looking out the window the heading looked pretty good. Of course it's hard to tell, [because] you've only got mountains out there in the distance as reference." He touched down and skidded across the lakebed. After coming to a complete stop, he remarked with a satisfied air over the radio, "How about that!"

Just over four weeks later, Milt was ready for his second flight. During that intervening month, the entire world changed.

Flight day was Wednesday, 27 November 1963. The previous Friday, at approximately 12:30 p.m. Central Standard Time, President John F. Kennedy was assassinated while riding in the back seat of a limousine being driven through downtown Dallas, Texas.

Kennedy had set the American course to the moon but had also recognized the importance of the research provided by the pilots flying the X-15. He met several of them on more than one occasion, usually to present aviation and space awards for their exceptional service to their country and to aerospace. His loss reverberates to this day. It is often lamented how there is never the same fire of political will to present the case for the exploration of space to the public as had been seen while Jack Kennedy was in office. Entire alternate histories have been written about what might have been if Lee Harvey Oswald had not pulled the trigger that afternoon. How far might America, and the world, have progressed by the early twenty-first century if we had not faltered so badly with the end of both Apollo and the X-15? How many follow-on programs died before even having a chance to fulfill their promise?

On that Wednesday afternoon at Edwards, as Thompson and the entire team prepared for the ninety-sixth x-15 mission, no one yet knew the full future ramifications of what had transpired. It can only be said with great certainty that the recent events in Dallas weighed heavily on the minds of everyone present. But they still had a job to do, a research program to advance, a rocket plane to fly. And on this day, just two days after Kennedy was laid to rest in a temporary grave site at Arlington National Cemetery, all those people with NASA and the U.S. Air Force did those jobs in their usual professional manner. Another successful mission went down in the history books, with Milt achieving Mach 4.94 at 89,800 feet.

By 19 February 1964 the terrible memories of Dallas had faded. A dry winter season permitted x-15 flights to continue relatively unabated, a rare occurrence throughout the program. Milt was back above Hidden Hills to drop away on his first flight to surpass Mach 5.

Depending on mission requirements and throttle setting, a normal LR-99 engine run could last anywhere from 65 to 137 seconds, or even longer on the few occasions where the external tanks were used. The average time was in the range of eighty to eighty-five seconds. On mission 3-26 the plan was to run to shutdown at ninety-three seconds, but the LR-99 had other plans, stopping ten seconds early. Milt had plenty of energy to return to Edwards, so there was little effect on the overall flight plan.

The same was not true with his next flight, 3-29, on 21 May, out of Silver Dry Lake. He again had a premature engine shutdown, only this time it happened at less than forty-three seconds on the clock. This time, there was no way for Milt to get all the way home.

Thompson launched himself from the pylon of B-52 no. 003 at 9:39 a.m., lit the LR-99, and rotated the nose angle upward to the desired point. Passing through Mach 2.90 the engine gave out, and he had to set up for an emergency landing at Cuddeback Dry Lake. The engineering report stated, "The engine malfunction was found to be caused by a pressure spike in the second-stage chamber pressure sensing line, a problem which has been present for several years. . . . The intensity of the spike was sufficient in this case to deform the switch mechanically, holding the contacts closed, and thus preventing engine restart."

Milt explained about planning the flight, saying, "We were trying to simulate a [supersonic transport] flight condition." The SST was supposed to cruise up around Mach 3, so NASA engineer Ed Saltzman laid out the profile for the X-15 to match that. "Normally, we didn't try to throttle back the engine until there was enough energy to get back to [Edwards]. . . . That meant we would have to go to about Mach 3.3 before I could pull the throttle back." That was too fast for the SST simulation data, so Saltzman and Thompson went to the director of flight operations, Joe Vensel, and asked, "This is really important. Could we keep down to this speed?" They convinced him to let the plan include the maneuver. "As it happened," Milt continued, "I got up to about Mach 3, right where we wanted to be, pulled the engine back—and it quit!"

Milt further explained how far the X-15 could travel for any given height, which greatly limited his landing options. "The lift-to-drag ratio of the X-15 was on the order of about four-and-a-half, and you can convert to see how far you can glide. At 45,000 feet—nine miles up—that's thirty-six miles."

The lakebed at Cuddeback proved to be softer than expected, creating higher-than-desired loads on the landing gear, although this was not the worst of the problems. Milt touched down a bit long on the marked runway, and before the slideout was complete, the aircraft crossed over a road that had been graded into the lakebed. This grading had driven piles of dirt about six inches high on either side of the road. At an X-15 symposium, Milt said of that moment: "I hit that road doing about 100 [miles an hour], just plowed through the banks that they had up on either side and bounced over the road and finally came to a stop about 500 feet beyond. . . . Well, I survived the flight, but I was sure sorry I had, because now I had to go back and face Joe Vensel."

Milt was once quoted as saying of an X-15 flight: "You knew you were going to be on the ground within eight to ten minutes, one way or another. You were either going to make a successful landing or come down on a parachute or make a smoking hole." For his one and only emergency landing in the aircraft, with minor damage to the landing gear, he had definitely accomplished the best of those three options.

The UFO "scare" concocted in conjunction with the X-15 conference in October 1965 had alien spaceships supposedly ready to chase down an X-15. Having one or more X-15s hunt down hostile flying saucers was a motif used for at least two motion pictures. One used simple stock footage, while the other brandished entire squadrons of the rocket planes to thwart an alien invasion.

In 1967, at a secret military decoding laboratory in Texas, a signal was received that warned: *Mars Needs Women*. Tommy Kirk starred as Dop, the Martian commander sent on a mission to replenish his planet's genetic material with willing—or not—Earth women. When our government refused to allow this harvesting of females, an x-15 was dispatched to shoot down the invading saucer. Considering the rocket plane had no weaponry, the outcome was a foregone conclusion. In the end Dr. Marjorie Bolen (Yvonne Craig from the campy *Batman* television series) fell in love with Dop. Together they saved the Red Planet for future generations, with no further need to waste the resources of the x-15 program bringing the Martians to heel.

A much larger and less comedic threat emerged in the 1959 storyline of *Uchu Daisenso*, which translates from Japanese as "The Great Space War." The evil alien race, Natal, launched attacks against Earth from their moon base. The United Nations Space Research Center responded with dozens of x-15s soaring into the upper atmosphere and deep space. Launches of the sleek, silver rocket planes—with sexy red racing stripes on the wings—occurred from vertical tower structures, others at a forty-five-degree angle, and even more out of silos like intercontinental ballistic missiles. The brave astropilots waged war against the aggressors, firing electrical rays from the x-15's Q-ball nose.

When released in the United States in the summer of 1960, the title had been dramatized to *Battle in Outer Space*. The movie was considered a welcome change from the normal Japanese monster flicks of the time. Instead of battling Godzilla, Rodan, or Mothra, there were no actors in rubber suits covered in scales and spikes. This time there were only spacesuits.

When Vice President Hubert H. Humphrey visited Edwards in April 1965 and told Joe Engle he would see about funding more squadrons of x-15s, maybe he recalled watching *Battle in Outer Space* and had seen how effective the rocket planes were against the alien invasion force.

Back on television, two shows used the x-15 to great effect. First was the pilot of *My Favorite Martian*, shown on 29 September 1963, exactly one month before Thompson's inaugural flight. Bill Bixby had the part of Tim O'Hara, a reporter for the fictional *Los Angeles Sun* newspaper. At Edwards a test flight of the x-15 was moved up a day due to the weather forecast. O'Hara was late getting there to cover the mission because he overslept.

x-15 no. 1 dropped from the b-52, lit the lr-99, and hit Mach 5.4 before the pilot, Summermeyer, had a close encounter. The radar operator saw two returns on his scope and called to the colonel in charge: "There's a blip on a collision course with the x-15—and gaining on it fast . . . whatever it is, it's going over 9,000 miles an hour!" Summermeyer excitedly called back to the base, "Something just went past me like I was standing still—a flying saucer!" The colonel grabbed the microphone and told the pilot, "Check your oxygen supply and stop babbling that nonsense about flying saucers. It's probably just a speck in your eye."

The military, not wanting a flap in the newspaper, shoved O'Hara back out the door the moment he arrived. Tim was dejected at first, but things picked up as he drove home, witnessing the crash landing of a one-man—or more accurately one-alien—craft. He rushed to the rescue to see if anyone was still alive, finding Ray Walston as the Martian, in his silver spacesuit. When Tim revived him, the first thing out of the alien was, "What are you waiting for me to say: 'Take me to your leader?'" The Martian, who was a professor of anthropology sent to study the primitive Earthlings, now was marooned on our planet with a broken spaceship. He lamented, "That idiot in your antique rocket plane almost ran into me. . . . It was lumbering along at barely 4,000 miles an hour. I had to strain my ship to get out of his way." Thus started a three-year relationship between the reporter and his "Uncle Martin."

The second-to-last episode of *The Outer Limits*, a well-regarded attempt to mimic the success of Rod Serling's *The Twilight Zone*, presented "The Premonition." It originally aired 9 January 1965, four days prior to Milt's tenth mission. In the story's scenario, x-15 no. 3 launched and was piloted by Jim Darcy from the Palmville Flight Test Center. His wife, Linda, arrived at the base and dropped off their daughter, Janie, at the day-care center before heading into the desert to see her husband land. Somewhere around Mach 6, the aircraft went out of control, and Darcy found himself in a crashed and smoking x-15. His wife was nearby in the family car, having had her own accident at the same moment. The problem was, they were the only two people running at normal speed. Everyone else had slowed down, their minuscule motions barely perceptible to the Darcys. They were in a time warp.

The reason all this happened was so they could save Janie from yet another accident about to occur. A parked truck lost its parking brake and was going to run over their daughter, who had gone outside the day care to ride her tricycle through the alley. The stars of the episode, Dewey Martin and Mary Murphy, were not well known, but both had full careers in primarily minor roles. The story was written by Ib Melchoir, famous for *Robinson Crusoe on Mars* and *Death Race* 2000.

During the course of the show, stock footage was used of the x-15 launching and in-flight. Most of the shots were of the no. 3 aircraft, but some from a distance did switch over to the no. 2, including the crash sequence. In reality, it was Scott Crossfield's third powered flight where the aircraft had an in-flight engine explosion and landed on Rosamond Dry Lake, breaking the fuselage behind the cockpit. The biggest gaff, from a technical standpoint, was when the pilot started the rocket engine as he dropped from the b-52 by throwing toggle switches for the eight chambers of the lr-11 engines. The exterior shots clearly show the lr-99 installed.

On the literary front the x-15 was featured in several nonfiction books of the time, but those dried up once the focus on space travel shifted from the California desert to the marshes of Florida. From the fictional side there were the eight books in the Mike Mars series, published between 1961 and 1964, which were aimed directly at young teenage boys. Donald A. Wollheim penned two titles specifically concerning the x-15: *Mike Mars Flies the X-15* and *Mike Mars, South Pole Spaceman*.

Mike's last name was actually Samson, but his full given name was Michael Alfred Robert Samson, whose initials spelled out M.A.R.S., thus the name by which he was best known. He and his best friend, Johnny Bluehawk, were part of the super secret Project Quicksilver, created to upstage the Soviets in the space race. Their boss, Colonel Drummond, explained their purpose by saying, "You men are our secret weapon. Without family ties, young and daring, ready for risks that cautious and older test pilots would not undertake, you are going to be moved ahead secretly, and the world is not going to hear of you until you have planted Old Glory on the Moon."

Mike felt landing on the lunar surface was only a small step toward his ultimate goal, being the first man on Mars, a destiny connected to his name. The author may have based his main character to some degree on the media legend of real-life rocket pilot Iven Kincheloe, who was originally supposed to be the air force's prime x-15 pilot before he died in the July 1958 crash of his F-104.

In the first book the x-15 was a target of sabotage, then shot down by a Sidewinder missile. For the second outing the x-15 was flown south to Antarctica, where it was used as the launcher for a Blue Scout rocket. The reasoning was that it must go into space through the polar hole in the radiation belts surrounding our planet. After the satellite launch from the x-15, a Russian plane dropped several bombs, creating a whiteout to force Mike to change his landing zone to a Soviet base on the ice. Of course, Mars foiled the attempted hijacking by landing, then destroying, the x-15 and burying it in a crevasse.

One relatively unknown movie appearance by the x-15 was in the Stanley Kubrick classic 2001: *A Space Odyssey.* A model of the rocket plane sat on the shelf in the Hilton hotel office of Space Station 5, behind the manager's desk. The model was modified to represent a possible Earth-to-orbit vehicle, giving an obscure nod to the spaceworthiness of the x-15, yet unnoticed by nearly everyone who saw the movie.

Milt Thompson returned to fleeting stardom in NASA's own production of *Research Project: X-15.* This publicity film by the agency featured Thompson "flying" the simulator and also had comments from numerous people associated with the program, including Harry A. Koch from Reaction Motors, Harrison Storms and Scott Crossfield from North American, and Paul Bikle and Joe Walker from NASA. The only air force representation was an appearance by Bob Rushworth as NASA 1 in the flight control room. Again, there was no mention of the navy's participation in the program. After talking of the Charles Bronson version of the x-15 story, Paul Bikle also gave me his succinct review of the government production: "NASA put out a thirty-minute movie, and I thought it wasn't very good, either."

Besides Nathan Scott's soaring score for the movie *X-15,* the rocket plane also got airplay on the radio through singer and songwriter Johnny Bond. The hit "Hot Rod Lincoln," by Charlie Ryan, inspired Bond to do him one better, releasing "X-15" in a similar vein to the race song in October 1960. It started, "Gather 'round you cats, and you'll hear about a race I had in the stratosphere. Ol' Joe had a slick jet, I mean, and I flew a souped-up x-15."

The lyrics had twin x-15s starting a race by launching at 80,000 feet from under the wing of a B-70. Bond's description of the rocket plane was surprisingly accurate, considering it was simply a fun parody song. About halfway through he sang, "Now this x-15 is a goin' thing, got two little stubs, they call it a wing. Won't help you much if the motor stalls. Hey, lookie below, Niagara Falls." Johnny and Joe continued their race, rocketing across the entire nation before heading out over the Atlantic Ocean. By the time they finished they left the planet completely, another recurring theme of taking the x-15 much higher than it was ever capable of achieving. "Well we circled the earth in three hours flat, yes I'm in orbit, no doubt about that. Here I sit, just circlin' in space; what's that? Don't ask me, I don't know who won the race."

With Hollywood and the "Top 40" continually disturbing the peace at Edwards, Paul Bikle was actually glad when the nation's focus turned away from the x-15 and went instead to the Mercury program. "I thought it was good. It got all of those reporters and everybody out of here. Oh, Jesus, it was terrible!"

Every flight of the x-15 was an experiment with inherent dangers. The idea was to eliminate as many of those potential threats as possible through proper engineering and planning. A good example: no one ever thought it an especially bright idea to fly an x-15 close to sunset. The rocket plane had no lights nor special instrument landing aids. The pilot required proper lighting conditions. Yet Milt ended up in the air near dark, something that never should have happened.

It was 29 October 1964. Milt recalled, "We were having gear trouble . . . so we ended up modifying the release system. It just so happened they got it all done late one day, and [no. 3] was scheduled for a flight the next day. They wanted to go up to try the modifications in-flight . . . so they needed a captive test."

With mission 3-39 scheduled the next day, Milt got the tap on his shoulder for test mission 3-C-58. "I happened to be the only x-15 pilot around, so they stuck me in the airplane and got aloft about, hell, it was 3:00 or 3:30 [p.m.]." The B-52 circled at high altitude to cold-soak the entire system, then Milt pulled the handle to deploy the nose gear and rear skids. They went down fine, but the late October sun was quickly fading. Milt said, "It was [almost] winter, so by the time we got all the way down, we landed in the dark." It was 4:53 p.m. as the B-52 touched down, ten minutes prior to local sunset. Even without the normal occurrence of cloud cover near the horizon at that time of year, there was very little light left in the sky.

As I sat and talked with both Milt and flight planner Jack Kolf, Jack interjected, "Some of these things, when you think back on it, weren't very smart. The theory is you never did anything—ever—that would preclude you from launching. You never took an airplane up unless you were convinced it was ready to fly, even if you weren't planning to." Milt finished Jack's thought: "Because something could happen. You may have to be dropped. If they had a fire on the B-52, the first thing they do is let you go." Milt returned to Edwards still safely embraced under the B-52's wing, clearing the mission to proceed.

Less than eighteen hours later, Milt was launched near Hidden Hills on flight 3-36. At 9:51 a.m. there was plenty of light left in the day, and except for an errant LR-99 fire-warning light fifty-four seconds after shutdown, the flight was flawless. NASA operations engineer John McTigue related his impression of Milt as an X-15 pilot: "Milt Thompson was very exceptional from the standpoint of being able to fly the airplane and get the precise positions [the engineers] wanted. He knew what the hell he was doing from a scientific standpoint."

Each pilot in the program gravitated toward certain types of flights. Joe Walker was remembered for his altitude work, while Bob White took the speed records. A very important aspect of hypersonic research was to study the effects of heat on the X-15's Inconel-X structure and control systems. This was where Milt excelled. He said, "I specialized in the low altitude flights and did a lot of the heating work. We had to stay at a fairly low altitude and go to high speed [to] try to stabilize in certain conditions so we could measure the heating rate of the airplane." More than half his fourteen X-15 flights were devoted to gathering heat data.

Milt talked of mission 3-39 on 13 January 1965, to 99,400 feet: "On this particular flight, I got up to flight condition, then throttled the engine way back, put the speed brakes out, pulled into a four g turn, and just sat there until the engine burned out. Now, because it was optimum condition for the fuel of the plane, the engine burned about eight seconds longer than we had anticipated, and that's a hell of a long time when you're sitting there in a four g turn and close to Mach 6. I was turning away from Edwards, and I thought, 'Goddamn, I'm probably halfway to Barstow by now!'"

He decided to push over, which shut the engine off. "When you hit zero g, and all of a sudden your fuel goes all over the tank, that's an automatic shutdown. I reversed the bank, but . . . [the aircraft] started gyrating all over the place!" Once again, if the x-15 started acting as it shouldn't, the best solution was usually to take your hands off the controls. As soon as Milt did this, "It kind of settled down. Generally, the airplane is stable. In this case, it was like a pilot-induced oscillation. So, although I wasn't feeding it, the simple fact I had my hand on the stick meant forces were feeding back, and I was resisting, so that amounts to a PIO." Thompson made a normal landing on the Rogers lakebed at 10:57 a.m.

Jack and Milt spoke more about the design of the x-15 in relation to heat. Jack started, "Part of heat dissipation was using the structure as a coolant. It absorbed [the heat] throughout the structure of the airplane. Then, to some extent, we used those two big [fuel and oxidizer] tanks, which were 90 percent of the body anyway, to help in cooling. Another reason is Inconel is a black metal; it wasn't paint. We did paint them occasionally, but then it burned off." Milt picked up from there, saying, "Inconel-X was good up to about 1,800 degrees [Fahrenheit], but it started losing a little bit of its strength at 1,300. So you didn't want any particular spot getting hotter than, say, 1,200 degrees. . . . If you had a hot spot on the leading edge, you wanted to release that heat as quickly as you could back into the rest of the structure." Milt mentioned that after nearly every flight they did have to paint the markings, such as emergency instructions, "No Step" signs, and things of that nature, back onto the x-15 fuselage and wings. Jack exclaimed, "Yeah, they'd melt off and make a mess!"

Heating flights were special, in a strange sort of way. Milt recalled, "When you got into the heating region, you could hear the pops and bangs, and you could feel the airplane twitch. It really heated up fast, and that's the problem. You didn't have a smooth onset to all parts of the airplane. You got certain parts that were hot real quick, and you'd have a cold part next to it, and that's how you'd get these buckles and pops and things. . . . That really got your attention!"

In the end the x-15 gathered so much data on so many areas of flight research that these data were often never utilized, especially with the lack of a follow-on effort once the rocket plane program ended. Jack was sad when he commented, "We've thrown away an awful lot of stuff that just sat around for tens of years. . . . Data, piles and piles of data. The reason we threw it away is because it really wasn't any use to us any more. After a while, you don't even know where it came from. You didn't know what flight it was or what parameters you were looking at."

For a set of four flights, lasting exactly three months, starting on 25 May 1965, Milt changed direction from heating to a short stint of altitude work. Each of these missions exceeded 100,000 feet and were the only times he did those types of flights. His last flight aboard the x-15 was mission 1-57 on 25 August, when he attained 214,100 feet. Milt said, "It was busy on the altitude flights. I really only got the chance to look out the window once, and that was my last flight. . . . I got up on top after I had everything fairly settled, then looked out the window. Gee, it was a really impressive view! You could see all the way from San Francisco to Baja California. All I could see over the nose was the ocean." He laughed and continued, "I thought, 'Oh, shit! Somebody didn't plan this flight right!' That was pretty impressive, you know. It was beautiful up there. You just have a few seconds to take a look."

As he reentered and headed into the Rogers' landing pattern, Milt found the x-15 was pitching up, and he was having a very difficult time bringing the nose back down. A combination of problems associated with aircraft trim and the center of gravity caused Milt's difficulty. He regained full control and laid the aircraft smoothly onto the lakebed for the final time at 10:03 a.m. Joe Engle at NASA 1 radioed his congratulations, "Very nice Milt."

This was not the end of Thompson's career in experimental aircraft. In fact, the problem was he had too many to fly. "I got out of it because I had two good programs: the x-15 and lifting body. My boss, Joe Walker, said to me, 'You've got too many programs, and you have to choose.' So I elected to go on with the lifting body. That's when Bill Dana picked up from there and replaced me in the x-15 program."

Eleven months after completing his time on the x-15, the first of what was called the heavyweight lifting bodies was ready for flight test. The Northrop M2-F2 was an aluminum-and-glass direct follow-on to the plywood and tubular-framework M2-F1. Milt flew the first five unpowered glide tests, starting on 12 July 1966, dropping from the same pylon on the B-52 mothership as used by the x-15. Several other pilots joined Thompson, including Bruce Peterson, Don Sorlie, and Jerry Gentry.

32. Milt Thompson talks with reporters following the first unpowered glide flight of the M2-F2 lifting body on 12 July 1966. On the same day, Pete Knight flew mission 1-64 in the X-15 to Mach 5.34. Courtesy of the Armstrong Flight Research Center.

Over the next several years, differently shaped lifting bodies were introduced to the program. Northrop came in with the HL-10, and Martin Marietta produced the X-24A, then X-24B.

Bruce Peterson became somewhat famous after the sixteenth glide test of the M2-F2 on 10 May 1967. He rolled the vehicle on landing. From all appearances it was a horrific crash, with the M2 flipping six times, but he escaped with treatable injuries. The only exception was the loss of his right eye, which actually occurred from an infection during recovery rather than in the accident.

Footage of the crash landing got science fiction writer Martin Caidin thinking, which led to his creation of the 1972 novel *Cyborg*. The premise was that the pilot was horribly maimed in the crash, allowing a government agency to rebuild him with many robotic parts, creating an American secret weapon.

In early spring 1973 the ABC network released a made-for-television movie adaptation of Caidin's book as *The Six Million Dollar Man*. It proved so popular ABC ordered a full series and ran the show for five seasons, with several additional TV movies, plus the spin-off, *The Bionic Woman*. During opening credits of each episode the HL-10 is shown being dropped from the B-52's wing, then cutting to the crash landing footage of Peterson in the M2-F2. Often seen in the corridors of the NASA facility at Edwards, Bruce wore his black eye patch, and people were always fascinated to meet the real man behind the fictional "Steve Austin."

Milt Thompson's final flight in the M2-F2 was on 2 September 1966. He never flew any of the other lifting bodies, including the rebuilt M2. The new designation was M2-F3, easily distinguished from the earlier version in that it now came equipped with a third vertical tail in the center of the aft upper body. The extra tail improved roll control to prevent a repeat of Peterson's accident.

Jack Kolf was a lifelong friend of Milt's. Every time I saw one of them, I saw them both. The first interview ever conducted for this book was with Thompson, arranged by Dr. Richard Hallion, then the head of the U.S. Air Force Flight Test Center History Office. Milt was relaxed and happy to talk that day, and it wasn't long into our conversation he invited Jack to come in too. The reason was to have Jack explain a research program idea he came up with to launch an M2 lifting body off the back of the YF-12A, a prototype version of the famous Lockheed SR-71 Blackbird.

Before he even yelled the invite into the next room, Milt started to laugh very hard. "Hey, Jack! Come here and tell them about your idea of launching the M2 off the YF-12." Milt continued to laugh as Jack sheepishly entered his office. He did not appear too particularly pleased, but I could tell he was having fun too. "It wasn't my idea!" Jack deadpanned. "Paul Bikle expressed some curiosity about what we could do if we launched from a much higher altitude and velocity, so, having an M2 simulator, we went down and set that up. The initial conditions were Mach 3 at 80,000 feet, which is about what the YF-12 can do. Then we proceeded to launch the airplane, lit all four chambers of the rocket engine—and decelerated all the way to burnout!" The configuration and drag inherent in the lifting body design could not be forced to go to such high velocities.

All of us in the room were laughing as Jack went on: "Needless to say, we reported the results to Bikle, and all I did was turn around and leave." I asked, "That was one of those programs that lasts five minutes?" Jack answered, "That's about all it was too. One shot! We weren't going any farther on that one." Jack turned to leave, and as he passed out of Thompson's doorway, he looked back a moment, growling, "Thanks, Milt."

Within a year of completing the glide flights in the M2-F2, Milt left piloting duties for management. He was the director of Research Projects, so he had "all the projects at that time," he stated. Thompson later used his X-15 expertise to help convince the Space Shuttle design team to delete any thought of using jet engines during the landing phase of the new reusable space plane. Years of flying the X-15 showed unpowered landings from hypersonic velocities not only were feasible but could be accomplished routinely. The X-15 came back safely from space 12 times, while the shuttle eventually did so on 133 occasions, without the need of costly and weight-prohibitive jet engines to provide power during the last few minutes of flight. These had been part of the original design of the shuttle and would have changed the program completely if Thompson and others had not convinced NASA to leave them on the drawing board.

Milt talked about that decision: "They asked us to do an evaluation of the entry controllability, so we put together a complete simulation here. . . . We didn't have the exotic simulator they have [at the Johnson Space Center]. We actually looked at reentry, then we made our comments and recommendations on the control system."

Thompson's career finally took him into the position as chief engineer at the renamed Hugh L. Dryden Flight Research Center at Edwards. When I first spoke with Milt, the dead giveaway of what his job entailed was on a shelf behind his desk: a proudly displayed blue-and-white-striped railroad engineer's cap. When asked what the chief engineer does at Dryden, he laughed easily, then said, "Anything I want it to! How's that?" Then he spoke more seriously of his job responsibilities: "I kind of look at all the programs that we have—the technical content—and make recommendations on them and what we ought to be doing, that kind of thing. And then I spend a lot of time on flight safety aspects of all the things that we use, and if we are doing it right and doing it safely."

Operations engineer Bill Albrecht said, "There were always a lot of shenanigans going on, people pulling tricks. We're so somber now; we don't do those sorts of things." He related his version of a now infamous story about Milt. "We had a wet lakebed, a lot of water, and somebody brought in a pair of water skis. He challenged Milt to get the skis on, go out on the lake, and get somebody to tow him. The only thing that could tow him would be a helicopter. . . . The helicopter was sitting on the ramp, pointing toward the lake, and down went Milt. He actually put on the skis, and he was going to go out to the ramp that went down into the water, then the helicopter was going to tow him. They had it all planned out."

Flying x-15s and lifting bodies was a dangerous business. Adding helicopter waterskiing to that resume didn't seem like such a good idea. Albrecht continued, "Bill Dana saw what was going on [and] he walked into Joe Vensel's office and said, 'Joe, you ought to look at what's going on outside.' Joe looked . . . then he ran down the stairs. The chopper's blades were already spinning. He ran out in front of the chopper and gave them a dirty look, and then went like that [motioned to cut the engines by slicing across his throat]. Whoever it was piloting the chopper shut it down. Joe Vensel stomped off and went back into his office." As soon as it came out that Dana was the one who turned Milt in, somebody presented him with the Benedict Arnold Award, equating the action somehow to treason. "They had this put in a frame and presented to Bill. . . . That picture floated around for years."

Jack McKay's son, John, worked alongside Milt for many years at Edwards, and he had a favorite story about Thompson. "They had one hell of a party and Milt was out all night. He came home and snuck back into his house [and] was sitting there taking off his shoes when [his wife] Terri woke up. She asked, 'Are you just getting home?' And he goes, 'Nope, sweetheart, just going to work.' He went to the pilots' locker room [at Edwards] and slept it off for a couple hours. And he confirmed that to me sitting in my office one day. . . . Milt didn't like to talk about himself, but if you brought it up, he'd want to clear the air."

10. Fastest Man Alive

Things were looking really good, and I was really enjoying the flight. All of a sudden the engine went "blurp" and quit.

William J. "Pete" Knight

The T-39 Sabreliner cruised eastward through the cold early December air at 40,000 feet. Aboard, a seven-year-old boy was having a great time, being allowed not only into the cockpit but even to handle the controls. Maybe this experience was part of his reasoning to later become a pilot himself. Certainly some of that decision also came from the fact his father was one of the greatest, and definitely the fastest, of the test pilots, Pete Knight. David was accompanied by his parents and brother that momentous day of 3 December 1968 and still recalls his time on the small jet as it traversed the air space between Edwards AFB and Washington DC. The purpose of their trip was for his dad to meet with President Lyndon B. Johnson and receive the Harmon International Aviator's Trophy. However, meeting a president was much lower on David's list of cool things than sitting at the controls of a jet. Many years later he had no memory of being in the White House or meeting Johnson. Besides the flight "my recollections are of playing on elevators in the large buildings in DC." These activities were more important to a young boy than any presidential pomp and ceremony. Photographs from the day clearly show David and his older brother William, uncomfortable in their suits and ties, next to the president.

The reason for the invitation from President Johnson, and the cross-country flight to receive the Harmon Trophy, had occurred more than a year previously, on 3 October 1967. Pete Knight flew the X-15 rocket plane to 4,520 miles an hour, or nearly seven times the speed of sound, that day in the rarefied desert air at just over 100,000 feet. In many ways Pete's flight was the culmination of the X-15 research program, with a speed record that has stood for more than five decades, with little chance of being surpassed in the foreseeable future.

Pete's journey toward that flight began on 18 November 1929 in Noblesville, Indiana. He attended Butler and Purdue Universities before enlisting in the air force in 1951, took a couple of years to finish his cadet and flight training, then almost immediately began making a name for himself in aviation by winning the Allison Trophy Race at the September 1954 Dayton Air Show.

Pete told the story of how he flew an F-89D Scorpion on a time-to-climb record in the skies near Vandalia, Ohio, that day. "I practiced quite a bit and developed a profile that would comply with the regulations and the rules of the race and still get to altitude in a minimum of time. . . . We didn't have very much fuel, [because] we didn't need any more than it would take to climb to altitude and land." Sitting on the runway with the engines burning used a lot of fuel, and Pete became concerned there would not be enough to get through the event.

The moment he was finally airborne, Pete got the gear up and firewalled the afterburners. "I went straight up [to] go through 10,000 feet . . . [then] shut one [engine] down and glided over to land. Just before the landing roll, I fired up the other [engine], so I could taxi across to the finish line with [both] engines running. We made it but didn't have any fuel indicated in the airplane. . . . I had beaten everybody by about fifteen or twenty seconds."

Knight excelled with each airplane he flew and was recognized for such as he moved upward in his career. By 1958 he graduated from the Air Force Institute of Technology at Wright-Patterson, then almost immediately went into the Test Pilot School at Edwards as a member of Class 58C, graduating on 24 April 1959. This location put him close to the aircraft for which he would ultimately be remembered: the X-15. The rocket plane was shipped to Edwards after its rollout at the North American plant during the time Knight was attending the school.

Pete thought a lot about aiming for the X-15 program but initially sidestepped the idea by becoming one of a number of pilots assigned to the consultant group on the X-20 Dyna-Soar. "I had always wanted to fly the x-15," he said, "but I was more interested in the Dyna-Soar program at the time." This was probably attributable to the promise of the x-20 to go well beyond the x-15 as a possible suborbital test vehicle and later, as the program expanded, into orbital flight with a military reconnaissance role in space.

Pete recalled, "There were five air force pilots assigned to the program. We worked hand-in-hand with Boeing in the development of the cockpit, subsystems, the procedures, and the profiles that would be used to demonstrate and test the x-20." This group helped define what it meant to be an astronaut by going through the same regimen as the x-15 and Mercury pilots. They participated in early centrifuge tests, as well as the same infamous medical testing at the Lovelace Clinic in New Mexico, and Brooks AFB in Texas. The Dyna-Soar was a viable spaceflight program for several years, and as Pete stated, "We had over 400 hours of simulator time in the Dyna-Soar."

The x-20 was canceled in December 1963 by Secretary of Defense Robert S. McNamara, and Knight thought that was definitely a wrong decision. "We were all very disappointed they canceled the program. At the time . . . we were pretty well on our way, with a number of systems that were qualified. We were ready to move on, and the schedules were looking good. We really wanted to go ahead and build the vehicle and at least do the air drops we had planned." Pete lamented that instead, it all was literally tossed away. "They canceled the program and scrapped everything. We had, as I remember, 5,000 parts laying around up there that were ready to go into the no. 1 vehicle. . . . It was an extension of the x-series airplanes. If we had proceeded with it, we would have been much further ahead in the space business, maybe with a more realistic Space Shuttle operation."

Once finished with the Dyna-Soar, Knight moved on quickly, returning to Edwards for the second half of what had now been renamed the Aerospace Research Pilot School. As he explained, "At that time, the test pilot school was a year-long course and they had split it up. The first six months was the test pilots' portion, and the second six months was the space portion. . . . [While] working on the Dyna-Soar, I never went through the space part of the school. I figured if I was going to get on another space program, I had better have that school, so I went back for [another] six months."

Graduating in mid-1964 Pete was ready for the next step. He thought of applying for the astronaut program with NASA but ultimately stayed with the air force and opted for the x-15. "After Dyna-Soar, I was happy to get on the x-15. . . . They were beginning work on the expansion of the envelope to Mach 8."

The longest-serving x-15 pilot, Robert Rushworth, recalled that bringing Pete Knight aboard was a choice initiated through him, as the replacement pilot he'd like to see in that role. Rushworth had been trying to depart the x-15 for some time, but his plans were frustrated when Joe Engle did not stay long enough to take over, jumping ship to the astronaut office in Houston to become part of the Apollo program.

Two weeks prior to Joe Engle's last x-15 flight, Pete Knight took his first. Robert Rushworth thus started his mentoring program all over for a second replacement pilot. This time he succeeded and was able to move on with his air force career.

On 30 September 1965, flight 1-60-99 was ready for the takeoff roll. Heading northeast from Edwards, B-52 no. 003 laid a ground track not far from Las Vegas, before turning back to line up for Knight's launch. After all his simulations, he felt he was ready for anything. Even then, as he explained after the flight, the shock of the drop was more than anticipated. "I couldn't tell you what happened at launch, other than it was all I could do to keep it right side up! When I hit that switch . . . it just blew me off of there. . . . It was probably a second before I could do anything. I had my hand on the throttle right from launch, but I didn't move it until I got the airplane squared away, and it kind of surprised me." Pete concentrated on his job of flying the profile, wanting to make sure he did everything right. He said later, "I think I was probably fighting it pretty hard to keep it straight."

Often, Pete talked of his experiences in the x-15 as if the airplane were a companion on his flights. Like Milt Thompson saying the x-15 was a bull to be ridden, the pilots often imbued the airplane with a soul. On this first flight for Pete, he trusted the x-15 to not let him down as the LR-99 accelerated him to above Mach 4. "I never looked out, from the time we launched until burnout. It felt like . . . we were coming right on up over on [our] back. I knew we weren't . . . but that's what it felt like. I finally figured it out that it was just an increase in g. . . . You don't think it'll be this much, [but] not looking out, you can't correct the sensation, so . . . I just flew the gauges and let it go."

The only purpose of this flight was to give Pete experience in the x-15, so

33. Three-view layout of the advanced x-15a-2 rocket plane with external fuel tanks. The core of this aircraft was the original x-15 no. 2, which rolled over on Mud Dry Lake on 9 November 1962. The rebuild consisted primarily of lengthening the fuselage by twenty-nine inches and adding the external tanks for increased performance. © Thommy Eriksson.

his speed barely topped 2,700 miles an hour, and he quickly pushed over into a flight peaking at 76,600 feet. He found Rogers Dry Lake and wanted to make his first x-15 landing perfect. The skids bit into the lakebed and the nose wheel slammed down. Pete deflected the horizontal tail surfaces to help push the aircraft downward, keeping it from jumping. Throwing up a dirty rooster tail of dust, he quickly slid to a halt. Ground personnel surrounded the aircraft, popped open the canopy, and Knight's first flight was a successful piece of history.

Each of the three x-15s was utilized for particular tasks. The no. 1 aircraft was always used for first flights of new pilots and also for experiments. No. 3 was primarily an experiment carrier but was also the one normally used for the higher altitude flights because of its adaptive control system. The no. 2 aircraft, now known as the x-15a-2, was the only one of the three at this phase to be used to expand basic flight research on the hypersonic airframe.

Following McKay's rollover accident in 1962, and the subsequent re-building of the aircraft into a version more than two feet longer and about 2,000 pounds heavier than the original, there was much to still be examined, especially where pushing the speed envelope was concerned. The addition of the external fuel tanks increased the burn time of the LR-99 by 50 percent. The hope was to increase the top speed by two full Mach numbers, or somewhere in the range of an extra 1,300 miles an hour.

Pete was still becoming familiar with flying the research vehicle and built up his experience with four X-15A-2 flights in the Mach 5 range within a bit more than five weeks.

The last flight before the technicians went back to preparing the external tanks for attachment, mission 2-49, involved a new experiment called the Maurer camera. The idea was to know if the hypersonic airflow around the X-15 might impede a high-resolution camera from getting clear photographs of ground targets. This may have been the first look at an imaging system for a possible new high-speed reconnaissance aircraft, which might someday supersede the Lockheed SR-71 Blackbird. That spy plane's Mach 3 speed was half that of the X-15, making the Blackbird increasingly vulnerable as new defense technologies emerged. Knight spoke of how the experiment was to work, saying, "There were some photo resolution panels laid out [on the ground] so we could evaluate the quality of the pictures we got back."

Everything checked out fine on the Mauer system, including its gimbaling mechanism; then, prior to taxiing the B-52/X-15 combination for takeoff, the gimbal no longer worked and the camera mount was locked into a forward-right viewing position by a thermal blanket that had come loose. The mission went ahead with the camera locked in that position. The experimenters still hoped for favorable conditions to photograph the primary resolution target, located on the surface of Racetrack Dry Lake at the northwestern end of Death Valley. Knight flew a good profile, passing over the edge of Racetrack. After the flight, the film was found to be scratched, but usable.

On an otherwise uneventful flight, at approximately 25,000 feet, Knight experienced one minor glitch. The parachute that was supposed to safely float the lower ventral down to the lakebed near the end of the mission deployed of its own volition. The ventral itself had not been jettisoned and was still firmly attached at this point. The parachute popped out of the canister on the rear of the fin and buffeted against the underside of the aircraft, creating minor damage to the aft area near the LR-99 engine exhaust. At the proper time and altitude prior to landing, Knight jettisoned the ventral fin without its parachute, to have it irreparably impact the lakebed.

Many of the jettisonable lower ventrals were lost over the early years of X-15 flight testing when the parachutes failed. However, this malfunction on flight 2-49 was the first where the parachute completely left the aircraft prior to planned jettison. There seemed little need for the jettisonable portion to be flown again, yet flight planners decided to use it one last time on the next flight of the A-2.

The external tanks were ready for flight the first week in October. Two air aborts occurred within two weeks, one of which was directly attributable to the ammonia external tank. There were also eight cancellations for various reasons prior to getting airborne, including a large rainstorm that left Rogers Dry Lake anything but dry for nearly a week in early November. It took five weeks before Knight was able to launch with the fully fueled tanks in place on 18 November. This was, coincidentally, Pete's thirty-seventh birthday.

If all had proceeded according to plan that day on mission 2-50, the X-15 should have notched up another record, in that both the no. 2 and no. 3 aircraft were ready to fly within hours of each other. Bill Dana was first up with no. 3; however, a malfunction kept the B-52 from starting its taxi. The flight was aborted, leaving the double-flight, single-day record unbroken.

For Pete, everything was fine with the A-2, so the B-52, with pilots Lt. Col. Fitz Fulton and Col. Joe Cotton at the controls, took off in the early afternoon. This became the first of two missions above Mach 6 for Knight, and the first such high-speed flight for the X-15 program in nearly three years. It was also the third of only four launches to use the external tanks. Primary objectives were to test the handling characteristics of the A-2 with the full external tanks and also to see how the rocket plane handled the jettison of the empty tanks while still under LR-99 rocket thrust, accelerating through Mach 2. These had been the same objectives that had failed to be realized on Rushworth's final flight on 1 July. The A-2 program was finally, and literally, ready to pick up speed.

The launch point approached, and the X-15 dropped on time. "The launch was no more difficult, nor no more surprising, than any other launch,"

Knight noted matter-of-factly. With the LR-99 at full throttle, he pulled up the A-2's nose to match the intended flight path, but the airplane was rolling somewhat to the left because of the extra 1,500-pound weight of the oxygen tank over the ammonia tank.

This tank asymmetry, and the resultant roll, moved the ground track left of what was planned. A little over a minute into the flight, the two external tanks were going dry, and Pete reached toward the switch to punch them off. The mid-November afternoon launch meant that Knight was heading directly into the sun, which partially blinded him through the windows. He missed the switch the first time. "The sun was getting real bad at this point. . . . I hit the button and nothing happened . . . then when I moved my finger over and got it on the button, it went [at] about sixty-seven and a half seconds." Knight said of the jettison, "When those tanks go, it is the loudest bang and jolt I have had in a long time. . . . The airplane jumps right to zero g, and it jumps hard at first, and then steadies out on zero very easy."

What Knight did not realize at the time was he came very close to having the rocket engine shut down, which would have aborted the flight. This may help account for the jolting he experienced. As the tanks were being propelled away from the X-15, the engine had burped. With the ammonia tank going dry, and the slight delay in the ejection sequence, the fuel depleted momentarily before the switch over to the internal propellant tanks. The pumps got the fuel flowing fast enough so the engine thrust only slumped slightly before regaining full power. Pete said, "It sounded like the whole airplane blew up. I figured, well, it had to be the tanks [since] nothing else happened."

The two tanks tumbled, then the recovery systems deployed properly. A wind out of the west pushed both tanks east of the ground track into the Cactus Range, a ridge of volcanic mountains about three miles from the planned impact area. The extra weight of the oxygen tank brought it downrange from the jettison point, while the ammonia tank actually floated backward somewhat along the track. The inaccessibility of the recovery area hindered their removal. Three weeks later they were still awaiting a helicopter to fly uprange to bring them back to Edwards.

While the tanks floated earthward, Knight continued on his accelerating flight, achieving Mach 6.33, at 136 seconds after LR-99 ignition. As the aircraft heated, Knight experienced the expansion of the Inconel-X, saying,

"Just before we got to [Mach] 6 it was banging and popping quite a bit. I think three is the most I have heard on the other flights, and on this one there [were] probably six or eight bangs and pops."

Soon, he was lining up for landing on the Rogers lakebed. The lower ventral fin that had been destroyed on the previous flight was replaced with a new one. It was jettisoned, only to be damaged on impact when a peroxide fire ruined its parachute. The flight had finally proven the external tank system was a viable way to extend the burn time and thus the overall speed of the x-15.

With the tank system now qualified, there was just one element left to test before the final push for the envelope expansion: the scramjet. This small, supersonic combustion ramjet engine was envisioned as a way to sustain hypersonic flight for much longer periods than could be achieved through conventional rocket engines, such as the LR-99 used on the x-15. The problem was that the scramjet was behind schedule, so the x-15 was to have only an instrumented test shape mounted beneath the fixed portion of the lower vertical tail. From this, airflow characteristics, along with temperature and pressure readings, could be studied while the actual engine continued development. The shape was a stubby cylinder, about half the length of the ventral fin. It had a long, tapered shock cone on the front, doubling its overall length.

Six months after his previous mission, Knight made the flight with this scramjet model but without the external fuel tanks. In addition to the scramjet, ablative materials were installed in test sections, including around the ventral area.

A special two-piece cover, called an "eyelid," was fitted over the left-hand cockpit window to test its aerodynamic qualities. This was necessary to keep the window clear after expected sloughing off of the ablative planned to be installed for the buildup to the Mach 8 flights. As the x-15 slowed into the landing pattern, making a left-hand turn onto approach and touchdown, the eyelid was to be opened, giving the pilot much-needed visibility during this critical final portion of flight.

Once Knight flew the test, it was discovered that opening the eyelid created a canard effect, in essence, providing a tiny wing on the left side that pitched the x-15 upward, while rolling and yawing the aircraft slightly to the right. It was annoying, but something that could be lived with, considering the importance of being able to see the ground during landing.

Since the scramjet extended even further below the main fuselage of the x-15 than did the jettisonable lower ventral, the test shape also had to be ejected before the airplane could safely land. The ejection occurred as planned; however, the apparent jinx of being jettisoned from the x-15 went unabated when the cable between the parachute and scramjet detached, and the expensive model impacted the desert. Post-flight analysis showed that, although severely damaged, it was repairable. In all, it was considered a "satisfactory" test.

With the length of time required for the extensive engineering studies on the external tanks, the refurbishment of the scramjet model, and preparations for adding the ablative coating, Knight was temporarily given a respite from flight testing the x-15a-2.

This first mission for Pete following the initial external tank and scramjet tests was certainly a lot more dangerous than what had come before. As Bill Dana put it, Pete Knight "departed the atmosphere, reentered, and made a 180-degree turn, and landed without a single electron in the airplane." His description may be a slight oversimplification, but it does give a perspective to what Pete accomplished on the morning of 29 June 1967, on flight 1-73, the 184th of the x-15 program.

Knight's flight in x-15 no. 1 was planned as a repeat of the one Mike Adams had flown in June when the Western Test Range experiment malfunctioned. It had taken Adams three aborts before he launched successfully, but Knight was able to cut loose on his first attempt after becoming airborne. The b-52 took off at 10:22 a.m. from the Edwards runway.

Maj. William G. Reschke Jr. and Lt. Col. Emil T. "Ted" Sturmthal piloted b-52 no. 008 toward launch altitude, while heading nearly straight north for the Smith Ranch Dry Lake area in north-central Nevada. The launch aircraft, with Knight riding in the x-15 under the right wing, passed west of Smith Ranch, skirted the edge of Edwards Creek Valley Dry Lake, then made the 180-degree turn to head back south toward Smith Ranch. The b-52 lined up for the x-15 drop as they passed to the northeast of the launch lake.

As the airplane came into position, the launch panel operator, Jack Russell, topped off the x-15's liquid oxygen, while Knight pulsed the ballistic control system rockets, performed the propellant jettison checks, and readied himself to launch off the bomber's pylon. Just prior to 11:28 a.m., he counted down and the x-15 quickly dropped away.

The LR-99 engine kicked into life, pressing Knight back into his seat. NASA-I called out, "Rog, Pete, got a good light." Pete was busy setting up the flight profile, leaving the conversation one-sided coming from the ground, "Watch your heading. Track is looking good. Coming up on profile. . . . Beautiful track, Pete. Up to 80,000 [feet]. . . . Profile's looking real good." Then, at just under sixty-eight seconds, Pete said his first and only word recorded after launch, "Shutdown." NASA-I responded, "Understand shutdown, Pete. We've got a Grapevine [Dry Lake] time here. . . . How do you read, Pete?" No response. The airwaves from the x-15 were silent.

Johnny Armstrong, an x-15 flight planner, was in the control room monitoring the flight. He shared his thoughts, saying, "It was one of those situations where everything quit . . . all telemetry, radar data, we lost everything at that point in time." Armstrong believed the worst in those first moments. "With the history of Scott Crossfield blowing the ass end off of no. 3 out here [on the LR-99 test stand], to me, Pete just blew up."

No one in the control room or the chase aircraft truly knew what happened with Knight. Pete later explained what was transpiring, unbeknownst to his ground controllers: "As I went through about 100,000 feet, the engine quit. At the same time, every light in the cockpit came on. A few seconds later, all the lights went out again and it got extremely quiet. I'd had a double APU shutdown, and, of course, without the APUs there is no power in the airplane."

This type of total electrical failure had never occurred on an x-15. The best guess was that if such a situation arose, the pilot could not land safely. He must eject. Complicating matters was the exact time at which the electrical fault occurred. Post-flight data showed electric failure 67.6 seconds after ignition, almost exactly the breakpoint that determined the emergency lakebed location. At sixty-nine seconds the lakebed changed from Mud to Grapevine. Based on the actual shutdown moment, Pete's decision was to aim for Mud Dry Lake outside Tonopah. On the ground, nearly two seconds elapsed between Pete's "Shutdown" call and the controller's reaction. From their perspective on the ground, Pete should have gone to Grapevine, approximately twenty miles west of the Beatty tracking station.

With no radio, no one could ask Pete his intentions, not to mention the idea that maybe there was no Pete to answer. Based on the stopwatch, the chase planes were dispatched to where he was expected to try to bring the airplane down: Grapevine. Aboard the x-15, Knight was fighting to keep the x-15 under his command.

I was a little over Mach 4, and the airplane was wallowing around. I still had reaction [jets], so I maintained some control. At one time I was up about seventy degrees bank angle, and I looked down at Mono Lake, and I said, "You'd better enjoy the view because that's the last one you're going to see."

I really wasn't in a hurry to start the APUs because, as I remember, they have to spin down about twenty or twenty-five seconds before you could restart. I got the emergency battery on and was just trying to maintain control. . . . My thoughts were that this is going to hurt regardless of what I did. If I bailed out, we were over Mach 4 and that's the limit on the [ejection] seat. . . . So I thought, "Well, I'll stay with it and see what happens, see how long I can control it. Once I lose it, then I'll bail out."

As the aircraft went over the top, I started an APU, so I had one running. That gave me hydraulics to control the reentry. . . . I got the airplane level and started pulling the nose up. I didn't have any instruments, [and] there were no lights in the cockpit. . . . As I entered the atmosphere, and the recovery was taking effect . . . I turned around and pulled about six gs to go back to Mud. I didn't have any flaps. All I had was airspeed and altitude once I went subsonic, and the g meter.

Director of flight operations was Stan Butchart. He was in the control room as the test director, and thus in charge of all that was happening. Like everyone else, he was in the dark. "[Pete] was at one of these marginal areas when he disappeared. We didn't know where the devil he was going to go. Word went out: 'We lost him.'"

Knight said it was the familiarity of seeing Mud Dry Lake at the critical moment of flight that helped in his decision to stay with the airplane. He really did not like Grapevine and felt much more comfortable using the lakebed at Mud.

With the unique energy signature and speeds of the x-15, just being able to see a place to land does not necessarily mean it's possible. As Johnny Armstrong explained, "It was visible to him, and, of course, that's the hazard of the x-15—you can look down, but you can't necessarily land where you're looking, because you've got too much velocity built up. But in this case, [Pete] figured right."

Fellow x-15 pilot Bill Dana, who had been flying Chase 1 for launch, was heading south to return to Edwards, when he first saw Pete as a fast, black object approaching Mud. Mystery solved. Dana followed Pete onto the lakebed and called for the c-130 to support the emergency landing—the last hazardous landing in the x-15 program, as it turned out.

With only a single APU running, and the main electrical bus line dead, Knight had no flaps to help slow the x-15 and no dampers to smooth control stick inputs at touchdown, making the airplane "a little squirrelly," as he put it afterward. "I settled in and got it right down on the runway." The x-15 slid 9,050 feet before finally coming to a stop. Knight's travails were not quite yet over, as he explained, "I opened the canopy, and by this time Bill [Dana] had come by so I waved at him that everything was all right. I was in no hurry to get out and was kind of relieved to be on the ground. I was sitting there, and I heard this gurgling noise in the nose and couldn't figure out what it was, so I said, 'Well, maybe I better get out of this thing.'"

Quickly, Pete popped the canopy, but he was still firmly attached to the aircraft. "I had everything unstrapped [except] the leads . . . that supplied nitrogen and oxygen to the suit and helmet. I couldn't get those undone. In the suit things were beginning to get a little warm, and I still had this gurgling going on up in the nose." He fought harder to get disconnected. "I decided if I couldn't get those [leads] off I was going to blow [them off]. I pulled the emergency release handle, which also blows the headrest off the seat. The headrest went up and hit the open canopy. When it hit, it went forward—and my head was in the way! It just ricocheted off my head!"

Pete explained the problem with the headrest was all his own, saying, "The reason I couldn't get those leads off was that I hadn't turned the system off. There was still seventy pounds of pressure holding against that fitting. All I had to do was run off the pressure, and it would have come off real easy, but in all of the excitement I forgot all about that."

The c-130 support crew landed and saw Pete's head bloody from his boxing round with the seat headrest. They patched him up to get back to Edwards, where he was met on the ramp by Johnny Armstrong. Johnny said, "I was the only one to greet Pete. He got off, and kind of ignored me, then walked on into the building. He had blood dripping down his head from being clobbered by that seat. 'Pete, what happened?' [I asked.] 'Arrgh, ar-rgh, arrgh,' and he kept going." Pete went to the hospital, where they put in six stitches.

Everyone learned a lot about what the x-15 could do that day. Because of the difficulty Knight had in disconnecting from the cockpit seat by for-getting to dump pressure in the suit leads, it also highlighted that all x-15 pilots needed some remedial training in aircraft egress. Part of the official response to flight 1-73 was, "In the future all x-15 pilots will make an un-assisted practice exit from the x-15 on the ground with complete pressure suit connected and canopy closed."

For all that happened, Knight was awarded a Distinguished Flying Cross for conducting a flight profile in an electrically dead x-15, a feat no one ever thought could be done. He saved the aircraft from certain destruction if he had decided instead to bail out. Everyone's instincts said that's what Pete should have done. Bill Dana agreed, "That is truly impressive that you could have an airplane so redundantly designed that it would do all those things without any electrical system working."

Pete Knight said what all the pilots who flew it knew: "I think that in every flight there was some sort of emergency. You can probably list on one hand the number of flights that went without a hitch. . . . My first flight, I had an apu shutdown. Mike Adams, on his first flight, landed at Cud-deback. . . . There was always something that was a minor emergency, or sometimes a serious emergency. It didn't always go as planned, but that was one of the reasons the pilots were in there—to take care of those kinds of situations and recover the airplane."

The operations engineer for x-15 no. 1 at the time of the electrical fail-ure was Meryl DeGeer. Being in control of that vehicle, he was responsi-ble for all aspects of its configuration, maintenance, and handling. When the emergency landing occurred, and because he was fairly new to the pro-gram, he felt he should head up to Mud to see how field operations were handled in person. "I went up and rode back with the guys" he said. "And that was quite an experience. The ground crew had been doing that each time they had to bring an airplane back, [so] I said, 'Hey, I want to do it. I want to go with you.' They said, 'You're crazy, but here's a seat.' I found out what it was like to drive through Nevada . . . at 110 degrees [Fahren-heit], with no air-conditioning. I leaned over to talk to the guy beside me and laid my hand on the drinking water can and burned it!"

Getting the aircraft ready to transport meant verifying the safety of the vehicle and its systems only to the extent necessary, so as much could be preserved for review of the malfunction as possible. The upper vertical tail was removed, and a lot more work done before the x-15 could be hoisted by a large crane and loaded onto a long, flatbed trailer. "There were some people who got really good at . . . going up there and retrieving an airplane," Meryl said. He laughed when others told him how sometimes a crane operator not experienced in the x-15 did not understand how heavy the airplane was until he found out firsthand. "I'd sit there and watch it happen. A guy comes out and takes a look at the airplane, and he says, 'This'll be a piece of cake.' They get all hooked up to the sling, and he fires up—then practically kills his crane! He comes up against it, and it just doesn't budge. He's looking at it like it's an aluminum airplane, and it's going to be easy to lift onto the truck." After a few tries, the operator finally caught on that the x-15 was not a normal airplane.

Once the convoy rolled away from Mud, they had to cross the lakebed, then manage more than ten miles of back-country dirt road to get to pavement. Meryl continued, "We came out of Mud Lake and went to Tonopah to get to a highway. Then . . . down through a bunch of country that I've never seen since. They were making like eight miles an hour up some of those grades with the truck . . . driving on very narrow two-lane roads. If you met somebody, you practically had to get off in a ditch to keep from taking the top off of their vehicle. . . . I remember we got passed by one car, and we all noticed they had their windows rolled up. That meant they had air!"

About the time of sunset, the convoy was outside of the town of Baker, then, "we turned and went across the end of Silver [Dry Lake, and] went through the gate into Fort Irwin. . . . The sun was just setting, and there were all these tanks sitting and facing each other all down the road. . . . I had the feeling we were intruding on somebody's war! We pulled in there and parked. They brought so many [military police] out to guard the truck. We went into Barstow and got a motel, spent the night, got back out early the next morning to finish the trip home."

Once x-15 no. 1 was returned to Edwards, the job became one of finding the root cause of why the electrical system had failed. Stan Butchart assigned Vince Capasso to head the investigative group. "Vince and I had both [been to] a training class that [was] a problem-solving type of thing," Butchart recalled. "The basics they were trying to teach was that, so many times when you have an accident, you go out and you just shotgun; you look at everything. Well, they had a system of going back . . . far enough to find out where was the deviation . . . [and] what caused that deviation? I told Vince, 'This is a good time to put your training to use.'"

The initial focus was on the auxiliary power units. They had caused numerous problems over the years on the x-15, but nothing was found that pointed to that system, so the digging continued. Harry Shapiro and Gene Miller, North American Aviation electrical engineers, were asked by their boss to participate in the investigation. Harry said, "We flew up to Edwards every day, and we had them run different checks. We did that for a whole month. They'd fly us up in the morning, and they'd fly us back at six o'clock every night."

Even after NAA concluded its part of the investigation, engineers at NASA kept at it. More than a month after the emergency landing, a series of altitude tests were run where evidence emerged that the Western Test Range experiment, created to monitor a Vandenberg missile launch, was arcing and blowing fuses inside the x-15.

The altitude chamber testing showed the arcing happened at roughly 100,000 feet, almost exactly where Knight's engine shutdown occurred. The accident report stated, "Power transients [in the simulated altitude tests] resulted in the blowing of two fifteen ampere fast-blow fuses in the lab wall power." So, not only was it enough to shut down the experiment, but as Butchart noted for flight 1-73, "The bad thing there was it shorted everything in the airplane!"

Even after the experiment was returned to its manufacturer, Nortronics, and after the unit was supposedly brought back up to flight specifications, it continued to fail. The decision was made to ground the experiment until Nortronics could attest the problem was fixed and no longer a hazard to the research pilot or the x-15.

Several systems were redesigned to help make the airplane safer. The wiring was rebuilt for the APUs, the experiments' electrical requirements were moved off the main power bus, and important systems such as the flaps were connected to a new heavy-duty emergency battery. If this type of failure was to somehow happen again, the pilot would have more control and thus more options. For NASA personnel like DeGeer it was back to work and onto the next research flight. At NAA Harry and Gene earned a vacation. "We worked night and day, and my supervisor said, 'Hey . . . you worked hard for the last four weeks; I don't want to see you around here for the next four weeks.' He gave [us] four weeks off with pay. That's the type of organization we had."

Pete Knight spoke of the heat on the X-15: "The temperatures are about the same as [those experienced by the Space] Shuttle, but it's a matter of the impulse, how long you're there. The longer the impulse, the more problems you'll have. It's like running your finger through a flame—the temperature is there, but unless you hold it for a period of time, you don't get its full impact."

The idea, as far as the X-15A-2 was concerned, was to use the platform for aerodynamic research, as well as experiments at much higher Mach numbers. This also brought increased temperatures, possibly more than the original Inconel-X airframe design could withstand. The proposed scramjet required velocities in the Mach 7 to Mach 8 range for proper function. "The first [external] tank flights," Pete said, "weren't at those high Mach numbers long enough to give us a problem, but as we went to the higher Mach numbers—we were above Mach 6 for a longer period of time—the heat impulse was becoming longer and longer." For the airplane to survive, a method had to be devised to protect the Inconel-X from failure.

What was settled on was an ablator product from Martin, designated MA-25S. Covering the airplane in a protective cocoon, it was manufactured by mixing a resin base and a glass-bead powder with a catalyst. When completed it had the consistency and feel of a pencil eraser and, interestingly enough, also looked like that same material, right down to the coarse feel and pink color!

Taking this sort of material and applying it to a relatively smooth, hypersonic surface was also going to greatly increase drag, not to mention the weight penalty added to an already very heavy rocket plane. When all these factors were placed onto the performance curve, the maximum speed was limited to about Mach 7.4. This equated to approximately 5,000 miles an hour—nearly 800 miles an hour faster than an unprotected X-15. After the speed run on flight 2-50 to Mach 6.33, Knight said, "We decided that we'd better get the ablative on for the work after that. We put the airplane down and put the ablative on."

The Martin ablator was manufactured so the majority of it could be applied by spraying it onto the aircraft in a manner similar to a car getting a new paint job. The X-15A-2 was towed into a special application building, where it remained for five weeks as the process was accomplished. The surface was cleaned and prepped, then technicians in protective suits and masks carefully sprayed the material onto the outer skin to the calculated thickness for any given area. The thickness was a direct function of how high of a heat load was expected.

This was Knight's program; no other X-15 pilot was assigned to share the A-2's flight testing duties. He talked of watching the process unfold, saying, "The leading edges were a high-density material, formed, molded, and glued on all the leading edges of the airplane. The rest of it was sprayed on and sanded down to a predetermined thickness. It started out about three quarters of an inch, as I recall, on the leading edges and tapered down to probably a quarter of an inch on the trailing edges and the top of the fuselage."

There were two initial problems with the ablator. Pete explained, "The first thing we found was that this material, in the presence of LOX [liquid oxygen], if it got into a valve and the valve closed, it would detonate." Explosive sensitivity to oxygen, Pete understated, "wasn't acceptable in the fuel system." In addition to these filters, the team decided to paint a white wear layer over the pink ablative for protection against accidental detonation and to smooth out the coarse surface. In doing so, this white paint eliminated a major difficulty for Knight. He stated emphatically, "It was pink, and I wasn't going to fly a pink airplane."

The entire process making the X-15A-2 ready for flight took something on the order of 2,000 man-hours. This labor intensity showed right from the start how the whole idea of the reusability of a spray-on ablator was going to be called into question.

When the stark white rocket aircraft was rolled out of the paint shop, it was a huge contrast to the years of seeing the three deep black X-15s at the NASA center. Only a few bright red and yellow caution and rescue decals, with some black stenciled lettering for vents and jettison lines, were placed on the outer layer. All markings denoting the air force and NASA lineage, as well as identifying aircraft numbers, were deemed unnecessary. The bare metal of the Q-ball nose, the rear of the fairings and tail surfaces, and the engine itself were the only areas not covered by the pink ablative and snow-white paint. The external tanks were added for completion. On 4 August 1967 the aircraft was towed in front of the NASA hangar, where Pete Knight came out to take a look, gave his approval, and posed for some publicity photos. Running his hand over the surface of the pristine A-2 looked very much like he was patting down a favorite stallion.

Within a few days, the A-2 was ready to become airborne for the first time, if only for a captive flight to altitude on the B-52's wing. This was to cold-soak the x-15 structure at drop altitude and test all systems prior to the first attempted launch. All went well, so after landing the tanks were removed.

Two attempts at the first flight with the ablative were frustrated with minor problems, understandable considering the radical changes to the aircraft. Pete explained what happened, saying, "We'd changed not only the configuration of the airplane but also the concept of the airplane. The basic x-15 was designed to run as a hot structure, and that meant it absorbed the heat buildup due to the Mach number. Now we had changed that and made it a cold structure, and we didn't know if that would generate problems or not. So we flew one flight without the tanks and just the ablative [and dummy scramjet] to about Mach 5. There were no problems with it. It worked very well."

Later, it was shown this was not fully the case. Localized heating problems from hypersonic shockwaves were identified but not deemed serious enough to alter the flight research plan. One unexpected bonus was discovered after the 21 August 1967 flight of mission 2-52. The scramjet, even though it created a nose-down aerodynamic change, acted very much like the lower ventral that had been removed, in that it improved directional stability. As Knight approached the lakebed for landing, the scramjet was jettisoned and yet again sustained damage.

Post-flight, the ablative was examined and touched up, the scramjet refurbished once more, the external tanks mounted, and the advanced x-15a-2 was ready to be loaded onto the B-52 wing pylon. This first flight of a new research series was envisioned to extend the speed range into territory undreamed of by the original x-15 designers.

Flight planner Johnny Armstrong worked closely with Pete Knight to map

34. On the afternoon of 3 October 1967, Pete Knight prepared the fully loaded x-15a-2 (with external tanks and ablative coating) for the highest-speed flight of the rocket plane program. The vapor at the rear of the aircraft is from the jettison checks performed just prior to launch. Courtesy of the Armstrong Flight Research Center.

out flight 2-53. After the debacle of Rushworth's flight, when he had to jettison half-empty external tanks, this time around they went the other direction and tried to guess every possible scenario. When it was all done, Knight said, "All of those things added up to very significant emergency procedures . . . [so] I broke the flight down into sections. . . . It was a busy flight."

Knight climbed aboard the x-15a-2 on Tuesday, 3 October 1967 and was strapped into the cockpit. Soon, he was performing pre-flight checks and found his helmet radio was inoperative. A replacement was quickly brought out. Another problem was a "persistent dribble" coming from the ammonia fuel jettison valve. Pete joked in his post-flight report, "Well, from the time the whole operation started this morning I figured it would be a good captive flight. As time went on, I began to realize that everybody is getting serious about this thing, and I tried everything I could not to go."

Fixing these problems postponed the takeoff of b-52 no. 008 by ninety minutes, to 1:31 p.m. Col. Joseph Cotton and Lt. Col. William Reschke got off the tarmac and headed north to Mud Dry Lake. Almost exactly an hour later, more than 8 miles above Nevada, the countdown reached zero. Pete attempted to launch his rocket plane, but the x-15 stayed firmly attached. He explained in his report, "I reached up and hit the launch switch and immediately took my hand off to go back to the throttle and found I had not gone anywhere. . . . I reached up and hit it again, and it launched the second time. Launch was very smooth this time. It was one of the smoothest launches I have had."

The LR-99 accelerated the x-15 quickly, and Knight rotated upward on a flight path fifteen degrees above the horizon, only to find the airplane buffeted, so he flattened out slightly to thirteen degrees, continuing to accelerate. "I began to look around at other things," Pete's report continued, "and we came up on the tank jettison. . . . I went to internal [fuel] and got a good five degrees [angle of attack] for tank jettison, and hit the button. . . . The tanks went awfully hard. They came off immediately."

Except for the hard jettison, everything went according to plan as the external tanks ran dry. Underneath the ablative the x-15 skin remained unnaturally cool—as low as 34 and as high as 310 degrees Fahrenheit at various spots along the fuselage. Another minute passed, and Pete shut down the rocket engine. He continued to believe the flight was going according to plan.

Unbeknownst to him, the hypersonic shockwaves were playing havoc at the rear of the rocket plane. The scramjet created flow patterns that burned through the ablative coating on the leading edge of the ventral, to which it was attached. Bare Inconel metal was exposed to a boundary layer heated to 2,400 degrees. This was nearly double the original design temperatures expected at Mach 6, and the skin could not withstand it for any length of time. Burn-through occurred quickly, which then allowed the hot gases to start destroying the interior of the aircraft.

A peak velocity of Mach 6.7, or 4,520 miles an hour, was a record for the x-15 program. If sustained this speed meant the airplane could cross the entire United States in under thirty minutes. This flight also could have ended in tragedy if the velocity had been maintained for only a few more seconds.

More than thirty-five years later, the Space Shuttle *Columbia* experienced a similar problem when a hole was punched into the leading edge of its left wing by foam insulation soon after launch, allowing superheated gases to burn through the breach in the wing during reentry, dooming the vehicle and the seven astronauts aboard. Although the actual temperatures experienced by both craft were similar, the difference in the case of the x-15A-2 that afternoon was primarily that the heat did not last long enough to cause full structural failure of the vehicle. Coming back from orbit, *Columbia* was not so lucky. Just like with Rick Husband, commander of the ill-fated *Columbia* mission, Pete Knight was flying the planned profile on the A-2 with no knowledge of hazardous events happening elsewhere on his vehicle.

Knight was interviewed by the Smithsonian National Air and Space Museum for a television special on the x-15 and described the sensation of speed he encountered on this flight. "I think it's akin to driving a car. You can have a car run at 70, 80 miles an hour, and it's rather comfortable. If you push that same car to, say, 100 miles an hour, or 110, you know that you're pushing the limits of that car. I think the x-15 was in the same category that, as you begin to go above Mach 6, and you begin to push out farther and farther on the envelope, I think we were well aware of the fact that things are getting a little bit more critical."

At more than 4,500 miles an hour, Pete recalled, "You were aware of speed, and if you looked outside, as I had a chance to do fleetingly from time to time, you knew that you were moving across the ground. Even at 100,000 feet you were very much aware of the speed. . . . There was not much time for looking around or joyriding."

The LR-99 shut down after 140.7 seconds, the vehicle quickly decelerated, and the heat began to diminish. The damage, however, continued to grow. The full burn-through of the ventral occurred sometime approximately twenty-five seconds following shutdown. At that point, all telemetry from the scramjet ceased as the electrical connections were destroyed, along with the jettison lines for the fuel tanks, which were also in this area.

As Knight returned to Edwards, he tried to jettison the remaining fuel, but nothing happened. Chase pilots confirmed his situation. Pete attempted to reset the switches, but he did not realize he couldn't change the situation. His landing was going to be extremely heavy.

Inside the ventral, three of the explosive charges that released the scramjet were ignited, and the fourth failed mechanically under the added strain. As Knight entered the landing pattern to Rogers Dry Lake at 32,000 feet, the instrumented scramjet test shape fell away from the x-15, landing unseen, yet appropriately enough, in the Edwards bombing range.

"I didn't know I had lost the scramjet," Pete said years later. "Of course, I didn't know the condition in the lower ventral as to what the heat had done. . . . The flaps didn't work on that flight, because the temperature had gotten into the wing area and had gotten to the flap motors . . . but all that meant was that we kept the angle of attack down on landing and kept the speed up."

Once the A-2 slid to a stop, Pete quickly popped open the canopy, expecting the ground personnel to come and help him from the cockpit. Instead, no one was paying attention to him. "I didn't realize what had happened . . . until I got out, and I noticed everybody looking at the back end of the airplane. I walked around and looked at it and said, 'Oh, Holy Christ.' It was in bad shape."

Pete checked out the entire airplane and shook his head at the damage. "You could see where the bow wave off the nose intersected the wing and caused another hot spot. The shock wave interference that took place around the scramjet was where we had all the problems. . . . It looked like you had a blowtorch down there." Pete reiterated what might have happened: "I've often said, if I'd known it was the last flight, I would have gone to Mach 7. If we had, we may not have made it because of the temperatures that were getting up into the engine bay. I might have lost the airplane." Pete neglected to mention he might not have returned either.

While the damage was being inspected, one missing piece of the puzzle was, what happened to the scramjet? It was easy to tell by looking at the extensive ventral damage, that the engine model had not come off when expected. Pete's first thought was that somehow it had separated at the same time the tanks had been jettisoned. His flight planner, Johnny Armstrong, was the one who solved the mystery.

Sitting in his office at Edwards one afternoon, Armstrong explained his deductive process: "The clue [was] to go back and look at the data. . . . At about Mach 4, the instrumentation ceased. Our first thought was that it came off at high speed, but the more we got to looking into it, we finally concluded [differently]. . . . You could take the radar data, go back in time, and figure out where he was physically . . . and then kind of guesstimate on a map where that scramjet might be."

Johnny grabbed Bill Albrecht, who was the ops engineer for the aircraft,

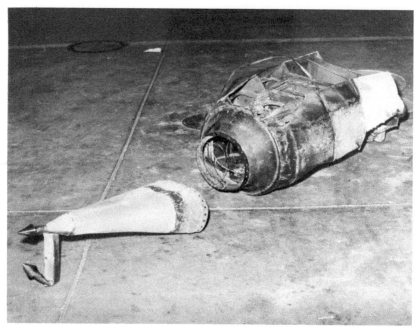

35. The instrumented scramjet model was burned off the lower ventral of the x-15a-2 during flight to Mach 6.7 on 3 October 1967. Johnny Armstrong led the team to find the wayward piece of equipment that fell in the Edwards bombing range. Armstrong kept the bent double spike seen at the lower left in this photo as a souvenir of the flight. Courtesy of the Armstrong Flight Research Center.

and headed out to the bombing range in a borrowed truck. "[We] went up a road, and I had marked where this thing might be. I had a checkpoint back toward the mountain range and back out toward the rocket base, and kind of swung that arc and said, 'Okay, Bill, you go that way to the north, and I'll go to the south.' I walked right up on the damn thing, half buried in the ground! I was just amazed and couldn't believe it."

After telling his story Johnny reached down to open the large bottom drawer on his government-issue desk and pulled out a bent slab of metal with two arrow-like cones at each end. He exclaimed, "Yeah! And there's the piece." I could see the pride still there after all the intervening years, as he handed over the battered piece of scramjet wreckage for inspection. "That was the little cone right there on the nose," he said, gesturing. "This is the ablative material that got charred up. This thing's supposed to have a ninety-degree angle, and now it's all bent. I just saved that right there in my desk."

Once the smashed scramjet was trucked back to NASA, it was immediately evident that, this time, the model was not going to be repairable. This actually made little difference, since that program was shelved before the x-15 itself ended.

Bill Albrecht, who had helped Armstrong retrieve the test shape from the desert, talked of what happened to the scramjet program. "They had a lot of development problems. . . . They got into laboratory work, and I think they demonstrated in some experiments that supersonic combustion was at least feasible. However, the prospect of getting a flight-rated engine to go on the x-15 grew very, very dim as we got further on."

Besides the scramjet, the idea of an ablator that could simply be sprayed on any surface was a fascinating prospect for a future craft returning from space. The idea was that whatever amount charred away during reentry would just be filled in, sanded down, and the next launch would be ready to go. Johnny Armstrong pointed out, "It did the job it was supposed to do. [However,] it was a real nightmare to try and maintain. I don't think anybody fooled themselves that it was going to be easy. It works, but you sure would like to have a different way to protect an airplane from hot temperatures."

When the Space Shuttle was designed, and the aluminum airframe was covered with tens of thousands of silica tiles glued in place, many people lambasted NASA for coming up with this supposedly bad idea. These critics of the tile system had not read the history of the x-15 and seen what a nightmare the alternative might have been. In the end, even with the maintenance headaches of the tiles, NASA had obviously made a wise choice, using the technology available at the time. Covering a fifty-two-foot-long aircraft with spray-on ablative for two flights was one thing, but worrying about dropping a wrench onto an airliner-sized spacecraft with an oxygen-sensitive ablator during 135 flights of the Space Shuttle would have been another disaster waiting to happen on an already fragile vehicle.

Immediately after the high-speed flight of the x-15A-2, no one had any idea this would be its final flight. With great hindsight, many of the engineers involved were kicking themselves over how they could have missed something that appeared so obvious as the shockwave impingements. Everything exposed to the hypersonic flow was going to produce its own individual shockwave. This included the antennae, speed brakes, and interaction of the scramjet shape with the lower ventral. It is interesting to note that on the test flight just prior to the Mach 6.7 high-speed flight, scorch marks on the white ablative pointed to exactly where problems would occur, yet the warning went unheeded.

The way the ablative and scramjet flights were planned was very different than the way the original research program had been conducted. During the initial buildup of speed and altitude, each incremental step was taken carefully. For example, as the speed envelope increased, the plan was to do so approximately one-half Mach number at a time, to not go too far too fast and thus encounter a defect that could not be corrected. This lesson served well for the safety record of the X-15, but then appeared to be abandoned late in the program.

When the A-2 was first conceived, it was simply considered an upgrade, so the airframe was cleared for its full regime almost immediately. Flight 2-50 set an unofficial speed record of Mach 6.33, which then led to adding the scramjet test article and ablative coating. The scramjet test flew to Mach 4.75 without difficulty, then the ablative was added and flown to a similar speed of Mach 4.94. As mentioned, impingement scoring on the ablative was noted after this flight, but not enough weight was given to the evidence. So, on the very next flight, instead of using the previous X-15 strategy of moving up half a Mach number to Mach 5.5, with ablative and scramjet in place, the decision was to increase speed 25 percent to more than 4,500 miles an hour. It should have taken three or four tests with this configuration to get to that point, yet no one apparently thought what they were planning was unwise.

No explanation has ever surfaced for this casting aside of safety measures. The hubris shown at this point can easily be compared to that exhibited in 1986 and again in 2003, with the losses of the *Challenger* and *Columbia* Space Shuttles. The process is called "normalizing deviation" and means that since nothing has gone wrong previously, everyone and everything is safe from harm in the future. It is a hard lesson that went unheard, and the price was paid decades later on the Space Shuttle, but also not long after on the X-15 itself.

Even with the problems that occurred on that final flight of the A-2, most of those involved with the X-15 program felt there was still a lot left to do. Knight said, "I would have liked to have seen it go to a higher altitude just for the sheer pleasure of it, although there probably wouldn't have been any

36. Pete Knight receives the Harmon Trophy for his Mach 6.7 flight from President Lyndon B. Johnson. The ceremony was held at the White House on 3 December 1968. Standing with Pete are his wife, Helena, and two sons, William and David. Courtesy of North American Aviation.

technical justification for it. And, of course, I would have liked to have seen it pushed out to Mach 8, but, again, we didn't have a requirement. If you can't generate a requirement, then the risks aren't worth it."

Knight understood that once the x-15 program drew to a close he became eligible for military active duty rotation. For a pilot, at that time, this almost certainly meant he would end up in Vietnam flying combat missions. His orders to do just that came through fairly quickly (by Knight's account, just forty-eight hours after the x-15 was canceled). As orders in the military can often take a while to process, he had some time to accomplish other tasks before he went to join the conflict in Southeast Asia. One duty was to talk about what the x-15 had accomplished. In this role Knight traveled and gave speeches, ending up in places as far as England and Sweden. While in England he met up with a friend, Bill Bedford, and Knight mentioned he didn't want to leave without getting a proper English bowler hat. "So," Pete said, "the next day I came back to the hotel and there was a box, and in it was a beautiful derby they had given me, so I wore it. At that time, I was wearing a herringbone sport coat, and they decided they were going to make an Englishman out of me." He was even given a briefcase and umbrella to complete the look. "I had a lot of fun wearing that hat around. It was cold. It was in January, I believe. I had a black top coat that I was wearing also, and I'd wear this bowler and it was funny the reaction that you'd get, because you'd go up to the airline counter at Heathrow Airport . . . and you could see the difference in how I was treated." Soon, the fun was over, and he was told to prepare for his transfer to Vietnam.

Since Pete had been out of the active air force loop for several years while on the x-15, he was expecting orders returning to training before he could fly in the war. Knight, and other air force pilots in the x-15 program, had worried about this scenario and had even gone so far as to suggest that during the rainy season at Edwards, they should be allowed to go overseas and fulfill their requirements for combat flying a few months out of each year. It was felt this would be too disruptive to both the x-15 program and to the commanders they would have to work under in Vietnam, so the idea was quickly shelved as unworkable.

Knight also felt that sending him back to school was a waste of time. He told his superiors, "I'll go in any airplane that you want me to go in, from a Gooney Bird [c-47 cargo airplane] on down, or up, but I'm not going to go to school for six months to learn how to fly an airplane. So they came back and said, 'How about the f-100?' And I said, 'Fine.'" Soon after, Knight reported to Luke AFB, west of Phoenix, Arizona. While there he took eleven check rides in the f-100 Super Sabre to get recurrent in the airplane and thirty days later was in Vietnam.

Knight's son, David, was not yet nine when his father was sent overseas for a year of combat flying, starting in September 1969. "I remember missing him a lot," David told me. He went on to say that once his dad left, his family moved to Vermont and rented a house, "to be with my mother's family. We would send each other [audio] tapes that we recorded. We would sit at the kitchen table and all take turns talking to Dad into the microphone. Then my mother would send us upstairs and she would finish the tapes. When we got a tape [from him], we would listen for a while and then my mother would send us [back] upstairs, and she would finish listening."

Pete Knight flew 253 combat missions before he returned stateside. David recalled running up the ramp to meet his father at the airport in Burlington, Vermont. "My kid brother was in my uncle's arms and was afraid of my dad at first because he was so young when my father left for the war." Many children have had similar experiences with their fathers away in the military. Even stateside assignments meant that they rarely would be home every night. For test pilots assigned to the x-15, their time at Edwards may have been the most stable period of family life.

Once Knight had returned from the war, he was assigned to Wright-Patterson in Ohio, attended the Industrial College of the Armed Forces in Washington DC, then went back to Wright-Patterson for a period before finally returning to Edwards as the vice commander.

During his final years in the air force, Knight delighted in still having the ability to fly the latest aircraft, at a time in his career when most officers were glued to their desks. Pete said, "It was an interesting time, in particular working for [Maj. Gen. Philip J. Conley Jr.], because he believed that people like me in command positions had to be flying. I [stayed] current in the f-16 and the t-38 . . . , flew the b-1, and got a couple flights in the a-10 and f-15. I was flying test cards, pulling my nine gs, along with the rest of them. I had a few emergencies in the f-16, but that was fun. I think the fact I was still flying made it a good job."

After retiring in October 1982, Pete was anxious to keep flying. One such adventure had the fastest man alive working as a consultant with a start-up company, Austin Aerospace, where they modified an ultralight aircraft into military configuration for sale to the U.S. Army. Pete stated the aircraft's mission was "harassment, recon, border patrol, those kinds of things." Although two airplanes were equipped with rockets and guns for test flights out of Mojave Airport and a demonstration series at Quantico, Virginia, the idea never got anywhere, possibly because of a crash of one of the test aircraft on 22 June 1983. Knight was quoted after the accident as saying, "One thing I've learned and that's that flight testing is the same whether you're flying forty knots, 400 knots, or 4,000 knots." This accident may have been the final instigation for Pete to set aside his love of flight testing to head onto a completely different endeavor: politics.

Less than eighteen months after leaving the military, Knight was running as one of nine candidates to vie for three seats on the Palmdale, California, city council. With his name recognition among a town still primarily centered on aviation, he handily won his first election. Just four years later, in 1988, he became the first elected mayor of the city. He continued to move up the political ladder in California, first with the State Assembly, and finally as a state senator. It was at this point where he left his most indelible political mark, not only in California but as a stalwart conservative opposing an early wave of social change throughout the country. Interestingly, his views did not seem to stem from his party but rather from his own family.

At the same time Knight was running for his first elected office, his son, David, was a junior in the U.S. Air Force Academy in Colorado Springs. As a father who was watching his son follow in his footsteps, Pete was extremely proud of David and the career path he had chosen. David caught the flying bug and earned his pilot's license at seventeen, the same year his mother, Pete's first wife, Helena, passed away. He earned a sponsorship to the U.S. Air Force Academy and was doing exemplary work, although his father worried that David's eyes might not hold up with regard to being a military pilot.

What Pete had no idea of at the time was that David was also struggling with the fact he was gay, which would make an air force career very difficult, if not dangerous. The "Don't Ask, Don't Tell" policy, finally repealed in 2011, was a military doctrine designed during the administration of President William J. Clinton. It allowed gay and lesbian military personnel to serve only if their orientations were not divulged. Because of the social and professional stigma that being gay carried, David tried not to acknowledge how he was born. As with so many people in this situation, the truth finally won out, and David accepted himself. The problem then came with the air force, and also with telling his father, which he finally did in 1996.

"When I was at the Air Force Academy, or when he was speaking at my pilot training graduation, or when I was returning from the Gulf War, where I flew fighters for my country, my father was very proud of me," said David. "His love for and pride in me, I assume, was because I was his son. I am the same son today. . . . I told him myself, and he was the first person I came out to. Our relationship changed from that moment on."

In addition to his son, Pete's brother, John, had also come out as gay. After seeing this twice within his own family, Pete decided he wanted to do something, and this feeling became Pete's imperative after John died of AIDS the year after David broke his own news to his father. Instead of showing support for his brother and son, Pete chose to use his position as a California state senator to mount a crusade against the gay, lesbian, and transgender community. David once cynically speculated in a 14 October 1999 editorial in the *Los Angeles Times* that his father felt this was a subject that could further his political ambitions.

What came out of this was what became known as Proposition 22, also sometimes referred to as the "Knight Initiative." Its purpose was to outlaw same-gender marriage in the state. As David wrote, "I believe, based on my experience, that his is a blind, uncaring, uninformed, knee-jerk reaction to a subject about which he knows nothing and wants to know nothing. . . . As far I can determine, he's made no attempt at understanding the issue. Therefore, I have a hard time with the fact that someone in his position is attempting to legislate discriminatory restrictions on a significant group of people that he has clearly rejected." It was difficult for David to say such things in public, but he felt he had no choice. In his editorial David continued, "My father seems to want things to remain as he has always known things to be—without change. He can't seem to understand that we, as a society, are growing and allowing more people the opportunity to share in the ultimate dream of happiness. . . . My father's idea of family values is very different from mine. He insists his are right and mine are wrong. . . . I must speak out publicly about what I perceive to be his willful, blind ignorance on this issue."

State Senator Pete Knight released a statement from his office in Sacramento the same day David's editorial appeared in print, saying, "I regret that my son felt he needed to force a private family matter into the public forum through an editorial. Although I don't believe he was fair in describing the true nature of our relationship, that is a subject which should remain between the two of us. I care deeply about my son." Pete's words ring hollow when he talked of this as being a "private family matter." Knight had made it all very public in the first place by deciding to move forward with Proposition 22 as a statewide ballot initiative against the community to which his son belonged.

Even after all the diatribes were exchanged, David never lost any love for his father, saying, "I have always, and still am, very proud of my father's accomplishments. I loved growing up as an air force brat. My father's actions later in life do not change how I [feel] about his accomplishments."

Pete apparently made no further public mention of his brother or son, but he did continue to push Proposition 22, which passed in the March 2000 California election.

The controversy did not stop after his statute became law; instead, the debate intensified. It eventually moved through the California court system, where the state supreme court struck down Knight's law as unconstitutional in May 2008. Although Pete had passed away several years previously, his wife Gail, stepmother to David, was undeterred by the court decision and took up the fight against same gender marriage. She helped spearhead a new ballot initiative, Proposition 8, to prevent marriage equality by amending the state constitution.

All across the country similar initiatives were passed. However, people started to fight back, beginning an upheaval that many have likened to a twenty-first-century civil rights movement in America. It is seen as a last stand for true equality in a country that has spoken of how all its peoples are created equal but has too often fallen short on that promise. Proposition 8, the direct legacy of Pete Knight and Proposition 22, was eventually found unconstitutional by the Supreme Court, thus securing marriage equality for all Americans.

Test piloting is a conservative institution by its very nature and would most likely attract like-minded people. When looked at from another perspective, it's interesting to think these pilots are actually often on the forefront of change, at least within the aviation and space community. These men and women push the boundaries of knowledge and rarely accept that something is impossible.

When the Bell xs-1 was first designed to punch through the speed of sound, a lot of reputable and scientifically knowledgeable people said it could not be done. The "sound barrier" was literally thought to be just that—until the U.S. Air Force and its primary test pilot, Chuck Yeager, put the idea to the test on 14 October 1947. Yeager and the Bell aircraft destroyed that fallacy forever. Pushing the edge is the bread and butter of aerospace designers, engineers, and pilots, but this does not always translate to life in general.

It is an unfortunate fact that for a younger generation who did not grow up with the x-15 and are unaware of Pete Knight, the fastest pilot ever to climb into its cockpit, he will now be remembered instead for his political policies toward the end of his life.

Prior to all this controversy, in April 1991, during a special event at the Rockwell plant in Palmdale for the rollout of the Space Shuttle *Endeavour*, my wife, Cherie, and I were speaking with Pete and his wife, Gail. With an impish grin, Cherie turned to Gail and asked, "How does it feel to be married to the fastest man alive?" Without missing a beat, Gail replied, "I caught him, didn't I?"

11. Chasing Experiments

There were some hare-brained things that went on in the x-15 program like there are in all programs. Capturing a micrometeorite was just one of them.

William H. Dana

It was a Thursday morning, as it was for a majority of x-15 flights. Lt. Col. Ted Sturmthal and Sq. Ldr. John Miller boarded b-52 no. 003 and successfully ran through their pre-flight checklist. In the life support van Bill Dana was helped into his a/p-22s pressure suit, one layer at a time, by NASA supervisor Roger Barniki. It was the 199th time Roger was there for an x-15 drop, and Bill was hoping it was about to become his 16th.

After being sealed into his silver spacesuit, Dana stood and picked up his portable air conditioning unit before stepping out the door of the van. He walked toward the yellow set of portable stairs that had been sidled comfortably into position next to x-15 no. 1's open cockpit, just in front of the right wing of the mothership. The rocket plane had been mated to the b-52's pylon the previous day, and like the pre-flight by Sturmthal and Miller, the x-15 proved itself ready to take flight as technicians prepared the way for Dana.

Bill climbed the stairs, then stepped over the sill and down into the rocket plane's cockpit, hunkering into the pilot's seat. It took about fifteen minutes to be secured by straps and make sure his suit was properly connected to the life support system. Bill then ran through his own checklist, verifying each switch was properly positioned. The ground crew hadn't missed a thing; it was time to go.

The hatch was snapped shut and secured in place. Everyone cleared the area of the last vestiges of equipment, then moved away themselves to a safe distance. Sturmthal called for verification of readiness, then spooled up the eight jet engines before starting his taxi toward the Edwards runway.

Three miles and another fifteen minutes passed as the combination lumbered forward. On the B-52's wing, Dana bounced and jostled with each imperfection of the taxiway. Finally, Sturmthal turned the plane into the wind to line up for the takeoff roll.

The vast expanse of the desert lay before them as everyone agreed all was in order. Ted pushed the throttles forward, the engines whined with power, and they slowly started moving down the runway. Soon, it was racing by as the bomber's wingtips bent gracefully up like an albatross ready for the downstroke that was to push it skyward. The wheels rose from the tarmac as the rest of the B-52 produced enough speed and lift for both itself and its 33,000-pound cargo.

After becoming airborne, Sturmthal turned toward the north and began his long climb to 45,000 feet, destined for the Smith Ranch Dry Lake area of north-central Nevada. Riding alone in the X-15, slung beneath the pylon, Dana continued his checks.

It was 8:56 a.m. on 24 October 1968, and there were sixty-six minutes before Dana was to launch on mission 1-81-141, the last of the X-15 program.

At the same time, 150 miles west of Edwards, at Vandenberg AFB on the central California coast, another countdown was taking place.

William Harvey Dana was born and raised a California boy, one of three from this state who flew the X-15. He was also one of three born in 1930, the second youngest pilot on the program after Joe Engle. Bill started life on 3 November of that year in Pasadena, the city that hosts the Tournament of Roses Parade and Rose Bowl football game each New Year's Day for a worldwide audience.

His family moved to Bakersfield, where Dana first got excited about flight, as he saw test missions of World War II bombers and fighters over California's Central Valley. Bill recalled being especially impressed when he once witnessed a flyover of Northrop's giant B-35 flying wing. This and other exotic aircraft were then being tested at Muroc Field, the forerunner of Edwards, where he was eventually to spend his flight test career. By the early 1940s, usually after homework was finished, Bill was often at work constructing balsa wood models of his favorite planes.

332 | CHASING EXPERIMENTS

Upon Bill's high school graduation in 1948, he earned an appointment to the U.S. Military Academy in New York, becoming the second x-15 pilot to undergo the rigors of a military education. Dana finished the West Point curriculum in 1952. His class also had the honor of celebrating the sesquicentennial, or 150th anniversary, of the academy, which saw its first cadets in 1802. Two notable classmates of Dana's were Ed White and Michael Collins. White became the first American to leave his spacecraft, performing a twenty-one minute extra-vehicular activity on the *Gemini* 4 mission in 1965. Michael Collins served as command module pilot on *Apollo* 11 in 1969, with Bill's fellow x-15 alumnus Neil Armstrong.

Dana then elected to join the U.S. Air Force, a path chosen by a quarter of the class of 1952 and a common occurrence before the establishment of the separate academy in Colorado Springs in 1955. He spent the next year in flight training, starting in Marana, Arizona, where he first flew in a propeller-driven North American T-6 Texan. Soon after he was transferred to Laughlin AFB in Del Rio, Texas, for the first half of advanced training. This brought him to an inaugural jet flight in a Lockheed T-33 Shooting Star. Bill was quoted as saying, "From then on there was no question about it, I wasn't going into the Corps of Engineers." He returned to Arizona, where he finished up at Luke AFB in the summer of 1953, an aviation goal now firmly in mind.

Bill's pilot wings were presented within a few days of the signing of the armistice between North Korea and the United Nations. However, even with the drawdown of military action after the war, he ended up being assigned overseas, flying Republic RF-84G Thunderflash reconnaissance missions in Korea throughout 1954.

Four years was enough of active military duty for Dana. At the end of that commitment in 1956, he decided to return to civilian life. His primary goal at that point was to get to the edge of flight research, as he had always felt there was too much lag between himself and the frontline aircraft. Becoming a test pilot was his way to rectify this situation, so Bill entered the University of Southern California to embark on a master's degree program in aeronautical engineering, achieving this in June 1958.

By September Dana was knocking on the doors of various airplane contractors at Edwards, looking for a job. Between interviews, he made a serendipitous stop at the NACA facility on base, which ended with Bill getting an offer he couldn't refuse, as a stability and control engineer—and the promise of moving into the pilots' office once he obtained his civilian license. With all the new, cutting-edge aircraft in the hangar, he jumped at the chance, setting his hire date as 1 October. The day he reported for work also coincided with the date when NACA restructured into the new space agency, NASA. In effect, as Bill Dana liked to say, "I was probably the first NASA employee."

Bill's arrival at the NASA High Speed Flight Station also coincided with that of the X-15, which rolled out of the North American plant just two weeks later. "The very first program I ever worked on, strangely enough," Dana recalled, "was a performance program on the X-15 simulator. That was in the fall of 1958. . . . I was working on a performance program on the [LR-11] small engines. Wendell Stillwell was my boss at that time, and I was mainly flying the simulator and helping him reduce the data. I spent a year in stability and control, then came down to the pilots' office, which had been my intent all along."

On 8 June 1959 Dana was out on the Rogers lakebed in support of the first X-15 flight. Scott Crossfield dropped from the B-52 and was making a very fast unpowered glide. There was little room for error, and Bill's role was to remove at least one unknown factor as Scott descended. He said, "I had an International pickup truck and a whole bunch of smoke grenades." His job was to light off smoke near the touchdown point, giving the pilot an idea of wind direction and velocity. Considering the X-15 had no ability to go around the field in case something was not right on approach, I asked Bill if he was worried about being too near at touchdown. He smiled with assurance, saying, "Scotty had to slip a mile farther to get to me."

Dana's primary project was a specialized F-100 Super Sabre. On board was an analog flight computer that could be programmed to behave as if it was a completely different aircraft. Like using hypnosis to make a person think he or she is a dog, this computer could have the F-100 react in flight as a giant cargo plane instead of a nimble fighter. The idea was to test control system design without risking a much larger and costlier prototype. Today, these sorts of tests are performed within the circuitry and microchips of a supercomputer, but in the late 1950s and early 1960s that was a luxury not yet invented. Pilots, with the help of rudimentary computational ability, were the supercomputers of this age.

While Crossfield, Walker, and White picked up the pace on the x-15, Dana had his sight set higher, on what was proposed to be the next step after the x-15, the Boeing x-20 Dyna-Soar. Dana's plan didn't pan out, as he explained, "That should have been our next logical step, but I didn't stay very long in the x-20. It was canceled by [Secretary of Defense Robert S.] McNamara, basically to fight the [Vietnam] war." Bill was pragmatic of the decision. "That's their job, to fight wars. The average guy who gets to the top of the air force isn't very concerned about [research and development]. He wants to go win wars, and it didn't look to McNamara like the Dyna-Soar was going to [do that], so he canceled it."

Dana's pragmatism carried only so far. He thought the cancellation neglected the long-term advancement of aerospace research just to satisfy short-term goals. "The x-20 is a program we all should have been allowed to fly."

One lasting thing Dana did get out of the Dyna-Soar was his wife. Judy Miller was working as a mathematician at Boeing, and they found each other in 1961 on Bill's frequent visits to the company's Seattle offices. A year later they eloped to Tucson, Arizona. Their marriage remained solid, raising two boys and two girls: Sidney, Matt, Janet, and Leslie (who became better known by her nickname "Cricket").

As early as 19 April 1962, on mission 1-26 with Joe Walker at the controls, Dana accomplished his first flight duties supporting the x-15 as Chase 1. Bob Rushworth flew alongside in Chase 2, with Maj. Walter F. Daniel and future x-15 colleague Pete Knight taking the positions of roving and landing chase, also known as Chase 3 and 4. By the end of the program, Bill flew in one of these support slots forty-five times.

While we sat early one morning discussing the x-15, the desert vista of Edwards expansive outside his large office windows, Bill energetically talked of what he considered one of his most significant accomplishments. "We moved down from the other wing of the [NASA] building in 1963, and I raced down here and got a corner window office! That was the very first thing I did. I moved down here and have enjoyed every single day since."

Dana continued occasional flights chasing the x-15, while Milt Thompson was named as the next NASA pilot, replacing Neil Armstrong, who left for the astronaut office in Houston.

Bill worked on a program around that time on the North American A-5A Vigilante. This supersonic bomber for the U.S. Navy had an unusual method of bomb delivery: the weapon was deployed aft from a tube between the two jet engines, ensuring a clean separation with a volatile nuclear munition. For NASA research purposes, the Vigilante was designated to stand in for a future American civilian supersonic transport, or SST, anticipated for service in the 1970s. Dana accomplished flight profiles in conjunction with a NASA and Federal Aviation Administration feasibility study of air traffic control handling that might be encountered during commercial supersonic operations.

The SST was never built in America because of alleged environmental concerns over interaction of the engine exhaust with the ozone layer. These ideas were first propagated in March 1971 by James McDonald from the University of Arizona, when he testified before the U.S. Congress on the matter. His theory was grabbed by the media. Unfortunately, pertinent facts went unnoticed, such as that McDonald had previously testified about unidentified alien craft visiting Earth on a regular basis, and were responsible for the giant electrical blackout affecting thirty million people in New England and eastern Canada on 9 November 1965. When he did speak of the blackout McDonald neglected to mention how human error caused the mishap when a protective relay had been incorrectly set. Later scientific research into possible ozone destruction created by SST flights showed the link probably did not exist. The profiles and procedures pioneered by Dana during the Vigilante tests were never utilized.

Meanwhile, Milt Thompson became more and more enamored with the possibilities offered by the lifting bodies and made the choice to jump full-time to that program, leaving a slot open for a NASA X-15 pilot. Bill recalled the situation: "Had [Milt] stayed in the program—which most guys would have—I probably would never have flown the airplane. Milt got all wrapped around these lifting bodies and wanted to fly the M2, so he faded off and I got his position in the [X-15] program." Bill Dana finally had his opportunity to enter the rocket plane business, and in May 1965 he was officially announced as Milt's replacement.

After six months of training and waiting, on 1 November Bill entered the X-15 cockpit with the aircraft still in the NASA hangar. This was a test to ver-

37. Bill Dana sits on the edge of the X-15 cockpit the day prior to his first flight in the rocket plane, which took place on 4 November 1965. The bandage on his nose has not been explained. Courtesy of the Armstrong Flight Research Center.

ify that while wearing his pressure suit he could squeeze into the tight spot and still be able to reach and activate all switches and controls unimpeded. Everything checked out once he was aboard, so Bill also ran through a simulated flight from the seat where he was to fly Mach 4 the next day. He finished up and went home, while technicians moved the X-15 into position. By 3:45 a.m. on 2 November the no 1. aircraft was mated to B-52 no. 008, and flight 1-62-103 was nearly ready.

There always seemed to be delays. Heading toward Hidden Hills Dry

Lake that morning, Dana's cabin pressure regulator malfunctioned and the telemetry link failed. With no way to monitor the x-15's systems in flight, the b-52 headed back to Edwards. Skipping to 4 November for the next attempt, everything went well—until Dana launched.

As he ran down the last few items on his checklist, everything appeared ready to go, but Bill needed a few extra moments. He told Bob Rushworth at NASA 1, "I'm going to launch ten seconds late." Rushworth acknowledged the small delay, then Dana finally radioed, "Okay Bob, we're going to go." Dana counted down from five, then confirmed, "Release." The rocket engine caught—then coughed and died. Dana said afterward he was taken aback at the moment he dropped from the shackles. "I just didn't expect anywhere near the violence of the launch maneuver that I encountered, but I didn't have any complaints about the aircraft response."

The x-15 continued to drop as seconds passed. "I was very reluctant to attempt an engine start at negative gs or zero g. . . . It took me a long while to accept the fact that I was going to have to keep the . . . nose down while I was going through my restarts [which put some positive g-loading on the aircraft]. This would have been obvious had I thought about it, but it took me quite by surprise because this is something you don't realize on the simulator."

Standard practice at this point in the program was to push the LR-99 throttle all the way to 100 percent immediately after launch, then drop back to the desired thrust level. With Dana a bit shaken, he went only to 50 percent power. Compounding the problem, a sticky fuel metering valve on the engine didn't completely open, something that probably would not have occurred at full throttle.

At six seconds after launch, Bill's reactions continued to be off-kilter. He said, "I turned the throttle off, [then] went through an incorrect re-light. . . . [So, I] turned the throttle off again and went through a proper air start procedure." Rushworth walked him through from NASA 1: "Go to reset and prime, Bill. Throttle off and reset. . . . Get an ignition ready light, and bring the throttle on."

It took more than twenty-four seconds, dropping 10,000 feet, before the LR-99 caught hold and started to drive Dana and the x-15 upward. Bill confirmed to NASA 1, "Got a good light, Bob." Dana later admitted in his post-flight report, "In the simulator, I was able to manhandle it quite nicely.

. . . Maybe I was being a little gingerly with it today because I didn't know quite what to anticipate."

Bill recalled, "I'd already looked and I had Hidden Hills down there, so I knew where I was going if it didn't light. Of course, I was quite happy when it did." He expressed some reproach for not being properly warned of the sensations that hit him at launch. He said, "Basically, that was the shortcoming of the pilots that had flown before. I think there was a sentiment in those days that the old pilots didn't really like to help the young ones. I don't know where they got that kind of macho outlook, but I really didn't get a lot of help in checking out at all. I went out and did it. I was absolutely shocked—it felt like you were shot off the shackles rather than fell off." Then he got hit in the back by the rocket engine firing up. "Yeah, that was the next surprise," Dana admitted, "but that was kind of a good feeling . . . that engine lighting up. I was down to my last try."

Bill talked of his initial worries: "Prior to my first flight, I thought landing would be the tough part. Don't know why I felt that way, but I thought at Mach 4 you're not going to hit anything. It's when you have to get it back on the ground that's going to be the work. . . . I got below Mach 2, and, hell, I'd been there every day practicing in an F-104. I said to myself, 'I got it made! I'm going to live through this!'"

Coming into the landing pattern, Dana regained confidence. "I was so overtrained for the landing, and I stayed that way throughout my career. I always shot fifty landings between each flight—maybe more if I had time—because I always felt that, if I ever had an emergency, I wanted the landing pattern to be instinctive. I didn't want to have to give any thought at all to getting the airplane on the ground."

As the skids touched down and the aircraft slid to a stop on the lakebed, Rushworth radioed the news to Dana that he was no longer a rookie: "Very good Bill. That was a good indoctrination wasn't it?" Bill responded, "That wasn't quite like the simulator, was it?" Bob joked of how, when pilots screwed up in the "Iron Bird" on the ground, they could go again with no consequences, saying, "You should reset there and go back on the launch!"

In a magazine article, Dana was quoted as saying of his first impression on seeing the aircraft: "I got to see [the x-15] the day after [the rollout on 15 October 1958], and I thought it was the ugliest airplane I'd ever seen. . . . I wasn't too impressed with it until they put the big engine in it, and then it had to command your awe. It was a 33,000 pound airplane with 60,000 pounds of thrust, and it really left the scene immediately when you lit that engine." Now that he was a pilot on the program and had seen and felt firsthand what the black beast was capable of, Bill knew he was in the right place.

The third time seemed to be the charm for Dana on his second flight as well as his first, but the three attempts at mission 3-52 took a lot longer to materialize this time around. He remembered, "I had a hell of a time getting my second flight. It took nine months. . . . Winter came and we had a lot of troubles, [but] after that things went a little better."

Originally, Bill was supposed to have flown a week after his first mission in November, but this flight was postponed because of rain. The problem was exacerbated as the wet season closed in, so it was eventually decided to take the time to do major wiring modifications to aircraft no. 3. Some of this was in support of a change to the standard instrument panel, composed primarily of dial gauges, to one that used rectangular strips to show status.

As technicians removed panels and delved deeply into the aircraft, cracks were found, necessitating replacement of inner-skin panels. Then the wiring was replaced throughout the aircraft, additional modifications were made to accept wingtip pods as on aircraft no. 1, a third skid was mounted at the base of the lower ventral, power-on checks were run, and the LR-99 installed. All was finally ready for mating to the B-52 on 17 June 1966. It was a protracted road that had begun because of a wet lakebed that had long since dried out.

Twice Bill returned to Hidden Hills for launch, on 20 June and 13 July, only to be thwarted by malfunctions in the X-15's inertial system. The second attempt had double trouble when the telemetry was unable to be read on the ground, a problem similar to what was encountered before his first flight. The third try on 18 July went as planned, with Dana launching at 11:38 a.m. on a mission to 96,100 feet at Mach 4.71.

How did the new instrument panel work? Bill shared his opinion, saying, "The basic X-15 had round dials, like all airplanes of that era. They came up with these vertical tapes they thought would cut down pilot workload. I flew ten flights . . . on no. 3 with those tapes. I liked the tapes at the time, but then I had to go back to flying no. 1 after we lost no. 3, and I liked the round dials just as well. Those vertical tapes were a 'gee whiz' item, but they really weren't any help to fly." Pioneered in the x-15 by Bill Dana, these types of cockpit displays are now the norm in most modern air and space craft.

Wade Martin worked quality control on the air force side of the x-15 house. He spoke of a typical day on the x-15. "Most all of us at that time carpooled and . . . I happened to be carpooling with the instrumentation guy and a couple of [propulsion system] test stand mechanics. We'd probably leave Lancaster in the neighborhood of 6:30 a.m. and try to get [to Edwards] somewhere around 7:30 a.m. At that time there wasn't any problem about anybody being late or not being there, because we would all show up—even if we were dying!" Being so enamored of their job that they came to work sick caused its own difficulties. Wade said, "I think one of the problems we had is that we would be finding ourselves sick on the job because somebody came in and gave it to us."

With an entire inventory of engines to keep track of, a large responsibility was ensuring every one of them was treated exactly the same. As Wade put it, "Otherwise, some of the engines being worked on would have one thing done to them and the others wouldn't." And, as with any government program, "the paperwork was a big thing because, not only did we have to keep track of the engines themselves, but I kept track of every valve that ever went on it."

Engine test runs actually gave a break to the routine. "If there was going to be an engine run that day, the maintenance shop had to shut down prior to going out for the test. That gave everybody a little bit of a rest, you might say, because the engine run would last a couple of hours, from the time you got it all loaded and cleared the area out, fired, then got back into the shop."

Operations engineer Vince Capasso oversaw the entire aircraft. He said, "As far as the airplane itself, a lot of the modifications were for aircraft improvements. You'd have a long schedule, and you'd be glad when it got done." It also required a lot of coordination between the various shops, especially when working inside the tight confines of the x-15. Installing new instrumentation and experiments often involved getting into the bay behind the cockpit, which had a simple elevator to ease access to the components. Vince explained, "The problem was, when you had a big [modification] going on, only one technician could really get in there and work; then you maybe had one or two guys hanging over the side [of the aircraft].

You were limited on how many people were in there. It took careful planning to lay the sequence out, so you didn't have a bottleneck of too many people trying to do things at the same time in that limited area."

The original purpose of the x-15 was to provide testing of new aerodynamic concepts in a hypersonic vehicle and also to verify procedures and ideas on controlled flight in a vacuum and reentry into the atmosphere. By the fourth year of the program, those objectives had been achieved, yet there were three very robust aircraft with a lot of life left in them. The question then became: what next?

New programs that were to go even higher and faster were on the chopping block. Most people at the Flight Research Center believed this to be a temporary situation, although, in the long run, that proved to be optimistic. Regardless, the question remained of either finding new uses for the x-15s or canceling the program.

Even before the first flight in 1959 the proposal had been made that the x-15 would make an excellent platform for other experiments. By August 1961 the x-15 Joint Program Coordinating Committee was formed to do just that, eventually suggesting more than forty proposals. These involved astronomical applications, support for the Apollo lunar program, follow-on aerodynamic research studies such as the scramjet concept, and even a delta-winged version of the x-15 with a higher-powered LR-99. On 13 April 1962, eight months after the exploratory committee set to work, NASA headquarters in Washington DC held a press conference announcing officially what was called the x-15 Follow-On Program.

First up was a proposal from Washburn Observatory at the University of Wisconsin. Their concept was to accomplish ultraviolet photography of bright stars, something impossible to do below the absorptive properties of the atmosphere—in this case, the ozone layer. Doing so in the x-15 also meant less cost than attaching an experiment to a space-bound rocket. Bob White flew the first attempt at gathering ultraviolet data on 14 December, his final mission in the program. The experiment failed to operate, which proved to be an ongoing problem with many of these packages throughout the years aboard the x-15. However, these were new and novel ideas, and they gave researchers first crack at something that might never have left the ground under a different, and more controlled, government space program.

As the x-15 matured, and new experiments were added, a great deal of the work supervised by engineers such as Vince Capasso involved their installation. Many other experiments followed, some successful, others not, including atmospheric density measurements, photography of the sun's spectrum, thermal insulation tests for the Saturn booster, Earth resources, tracking the exhaust plume of an intercontinental ballistic missile, infrared scanning, and capturing the dust left by micrometeorites as they disintegrated in the upper atmosphere. Within a year of the Washburn flight, more than 65 percent of data returned from the x-15 came from experiments outside the scope of the original reasons for which the aircraft was constructed. The percentage grew each year through the remainder of the life of the x-15.

On 4 August 1966 Dana exceeded both Mach 5 and 100,000 feet for the first time, then jumped his altitude to 178,000 feet on 19 August. Less than a month later, on 14 September with mission 3-55, Bill flew to 254,200 feet, coming in less than 10,000 feet below the magic number needed to qualify as an astronaut. This flight also featured the first use of the wingtip pods on x-15 no. 3. In the rear bay of the left pod was the solar spectrum experiment from NASA's Jet Propulsion Laboratory. During the mission it did not operate properly. The first attempt at the collection of micrometeorite dust didn't even get that far, as it malfunctioned during check out and was removed prior to flight.

It had been more than three years since Joe Walker attained the highest altitude of the x-15 program at 354,200 feet. No pilot had gone higher than 300,000 feet in a winged aircraft since that day. However, Dana was about to do so—inadvertently.

On 1 November's flight 3-56, instead of an early morning takeoff, a bad O-ring in a flow valve caused a halt, and Dana left the aircraft while the ring was replaced. When Bill was again secured inside the cockpit, Major Doryland and Major Reschke taxied out for a 12:22 p.m. takeoff in B-52 no. 003. One hour and two minutes later, Dana made a clean launch three hundred miles uprange near Smith Ranch, getting the LR-99 up to maximum thrust in slightly more than one second.

Several experiments rode along with Dana that day, including what was scheduled to be the last flight of JPL's solar spectrum instrument, an air density measurement to look for seasonal variations in the atmosphere, an optical background study to test the idea of high-altitude laser surveillance, daytime sky brightness checks for eventual use in star trackers for air and space craft navigation, and the failure-prone micrometeorite collection system in the tip pod.

For Bill, the flight was especially sweet in that he was cleared for a run at astronaut status, the sixth pilot to do so in the x-15. The experiments were secondary to flying a good profile. He explained what happened after launch: "Oh, yeah! I had one marvelous moment. . . . I'd already been to 250,000 feet, and they were going to let me go to 270,000 [feet] to get my astronaut rating. The way you flew these profiles was, after launch, you pulled up to some pitch attitude and you had a vernier indicator that allowed you to precisely fly that attitude. I think I was to go out at thirty-seven degrees . . . but it stuck and I just faded on up to forty-two degrees. So here I was five degrees steep, which is really steep as hell."

The LR-99 also ran for an extra two seconds, and Bill had to shut it down manually. Between those two factors, his altitude soared to 306,900 feet. Flight planner Jack Kolf stated how small errors quickly added up on the x-15: "It was really an interesting airplane in that if you let the engine burn a second too long, you'd go like 8,000 feet too high. And if your angle of attack was one degree too high, it was again worth 8,000 or 10,000 [feet]. Bill accumulated enough to take him about 40,000 feet too high." Dana was adamant: "I was not going to shut down early because I didn't want to miss my target altitude."

Pete Knight served as the NASA 1 controller that day. He called on the radio, "Right on the track, Bill. . . . Track is real good. We have you peaking out around 310,000 [feet]. . . . You're going to be in good shape for Eddy." Pete was using a common pilot slang for Edwards AFB.

As Knight called the approximate peak altitude, Dana saw the same on his cockpit instruments. Because of this, he knew he was in for a ribbing by Jack McKay. Bill recalled, "Jack had overshot his altitude by 40,000 feet several months before, and I'd been kind of hard on him for missing his altitude by that much." Now Bill was in the same boat and was ready for the consequences. As he acknowledged Pete's altitude call, Bill added, "Roger. [Is] Jack McKay sending in congratulations?"

This was not going to be the end of the foibles for the mission. In fact, as Bill shut the engine down, he created one of the classic moments in x-15 comic history. He smiled as he recalled, "As I pulled the throttle back, my elbow hit the quick release on my checklist, and there were pages floating around the cabin in zero-g conditions until I got gravity back on the airplane—which was three minutes later. I got a good chance to repent my sins that day. . . . That binder was perverse. You could never unlock it when you wanted to, and yet I knocked it loose accidentally."

The incident was caught clearly on the data camera, attached to the canopy behind Dana's right shoulder. "The [pages] were all floating around there in zero gravity, and you could see my hand up there pawing this shit out of the way!" It was an instant classic, a piece of film that was to be viewed countless times to inject levity into a meeting or conference. Bill expressed his predicament eloquently in his debriefing. "It was a great deal like trying to read Shakespeare sitting under a maple tree in October in a high wind. I only saw one instrument at a time for the remainder of the ballistic portion. . . . The alligator [clip] was locked tight and it took me ten to fifteen minutes to get it open, so I don't know how these pages came out . . . but there they were, and these will be in the camera film, which I think we can probably sell to Walt Disney for a great deal."

On a more serious note, Dana had just flown the last x-15 flight ever to surpass 300,000 feet. Speaking of his experience, Bill said, "The view was excellent. Unparalleled. . . . I looked out the right window and there was San Luis Obispo, and I looked out the front of the airplane and I could see the Gulf of California. Those two are a long way apart!" He laughed at the great memory. "It was just the highlight of my whole career, both literally and figuratively. That was far and away my favorite flight of all time. Kind of the high point of my life."

I asked him if the small windows were a detriment to how much he could see. "The view out the x-15 wasn't as bad as it looked. They were small windows, but your head was right up against them. The only place your vision was inadequate was aft on the left side where you always flew a left-hand traffic pattern. . . . You only had about a 100 degree field-of-view to one side, and that wasn't enough to see your runway. You had to have a landmark out on the desert to turn base leg on, until you could turn enough to get the runway in sight again."

After the annoyance of swatting twenty-two pages of checklist out of his

38. The x-15 simulator, nicknamed the "Iron Bird," was a bare-bones x-15 cockpit connected by hydraulics and wiring to representative flight controls. Here Bill Dana practices on 17 August 1966 for mission 3-54, which in two days will take him to 3,607 miles an hour at 178,000 feet. Courtesy of the Armstrong Flight Research Center.

way, gravity returned during the 3,750-mile-an-hour reentry, and they finally settled on the floor. In his report Dana said, "The reentry was very pleasant. I had forgotten what a fine-handling aircraft this is. It just exceeds belief. It is a whole lot better than the simulator." The landing was nominal, and from NASA 1, Pete radioed congratulations to the newest American astronaut: "Beautiful flight, Bill." To which Dana replied, "Thank you, Pete. Since my page sixteen is somewhere down on the bottom of the floor, maybe you could go over the checklist with me."

Roger Barniki was in charge of putting the checklists together and adamantly stated, "From that flight on, I'd get the checklist, put it in there, and I'd solder it closed! When they wanted a change, they had to come down to me to unsolder it and leave it that way until they flew again."

Another spate of bad weather prevented Dana's next flight for almost six months. Aircraft no. 3 was prepared with a group of seven experiments, including a coldwall heat transfer test. The idea was to cover a section of the upper vertical tail, a foot and a half wide by four feet, with an insulating piece of metal. A cork backing to the plate ensured the underlying skin kept relatively cool as the aircraft was pushed to above Mach 5. The metal sheet was then jettisoned using explosive bolts, exposing the cold skin to instantaneous heat from the hypersonic boundary layer air. On flight 3-58 there was no chance to check the operation of the system because Dana never made it to Mach 5.

As launch time approached on 26 April 1967, the Q-ball at the nose of the x-15 was not responding. Once the auxiliary power units were up and running, the ball nose should have been doing the same, but Dana got no feedback from the system. Analysis said there might be some frozen fluid preventing its movement. Mike Adams as NASA 1 radioed to Bill, "Things aren't looking too good." Dana replied, "I'm with you."

Troubleshooting continued, and it was decided to keep the pilot on B-52 oxygen instead of going to the internal x-15 supply. That gave everyone time to work the problem. Colonel Cotton made the two minutes to launch call, and moments later Bill exclaimed to Adams, "Holy barracuda, my ball nose just came on, Mike!" Things were looking up. Dana said, "Sunshade is up and I'm standing by for your one minute call, and confirmation that we will be able to go now." But the radio was also giving them problems, so doubt for the mission remained. Bill was anxious to move forward, saying, "Okay, I'm reading you now, Mike. Confirm that this is a launch situation?" Adams passed on the bad news: "This is an abort, Bill, and we're still having trouble." Reluctantly, Dana radioed his acceptance: "Understand, Mike, this is an abort."

As the issues sorted themselves out, they decided to make one more try at launch. B-52 no. 008 made a wide, circling turn back into position as the clock was recycled to the ten-minute point. They ran through radio and systems checks, then verified the chase planes all had enough fuel for a second attempt. By three minutes Mike radioed, "We're up to speed now, Bill." The Q-ball checked out and the radio continued to cooperate. When Adams asked for a radio status, Bill enthusiastically called, "Loud and clear."

The countdown ran to the end, and Dana flipped the launch switch to release from the B-52. It was 11:20 a.m. when the LR-99 lit, pushing the x-15 supersonic. Dana soon noticed a low fuel light, and at just past twenty-three seconds on the mission clock, and Mach 1.8, he made the decision there was no choice but to shut down the engine. "Got to shut her down, Mike, my pump inlets are running way low. Got Silver [Dry Lake] in sight, going to jettison." Adams confirmed, "Shut her down and go into Silver." Dana told his chase pilots what to expect: "I'll be landing to the southeast, left-hand pattern."

Maj. Jerauld R. Gentry was in a T-38 Talon as Chase 1, with NASA pilot John A. Manke in an F-104 as Chase 2. Pete Knight was waiting in vain, flying another F-104 near Edwards as Chase 3. Manke came in close to verify x-15 no. 3's condition. He called, "Okay, everything looks real good from this position, Bill. Looks like you got it wired here." Dana replied, "I got it, John. Sure glad I chose you for Chase 2 today."

Approaching the lakebed, with only light winds to contend with, Dana was lining up perfectly. Mike Adams had experienced an emergency of his own less than seven months and five flights previously, when a fuel tank ruptured and he had to cut his flight short and go into Cuddeback. As NASA 1 for Dana that morning, Adams joked over the radio network, "We can welcome him to the 'Strange Lake Club!'" Dana laughed as he agreed, saying, "I always wanted this thing [to go] cross-country!" At 310 seconds following launch, Dana made a perfect touchdown and brought no. 3 to a stop with 40 percent of the runway still in front of him. For an emergency, the tenth in the program, it was about as perfect as could be expected. From the B-52 Cotton passed along to Edwards and all planes on the network: "Bill says he's okay."

What happened on this mission? As Bill sat, relaxed in his corner office, he recalled, "I had an indication of a fuel line freeze-up at launch. . . . When the engine lit, I thought I had low inlet pressure, which meant I potentially had the possibility of burning through the [rocket engine] nozzle due to insufficient cooling." During post-flight inspection, the sensor was found to be the culprit and fuel had been flowing just fine to the LR-99. With the cockpit indication he had, and under the circumstances, electing to shut down and go into Silver was the only safe thing to do.

"The landing was just an absolute piece of cake because I'd practiced so much," Bill said. "I did that throughout the program. I never had the bad emergency I was really looking for. Had I had one, I think I probably could've coped with it. I could land that airplane where I wanted to. That was the only time I went into a strange lake in the x-15."

Back over Silver on 17 May, mission 3-59 went exactly as planned. "It was about as smooth as any I've ever seen," Dana was quoted as saying after the flight. He launched and rotated upward, then leveled out at approximately 70,000 feet to make his speed run. At Mach 4.5 he punched off the cold-wall insulation panel, making a clean break to expose the skin.

Flight planner Johnny Armstrong talked about his thoughts on this experiment: "We were trying to get kind of a quasi-steady Mach number by throttling the engine back, putting out the speed brakes to slow down the acceleration, [and] maintain[ing] a certain [dynamic pressure] level and angle of attack. . . . So it was a real challenge." With the coldwall experiment Johnny explained that the sequence was simple: "He pushes the button, and off goes this panel. You talk about being able to do the things then, and not being able to do them today, that's one of them. It's good we didn't get [Dana] fined for littering the desert, and I've often wondered where did that panel go? I was the one who made the predictions, but . . . I don't know that anybody went to Fort Irwin to say, 'Have you found anything on your range?'"

What Armstrong said of not being able to do things today that they could in the 1960s was certainly true. Dana reiterated, "It's very doubtful the x-15 could be sold in this day and age."

The mindset at the time of the x-15 was different. People looked at it as if they were on the frontier, which was literally true in many ways. Edwards was far from the beaten path, and even farther from NASA headquarters in Washington DC. Once the media moved on to cover Apollo, the men flying the rocket plane from the high desert of California were left relatively undisturbed. Support crews made do with what they had on hand, and the paperwork often was filled in after the fact. Research engineer Eldon Kordes spoke about the x-15 mentality: "People would come up to me and say, 'You can't do that.' And everybody would look at each other and say, 'Why can't you do it? . . . Let's sit down and figure out how we can.' There were a lot of things done that way."

Eldon went on to provide an example, saying, "I remember one incident where I wanted to do something on the airplane which had to do with some instrumentation. I was told it couldn't be done. So, I went down to the x-15 with a clipboard and started looking around. The guy that was the head of the shop at the time came out and asked what I was doing. [I told him,] 'We want this instrument mounted, but there isn't room to fasten it.' He said, 'Give me that.'" A few days later, Eldon was called back to the hangar, where the engineer showed off his handiwork. "He pointed and asked, 'What do you think about that?' He had engineered it, he did the machine work, and he had guys install it. I said, 'Wow. I didn't think you could do that.' He said, 'Now, cover my ass and go get me the paperwork I need.' I couldn't have even gotten in the hangar today without proper paperwork."

Dana continued to fly aircraft no. 3, with missions on 22 June and 20 July. Then, on 4 October with mission 3-63, Bill made the farthest flight on the program at 299.8 miles, which was to be tied nine missions later by Pete Knight on the penultimate x-15 mission. Mounted inside the nose of the left wingtip pod was Bill's least favorite experiment, the micrometeorite collection device.

When small rocks enter Earth's atmosphere, depending on how steep an angle at which they hit, several things could happen to them: they skip back into space, burn up completely, or explode into fragments. The meteor residue leaves a layer of microscopic extraterrestrial dust at altitudes above 150,000 feet. For the x-15 experiment, a set of surfaces were exposed when the nose cone of the pod opened at high altitude, to hopefully capture some of this dust and return it to a laboratory for analysis—a great idea in theory but very difficult to achieve in practice.

Meteor showers happen regularly many times during any given year. Each is associated with a parent comet, which leaves debris in its wake as it orbits the sun. That debris then burns in the upper atmosphere as Earth passes through the particle swarm. In early October the shower is known as the Draconids. The source is a periodic comet that orbits the sun every six years, codiscovered in the early 1900s by Michel Giacobini from France and Ernst Zinner of Germany.

When I spoke with Bill about his 4 October 1967 flight with the collector experiment, his first words were "Oh, Jesus Christ!" Then he talked of what a hare-brained scheme the whole idea was to fly a rocket plane into the mesosphere right in the middle of a meteor shower.

"I remember the same day that Pete set his speed record there was a big meteor shower," Bill recalled. "Well, to get out here [to Edwards] for that early flight, I had to drive to work in the dark, and I have never, ever, seen so many shooting stars in my life. There must have been a thousand shooting stars as I came in from Lancaster. And those were the micrometeorites we were going to try to capture! I always thought what a classically dumb idea it was in the first place." What of the possibility of capturing one through the x-15 cockpit rather than the experiment on the left tip pod? Bill stated, hopefully, "Well, I don't suppose it would have hurt us."

Johnny Armstrong said of the collector: "That's a phenomenal thing to try to do. . . . Most of these experiments, we went out and flew the airplane, and you did the job and said, 'Okay, experimenter, here's your data.' How many of those things were successful, I personally have never chased it down to see what they got out of it." Six missions of the x-15 successfully brought back small dust samples from the upper atmosphere, but in the end researchers found what had been returned was too contaminated by residuals from the firing of the ballistic control jets to be of any use.

After ten flights in no. 3, this mission became the last Dana was to fly in that aircraft. Forty-two days later, x-15 tail no. 66672 came apart in a hypersonic plunge to the ground, killing pilot Mike Adams.

After a hiatus of several months, x-15 no. 1 was put back to work. The no. 3 aircraft was now gone after the fatal crash of 15 November, and the x-15A-2 was still being repaired following the burn-throughs that happened on Knight's record speed flight of 3 October. The A-2 was destined to never return to flight status, as the program wound down in its final year, so x-15 tail no. 66670 was on its own.

John McTigue spoke about what it took to get an x-15 up and running on launch day, and a game he called "abort chicken." He said, "If nothing major went wrong, you could probably turn [the aircraft] around in little over a week. But most of the time it was more like two weeks." On the day of the flight, work started very early. "We'd start [liquid oxygen] servicing at about 1:30 a.m. for a flight that took off about 8:00 a.m. The crew would all get together for breakfast a little after midnight, something like that." Then they started running liquid nitrogen to cool the systems down. This was done right up to launch time and had the added benefit on hot summer days on the Edwards tarmac of cooling down the surrounding area where everyone was working. The liquid oxygen, which was begun so early in the morning, constantly vented off as servicing progressed, so it was also replenished throughout the process. Later, in captive flight, a liquid oxygen tank inside the B-52's bomb bay made sure the X-15's oxidizer was full when it came time to drop.

Back on the ground, as the flight preparation continued for departure, peroxide was loaded, and nitrogen and helium were brought up to pressure. John said, "We had our inspectors out there looking for little tiny leaks, because you had to be sure you're going to have enough source [pressure] when you got to the drop point. . . . We could stand a certain amount, so we'd monitor it for the hours we had between the time we found the leak and the time we were going to take off." There was so much effort invested in any particular flight that it was worth a lot to salvage the manpower and resources. "You waited until you were just about to put the pilot in and you'd say, 'Okay, we can make it' or 'We can't make it' based on the [leak] rate."

There were times when things just didn't go right, and the work was in vain. McTigue said, "We played a game of chicken to see who was going to abort first. Whitey Whiteside would be uprange looking at the weather, and he'd be calling to ask, 'How's the airplane?' We were asking him, 'How's the weather?' This was to see who was going to make the first move for abort, if needed." As an operations engineer, it was a matter of pride for John that a flight not be canceled due to the aircraft for which he was responsible. "A lot of time an abort was based on weather, when, in fact, if the weather had been good, it was going to be based on something with the aircraft. It's human nature to do that. We'd try to not put too much blame upon our airplane."

During the run up to Dana's thirteenth flight on mission 1-75, the game of abort chicken was lost once by the aircraft crew and once by bad weather at the Delamar launch lake. On the flight day of 4 April 1968, a third abort came close to occurring, which was the fault of neither. A helicopter being used in support of the flight was delayed by thirty minutes because it was involved in a rescue operation elsewhere, but once all the players were in place, the launch was ready to proceed.

Dropping away from the pylon, at 10:02 a.m., Dana headed to 187,500

39. While in space the x-15 deployed experiments to take measurements for researchers back on the ground. In this image the large structure behind the cockpit shows the deployment of the Western Test Range experiment, which was to record the signature of an intercontinental ballistic missile launch from Vandenberg AFB on the California coast. The box deployed from the left wingtip pod is the micrometeorite collection device. © Thommy Eriksson.

feet. He said after the flight, "I was just pleased to death to be able to get off the shackles today, and I figured on just flying the profile until shutdown." Everything went so perfectly, he said, "I think it is the first time in my life that I ever found time to look at the clock. I looked up and it said seventy-nine [seconds], so shazaam, off came the throttle!" Shutdown was right on the money at just past Mach 5.

A primary purpose of this flight was to accomplish a functional check of the Western Test Range (WTR) experiment, sometimes known as the Pacific Missile Range Monitor. The WTR package was to view a missile launch from Vandenberg AFB to test the idea of detecting ascent through the ultraviolet signature of the exhaust plume. Coordinating a Minuteman II intercontinental ballistic missile (or ICBM) launch with that of an x-15 proved more problematic than chasing down fragments of micrometeorites. Dana's only task with the system this time was to cycle it through a checkout to see if the machinery worked.

As Dana passed through 160,000 feet on his way up, John Manke, at NASA 1, called the x-15 pilot to open the experiment. A double door directly behind the cockpit, known as the skylight hatch, popped open, and a directional mirror system extended upward on two support legs. The primary electronics, consisting of a video camera and optical tracking system, stayed inside the x-15. Everything appeared to work perfectly, until readings were received on the ground through telemetry channels that differed from those onboard the x-15.

Manke radioed the order to retract the mechanism, but Dana thought it had already done so. That proved incorrect and as the x-15 was descending back through 170,000 feet, Manke told Bill, "Hit emergency on experiment." If it did not retract, the aircraft might be in trouble as it reentered. The opening was on top of the fuselage, so the x-15 most likely would have survived, but it would have been a hairy situation.

The x-15's emergency system worked—the mirror refolded itself as the legs dropped back into the fuselage and the skylight closed behind it. Talking with his debriefers later, Dana confirmed there was no doubt it was safely secured. "I did hear it clank in. . . . I think I heard both the retract and the doors slamming, if memory serves me correctly."

By the time the program was nearing completion, twelve pilots had flown the rocket plane, but there was another who was nearly ready to join their ranks: John Manke, who served as NASA 1 flight controller for the just-completed mission. Bill related, "John was within weeks of checking out in the x-15 when Mike [Adams] was killed. Paul [Bikle] just wouldn't hear of checking him out. I was in Paul's office a couple of times, begging him to let John fly, because I realized what a big thing that was going to be for him, but it didn't happen. . . . After Mike was killed, Pete and I were the only two pilots left on the program, so any time Pete was traveling or doing something else, I had John serve as my controller. John was in the x-15 program; he just never got to fly the airplane."

Dana also experienced frustration in that he never had the opportunity to work on the x-15A-2. "I was in line to fly it in the spring of 1969. And, of course, we never got there. . . . They were planning to phase me into no. 2, but we hadn't gotten as far as specifics. That was probably the biggest disappointment I had in the program."

For his last three flights, Dana again started moving up in altitude, although he never approached the 300,000-foot mark from his first astronaut qualification flight.

Attempting to get mission 1-77 off the ground proved troublesome, to say the least. This was partially due to the coordination required to get both the x-15 and Minuteman II flying at the same time. A rundown of the various schedule changes shows the first time was on 12 May, with a slip to 22 May because of a delay in the missile launch. On 13 May another slip occurred to 23 May. Then the missile launch was scrubbed altogether, so the decision was to fly the x-15 on 22 May, after all, because of other priorities. After being mated to b-52 no. 008 the 22 May flight was canceled due to bad weather at Smith Ranch. Going the next day proved just as fruitless, as weather again moved in, although this time the b-52 with Dana and x-15 no. 1 at least had the opportunity to get airborne. An engine igniter malfunction occurred after the abort, so either way the flight was not going to happen. The ICBM launch came back on the schedule at Vandenberg for 11 June, but weather again forced the x-15 to stay grounded. On the following day, Wednesday, 12 June, after a delay due to the XB-70 taxiing and a radar problem in the air, everything appeared to finally come together. Dana accomplished his launch at 8:31 a.m.

He was not happy with all the delays, especially with hardware such as the radar, and made his feelings known at the post-flight debriefing. "It seems to me," Bill began, "that if we can spend $600,000 for an x-15 flight, we could probably spend $20,000 for an S-band [radar] tracking capability. . . . By God . . . it costs us $100,000 to abort, so if we are going to fly this program let's fly it, and if we are not, let's cancel it."

Part of his attitude was from knowing the decision had already been made to fly out the rest of the year of 1968, then terminate the x-15 program. Dana thought that was a bad idea, and he certainly wanted to at least get in as many flights as they could before the 31 December deadline. Every delay meant one less flying day of the year.

Even before mission 1-77 left the shackles, the Vandenberg ICBM launch had once again been scrubbed. After five delays and one in-flight abort, the team was determined to not have the x-15 go home again on the wing of the

B-52, as there was other research business to accomplish. Once launched on his profile, the only action left for Dana with regard to the WTR experiment was to cycle it through its paces. This began on the upward arc, heading to a peak altitude of 220,100 feet.

Coming into reentry, Bill initiated the WTR retract sequence. Once again, the mechanism failed to respond, necessitating use of the backup system. Dana said, "I am pointing out that [on] my last two flights I hit the emergency retract, and that was followed immediately by clunks and amber lights going out. And that, of course, is all I know about what is going on. It doesn't concern me a bit, because I have been able to get the experiment down both times. . . . We have been allowing more than the normal ten seconds, or whatever was allowed, before we hit this emergency [retraction], and the experiment has not been coming down in the time it was supposed to." James R. Welch, head of the X-15 Research Project Office, wrote in his response report of the incident: "Post-flight investigations failed to explain the delay in the retract sequence. Investigation is still proceeding, and an extension of the time pad is planned for the next flight."

Dana's lack of concern for the difficulties encountered was shown in his post-flight statement: "My tracking, if it wasn't quite what you expected of me, was because I was gazing fondly out of the right-hand window where I saw nothing but blue skies."

Another attempt was made to verify the experiment's electronics on 21 August with mission 1-79. This time the overall mechanism was working relatively well but still experienced some problems because X-15 no. 1 was not within the requisite flight path angle. With the aircraft being out of position, this drove the mirror control motors to their stops. However, when it came time to put the package away in its holding bay below the skylight hatch, for the first time it retracted exactly as designed. Apparently the bugs had been worked out of the system, and a new try at recording a real Minuteman II launch began immediately.

Flight planning for mission 1-79 called for 250,000 feet, ideal for missile tracking and the several other experiments on board. However, Dana slightly overshot his altitude, attaining his second astronaut flight above 50 miles when he peaked at 267,500 feet. Bill became only the third X-15 pilot to surpass the embarkation line to space on multiple occasions. Eight pilots achieved that status during the program. When Dana touched down at

Edwards at 9:14 a.m., this also marked the final flight anyone in a winged aircraft was to make into space until the Space Shuttle began flying nearly thirteen years later, on 12 April 1981.

With Lieutenant Colonel Sturmthal and Squadron Leader Miller again at the controls of B-52 no. 003, and Dana snug inside X-15 no. 1 on the bomber's right-wing pylon, the mothership made a perfect takeoff for mission 1-81-141 at 8:56 a.m. on Thursday, 24 October 1968. At Vandenberg, the Minuteman II countdown was also proceeding without a hitch. The two missions appeared ready to intertwine with each other in just over an hour.

Sturmthal made one giant circle over Rogers Dry Lake, attaining approximately 35,000 feet before heading north to Smith Ranch. The hour at high altitude, and the climb of the last 10,000 feet to launch position during the cruise, helped to cold-soak the X-15 systems. Radio reports updating the ICBM status continued to be favorable. All systems were definitely a go.

With about eight minutes left in the X-15 countdown, the mothership swept around 180 degrees to the right to get on the southerly heading required for Dana's flight. This direction was opposite the normal turn made for other missions, but was done to put the X-15 on a perfect path to capture the missile's trajectory.

The flight had been delayed by twelve days when technicians at Vandenberg found their missile was not trackable from the X-15. What that meant was never officially revealed, but the problem appeared solved. On 24 October everything was right where it should be, the weather was cooperating, and the missile and X-15 crews were both on track.

Pete Knight, the only other active X-15 pilot, sat in the control room serving as NASA I. He radioed to Dana, confirming the Vandenberg countdown was also proceeding normally: "Bill, all looks good here—and we got a missile." Dana swept his eyes over the switches and gauges. "Five seconds," he told Pete, the B-52 crew, and all six chase planes following the mission from various locations along his projected pathway back to Edwards. Dana launched at 10:02:47.3 a.m., within three seconds of the planned time before leaving the ground more than an hour before. Sturmthal called, "Got a good light, Bill." Pete confirmed, "Okay, Bill, you got a good light. Check your [angle of attack] and your heading. . . . Real good. . . . Coming up on the profile, on the track." Then Knight confirmed what the entire process had been leading up to: "And we got a missile up, Bill." It was hard to believe after so many snags. Bill said simply, "How about that."

Boosting with full thrust on the LR-99, Dana headed upward. At the same time, 375 miles away, the LGM-30F Minuteman II popped out of the underground silo at Launch Facility-04 on the Pacific coast, arcing on a trajectory that was to take it toward a reentry near the Kwajalein Atoll in the Marshall Islands, approximately 2,400 miles southwest of Hawaii.

Passing 180,000 feet, Pete called, "Okay, Bill, we're clear to extend the experiment. . . . Right on profile." Bill flipped the switch, the skylight door opened, and the mirror was supposed to stand at attention and rotate to lock onto the missile, hundreds of miles ahead. Dana radioed, "It's up, Pete, and I saw the missile."

The Minuteman's three first-stage solid-fuel motors burned out, then separated. The second stage cut in on time, and the flight continued unabated. Flight planner Jack Kolf wrote in his response report what happened next: "At the time the WTR extend switch was thrown, a current decrease was noticed on the . . . number two electrical system. WTR power dropped off, and the computer light [inside the cockpit] came on. This power loss lasted for 0.11 seconds, and then data appeared normal. Two and eight-tenth seconds afterwards there was a momentary increase in current and WTR lost power for the rest of the flight. It is believed that the relay in the WTR experiment burned out at this point." Dana laughed when he recollected years later, "I saw the missile, but nobody else did!" The Western Test Range experiment had failed for its final time.

Dana spoke at the debriefing, detailing what he saw from the pilot's viewpoint. "Everything looked good going across the top [at 255,000 feet] except the light that said the experiment was up, [but] wasn't on. . . . Sure enough, when I turned the experiment switch off, the 'doors open' light went out immediately, indicating that the experiment was not up, and I felt no clunk, so I personally am convinced that the experiment was not up. I suppose this will show up on my cockpit camera film."

Pete Knight spoke about the flight, saying, "The complexities of launching both of them was such that we never got them to work together." Johnny Armstrong was simply frustrated when he exclaimed, "Finally got the two up at the same time, and the daggum thing didn't come up out of the hatch!"

Reentry brought Bill's top speed to Mach 5.38, or 3,716 miles an hour. He said, "I got as many gs as I personally wanted, which was about 4.5." Even though he felt sure the experiment package had not worked properly, he was still happy with the mission, saying, "I had more time coming home from eighty miles out today than I have ever had in my life." Knight, at NASA 1, put the focus back on the next step: landing. He radioed, "Good shape, and you should be headed right at Edwards. It will probably be just off your left nose. . . . Got it in sight?" Dana acknowledged, "Rog, under the nose, Pete."

Lining up on the Rogers lakebed, Knight called, "We got you coming downwind. Looks real good, Bill." Dana replied, "Looks good to me." Then the exchange got a little silly, as Pete said, "That's the one I was worried about." Dana didn't catch on right away, asking, "The down wind?" Pete joked, "No, looking good to you." Dana got it now, saying, "Oh yeah, that's right. We intend to satisfy one person with this traffic pattern."

Touchdown occurred at 10:14:15.6 a.m., giving a total mission time of 688.3 seconds. Pete told the radio network, "Looks like a beauty." X-15 flight 199 was at an end.

A little melancholy years later but still giving a small laugh, Bill said, "That turned out to be the last flight. No one knew it was going to be the last flight. The program ended on kind of an ignominious note. I'd have liked to have seen a more Christian burial for the X-15, but Mr. Bikle was calling the shots, and he said we'd fly through calendar [year] 1968, and we did. The last flyable day we had an in-commission airplane was October 24th, so that was the last flight we got. It's about all the glory I ever got out of the program, so I'll take it!"

Even with the forewarning, the end of the X-15 program shook Dana's confidence in what flight test was all about. Since first joining the ranks on the first day of NASA's existence, Bill made it clear he wanted to be one of the select few who were constantly expanding knowledge in aerospace. He hoped to possibly set a few records himself along the way.

Bill said what he thought of the idea of letting other government agencies take the credit. "I never fully understood why NASA didn't take more of a stand at getting us our share of the records, but they didn't. I always considered that it was beneath my dignity to go ask for it, because that just wasn't me. There were a lot of guys in the program that really wanted records. I don't know if it was for their own personal glory or if it was for their agency's glory. . . . I wanted to do a good job on every flight I was given. If some of the records came my way, I was going to be happy."

The one record Dana received was being the pilot of the final x-15 mission. Was that one he would have rather put off two or three years? "Yeah, you couldn't have said it better," he agreed. "I thought there was more to do. I think a delta wing x-15 would have been nice data for the Space Shuttle. . . . I thought the delta wing was going right up to the day I was told it was canceled. I kind of felt stabbed by [Paul] Bikle that they decided to cancel it, [but] it was beyond recall then. We still had good work to do on x-15 no. 2 [with] the scramjet research. Those two were the big things we wanted to do, and they looked very exciting to me."

Just because the x-15 was relegated to the Smithsonian didn't mean Dana was through with rocket plane flights. There were several lifting bodies still in test at Edwards. It was thought at the time that this was the road to the eventual design for a reusable vehicle to fly into orbit and back, a way to make access to space routine and much less costly.

By April 1969 Dana was piloting the HL-10 and M2-F3 lifting bodies. Between then and December 1972 he averaged seven flights each year, making four glide and twenty-four powered sorties in the flying bathtubs, greatly exceeding his sixteen in the sleek x-15. However, the envelope was much different. These flights rarely went supersonic, and none ever exceeded Mach 2. All powered flights of the various lifting bodies used recycled LR-11 engines from the x-15 rocket plane.

The next year, 1973, one last lifting body came on line, the x-24B. It was sometimes known as the "Flying Flatiron" because of its long, tapered nose and perfect, horizontally level underside. Dropping from the modified B-52 pylon that had once securely carried Dana aloft for his flights in the x-15, Bill accomplished just two missions in the program during September 1975, bringing his rocket-powered total to forty-two missions.

"That wrapped me up in the rocket airplane business," Dana recalled. "I thought the lifting bodies were going to go further than they did. So I thought maybe I could have an even better adventure in lifting entry, but it didn't work out that way."

With rocket planes a thing of the past, he moved on to other aircraft programs. "I went into the F-15 and the . . . HIMAT [Highly Maneuverable Aircraft Technology]. I considered this kind of a run of bad luck. I don't think it was fair I was assigned to that program." The tiny red-and-white HiMAT vehicle was remotely piloted and looked into many advanced techniques for future generations of fighter aircraft. Not being in the cockpit himself grated on Bill. "I thought I was far too senior to be asked to fly model airplanes, but I was asked to, so I did it. It was just an episode in my life I'd like to forget. It took a lot of my attention away from what I really wanted to do." John McTigue remembered, "Oh, boy did he hate that thing!"

John McKay, son of x-15 pilot Jack McKay, worked at Edwards with Dana. When we talked John told me, "Bill ended up being project pilot on the [F-18] HARV [High Alpha Research Vehicle] when I was crewing that airplane. Probably the best thing that happened was when he had that heart attack and couldn't fly anymore. Bill was starting to forget stuff. As the ground crew guys, we were somewhat relieved, because we were just going into a phase of some really heavy inverted spins that Ed Schneider started." This is one of the most dangerous maneuvers a test pilot can accomplish. "It wasn't going to be long after that that Bill was going to get there. Bless his heart, age was just starting to catch up with him. You know, it's like a baseball player, you can't leave the diamond. That's been your life forever." Dana retired in 1998 after forty years at NASA.

During development of the Space Shuttle in the 1970s, Dana was involved peripherally with the program due to his knowledge of hypersonic gliding in the x-15, as well as his lifting body experience, although the latter did not affect the shuttle the way Bill had originally envisioned. His feeling was, "NASA made a lot of dumb decisions on the Space Shuttle program. A whole lot of them were done just to protect their own self-interest. But that's the way all bureaucracies work—their primary purpose in life is to continue their existence. . . . As any government bureaucracy, its first objective is to stay in business and defend its territory from people who would like to throw it a curve. Both NASA and the air force were directing their activities to monopolizing whatever turf they had."

With his rocket plane experience and two spaceflights in the x-15, why didn't he consider joining Neil Armstrong and Joe Engle in Houston and eventually on orbit? Bill explained, "I suppose, it's not simple to articulate my motives, but I wanted to fly more than once every year or two, or whatever it was. Joe Engle got out of the x-15 program just as I got in, and Joe didn't fly in space again until the second Space Shuttle flight, sixteen years later. That's what I wanted to avoid."

Going into space again was something Bill Dana definitely thought about. "I certainly would've been thrilled to have ever been able to orbit. . . . I admire those guys, but I've just never thought I'd trade places with them. . . . I never had fantasies about flying around in a capsule. The [Space Shuttle] orbiter, of course, is a different matter. That's basically an airplane once it hits the atmosphere, and that's the way it should be done."

12. In the Line of Duty

I'm in a spin.

Michael J. Adams

Freida Adams made a point of attending each flight of the x-15 that her husband, Mike, flew for the U.S. Air Force. She recalled the first one, saying, "It was scary, but exciting. It was like Flash Gordon. You'd just get caught up in it." Freida watched what preparations she could. "They would bring us out in a trailer and get us pretty close. You'd watch him walk across [to get into the x-15]." She waved, hoping he would see her but knowing that once in his pressure suit visibility was limited. "I was just always standing there wide-eyed. I can still see him walking out slowly. [Because of the suit] he had to walk kind of like a robot."

Once her husband entered the x-15, preparations continued. Eventually, the b-52 spooled up its engines, taxied away, then lifted off into the desert sky. While the lumbering bomber headed for the point on the map where her husband was dropped away to start his flight, Freida was taken back to the NASA control room, where she watched the bulk of the mission unfold. When the flight was over and the ground crew had helped Mike from the x-15 cockpit, she hitched a ride back out on the lakebed to greet him. "It seems to me I was out on the lakebed most of the time . . . nothing but big space and all the activity going on."

Mike Adams became the twelfth and final x-15 pilot in the program. He made seven research flights, matching NASA pilot Neil Armstrong. Adams served the shortest amount of time, at just over thirteen months from first flight to last.

Unlike his career in the air force and his short stint on the x-15 before his death, Mike's childhood could be called very stable. He was born Michael James Adams in California's capital city of Sacramento on 5 May 1930, staying in his hometown from grade school at Donner Elementary, through two years at Sacramento City College, before finally leaving to join the military. His younger brother, George, remembered, "When we were growing up, we did all the simple things in the neighborhood. In the summertime we played kick the can and mumble peg, drank water from a hose, and went to the matinee movie on Saturdays. It was a modest upbringing, but we enjoyed it."

Mike's first job came while he was attending Stanford Junior High School, delivering newspapers on his bicycle for the *Sacramento Bee*. Mike also spent time playing with his best friend, Charles Gerdel, who eventually recommended Mike to Frank Simms at Simms Hardware for a second job during their first year in high school.

Mike and Charles had been part of a kids' neighborhood group they self-importantly called "The 8th Avenue Gang." Charles laughingly stated, "We used to run around with rubber guns." Working at Simms changed Mike's taste for toy guns and children's war games into what became a life-long penchant for heading out into the mountains with real hunting weapons. Charles recalled, "Simms got us real interested in target shooting. We both had heavy-barrel .22 rifles. The barrel was big! . . . We used to ride the little Cushman over to the rifle range at the old yards where they kept all the city trucks. They had a range in one of the warehouse lofts. Mike got many medals."

With lots of wilderness area nearby, they soon took up deer hunting. A favorite spot was Castle Peak, in the Tahoe National Forest near Donner Summit. According to Charles, "That's a big mountain up there. We went up to the top where there was nice brush and short trees, [and] we knew some of the deer were up that high."

Another area where Mike excelled was school theater, which earned him a special mention in the Sacramento High yearbook. He won the lead in the drama club's senior play with the role of Judge Harry Wilkins in the romantic comedy *Dear Ruth*.

Outside of dating and hunting, he also enjoyed many other hobbies, such as fixing up old cars. George noted that in 1949 his brother had "acquired a '37 Chevy, and he rebuilt the engine in conjunction with one of his pals.

... It was a hot car. . . . He had worked on it for the better part of a year. It was prime."

Sports also caught Mike's attention. "He played a little football, [but] he was a better baseball player. He threw the javelin in college, and, I think, like myself he was a good jock but probably didn't excel at it."

When George talked about Mike, it was clear to see that he always looked up to his older brother. "He wasn't real talkative, but I think he was liked and respected by those who knew him. . . . Growing up, we slept in the same bedroom [in] twin beds. I'd scratch his back, and he'd tell me ghost stories."

At almost the same time Mike completed his second year at Sacramento City College, the Korean People's Army escalated several years of skirmishes with South Korea into a full-scale war. It was 25 June 1950, and just two days later President Harry S. Truman committed American troops to the conflict.

To meet the manpower required to back the promise, draft notices to eligible men quickly went out. Mike was a prime candidate to be swept into the army, so he made an end run, promptly enlisting in the air force. After completing basic training at Lackland AFB, outside San Antonio, Texas, Mike applied for Officer Candidate School. Before entering OCS, he returned home and convinced Charles to follow him back to Texas, telling his friend, "The air force isn't too bad. You should join." Charles went into ground maintenance rather than becoming a pilot, as was Mike's aspiration. "It got so crowded by the time I got down there," Charles said, "they had to move us off to Lubbock, Texas. We froze up there!"

Two years later, on 25 October 1952, Adams graduated from the pilot training school at Webb AFB, Texas. He was then shipped off to Korea in April 1953, where he flew the F-86 Sabre fighter-bombers for a total of forty-nine missions during the war, earning an Air Medal in the process.

While in Korea Mike had enough downtime between missions to find ways to relieve both the tension and boredom. One such hobby was teaching himself to play the accordion—an interesting choice and one of several musical instruments Mike dabbled in during his life.

It was his return to stateside duty that changed his personal life forever when he met Freida Beard, a self-professed "southern belle." Freida remembered, "He came back from Korea and went to England Air Force

Base in Alexandria, [Louisiana]. That's where I was living. We had a real nice base there. . . . My brother knew this man, Forrest Campbell. He and Mike were from Sacramento . . . [and] had gone to the same high school." Freida's brother enlisted Forrest into a conspiracy to match his sister with Mike, an eligible bachelor just back from the war. "So they would try to set up a party and get us to come. They did that, I bet, four or five times. It just got ridiculous, and I'd always say, 'No, I already have plans.'"

Although Freida tried to avoid the setups, Mike succumbed to the pressure and gave her a call. Freida reminisced, "He had this real deep voice . . . and he said [imitating a deep voice], 'I was just calling and this is Mike Adams. I was just wondering if you'd like to go to a movie on Wednesday night.' And he said, 'If we don't like each other, we don't have to waste the weekend.' I laughed so hard, I fell on the floor practically. I told my friend, 'This is a new twist.' I said, 'I'd love to go to the movie with you and not waste a weekend.'"

The date was anything but a waste for either of them. "I think it was less than two years, [but] I just knew I would get a ring. . . . I decided Mike might be the one, so I made up my mind, and I thought, at Christmastime he would give me a ring."

Mike was not as attuned as he should have been to Freida's thoughts on the subject, instead buying her a 20-gauge shotgun that holiday, hoping she would join him on his frequent hunting trips into the wilderness. "Well, that didn't fly too well," she said. Mike had expected her to be excited over his idea of togetherness, but her dejection quickly got through to him. Finally catching the hint, he proposed soon after the new year.

Mike's brother, George, was surprised by what was happening. "I don't think he was ever close to tying anything other than his shoes. As far as I know, that's the closest he ever came to getting married—and he did!"

"We married in January [1955]," said Freida, "and not long after he had to leave on a six month TDY [temporary duty] tour to Germany. When he returned, it was like a stranger had gotten off that airplane. [He] came back . . . [and] we had to get to know each other all over again." Any spare time he had while in Germany with his squadron, he spent shopping for his new wife. "I still have things he brought home. The clock from Germany, that clock has never quit running."

As evidenced by his earlier Christmas present to Freida, Mike always found time for hunting. Freida occasionally shared his trips and even enjoyed shooting, but not if it was toward an animal. "I'd grown up, my brother and I, shooting targets. . . . I was real accurate with target shooting. . . . We had a lot of fun. When you go with Mike, you went out in the boonies. You went to places where you could shoot, or you went with a picnic basket, and you went and found places where you were in nature. . . . I always said he liked flying, hunting, and family—almost in that order."

Once Adams had returned from Germany, he pursued his military career in earnest, taking advantage of the educational opportunities afforded GIs in post–Korean War America. First up was heading to the University of Oklahoma for his aerospace engineering degree. This was also where Mike and Freida's first child was born. His given name was Michael, the third generation to bear that first name. Father and son shared the full name so baby Michael became Mike Jr., or sometimes "Little Mike."

Next up for Adams was the Massachusetts Institute of Technology for advanced studies in astronautics, where his second son, Brent, came on the scene. He returned to air force duties at that point, being assigned to Chanute AFB, Illinois, as an instructor. Their third child, Liese, joined the family there.

Whenever school and military life permitted, Mike flew back to Sacramento to visit family and friends. George always cherished the times his brother returned, saying, "When Mike would come to McClellan or Mather [AFBS], he'd buzz us along 8th Avenue at a couple thousand feet. . . . 'Oh, Mike's home.'" George laughed. He also spoke of a special pet that Mike took home to the family on one of his trips. "When he was out here on a Christmas vacation, he'd flown into McClellan, and the family acquired their first dog. Mike had bought . . . a [Brittany] spaniel. He put it in a cardboard box, tucked it in behind the pilot's seat, and flew home." Brent explained further, "I remember my dad trying all the time to make [him into] a bird dog. My mom named him 'Tripod' because us three kids kept fighting over whose dog this was."

It was while instructing at Chanute that Adams was selected to move out to California's high desert, taking his family along to Edwards, which became his home for the remainder of his life. The reason for the move was to attend the Experimental Test Pilot School, where he was singled out for his flying excellence and as the best scholar in Class 62C, earning him the prestigious Honts Trophy. He then moved on to the Aerospace Research Pilot School as a member of Class IV. Adams graduated with honors in December 1963.

While still at ARPS, it was clear Adams's goal was to become involved with America's burgeoning space program. Less than three years had passed since Yuri Gagarin was launched into Earth's orbit from the Baikonur Cosmodrome in the Soviet Union. In the United States NASA was still trying to play catch-up to the Russian achievements. Six one-man Mercury spacecraft had been flown, the last by Gordon Cooper on 15 May 1963. Two groups of American astronauts were already in training for upcoming Gemini and Apollo flights, and more men were needed.

Mike and his good friend, Dave Scott, applied for the third group. For a while it appeared they might both make the cut. They had flown out to Brooks AFB, one of a ring of air force bases that used to surround San Antonio, to undergo their medical evaluations. Then Mike and Dave jetted to Houston for interviews at the newly opened Manned Spacecraft Center. The process was going smoothly.

At the time, Freida was less than enthusiastic about the idea of her husband being shot into space. This was a reasonable emotion, especially considering the close call Mike had after they had returned to Edwards in a two-seat F-104B with Scott. While practicing a low lift-to-drag ratio landing in the Starfighter, simulating the same type used in training X-15 pilots, their jet engine lost most of its thrust. Mike told Scott that if they hit, he would eject. Scott pulled the nose up to flare for landing, but soon after the back end of the F-104 struck the ground. Adams did as he said he would, safely punching away from the accident. In the seconds remaining, Scott stayed with the crippled craft.

In a serendipitous turn of events, both pilots made exactly the right decision. With the rear end angling downward at impact, the engine was displaced forward, smashing through the very spot where Adams's seat had been just moments before. Mike would have been crushed if he'd stayed with the aircraft. Scott saved his own life by choosing not to eject. The investigation showed his seat was damaged at the moment the fuselage hit the desert, and if he had pulled the handle to fire away, he would have been killed.

Even when the pilot survives an ejection, he often suffers some injury. George remembered, "I know Mike had some back stress . . . but he came through it nicely." Problem was, Mike was being considered for the new class of astronauts, and even a transient and minor back injury could pull him out of the running. When the announcement came on 17 October 1963, Dave Scott was among the fourteen pilots chosen to move to the astronaut office in Houston. Mike was not on the list. His back injury at just the wrong time released his spot to another candidate, Ted Freeman.

How did Mike react to the news? Freida recalled he was "very stoic. No comment." Even with no words spoken, she could tell how deeply the loss of the NASA slot affected him. It was at that moment her reservations disappeared about Mike's goals of riding a rocket. "I reacted just like a woman. 'Why?' I wanted to know. 'Why was he not chosen?' It was a disappointment for me; I'm sure it was for him—a bitter disappointment." After that, Freida made sure Mike understood she was completely behind him. "Now we're pushing for what he wanted. . . . So we got caught up in it."

Still assigned to Edwards, and putting the astronaut selection behind him, Adams moved on to his first assignment following ARPS graduation, working on the Lunar Landing Research Vehicle (LLRV). This ungainly contraption was all framework, plumbing, and engine, with a pilot's seat bolted out to the side. It was sometimes called the "Flying Bedstead" due to its four-poster appearance. This was his first direct involvement with the space program, as the LLRV was built to simulate lunar landings. The first vehicle arrived at Edwards from the Bell Aircraft plant in April 1964. Chief test pilot for the new program was Joe Walker, who had recently completed his time on the X-15.

Mike's desire was to work in space himself, not test equipment that others might use. Eventually, this led him to apply for a new all-military manned space program run by the U.S. Air Force, the Manned Orbiting Laboratory, or MOL. The spacecraft was a modified two-man Gemini, then nearing flight stage with NASA as a precursor to Apollo. The function of Gemini was to expand spaceflight to longer duration and to be able to work there productively. Mercury had literally been a man-in-the-can program for the most part. Cramped quarters and no way to open the hatch while in space meant the astronaut could do some experiments, but not much more. Gemini was very maneuverable, even being compared by some astronauts to a sports car. To this day, many people say that Gemini should be classified as the best spacecraft ever developed in America. MOL attached the Gemini-B to a dry upper rocket stage, outfitting it as a secret military reconnaissance platform in space.

Many historians later compared some of the imaginative medical tests used to weed out the early astronaut prospects as being closer to the methods of the Inquisition rather than to modern science. The doctors had no idea what to look for, so they went for broke, devising one bizarre test after another, hoping they would hit on the right combination of health and stamina. The most heinous of these tests had already been dismissed as irrelevant by the time Adams was first evaluated as part of the selection for the third group of NASA astronaut candidates. After his selection as a MOL astronaut, Mike faced another round of medical evaluations.

One area of space medicine still not fully understood involved a pilot's susceptibility to vertigo. A standard test had been put into place, but doctors weren't fully satisfied that what they were looking at would translate into an infallible indicator for any given individual.

Paul Bikle, NASA Flight Research Center director for most of the X-15 program, described what happened. "When Mike was in the MOL selection . . . they had screening tests for vertigo. They were in the process of developing a new [test], which they later decided was a better test. He went through [the screening], but as a guinea pig. . . . They were just trying to develop the experiment—anything that turned up in it wouldn't interfere with their selection." Adams felt it was in his best interest to cooperate as much as he could, so he agreed to undergo the new test. Bikle explained, "It turned out he became completely disoriented." However, as promised, that record was sealed and no mention was made outside the medical officials administering the evaluation.

Bob Hoey, a civilian X-15 flight planner working for the air force, recalled more specifics of the test. "They had a motion chair, and most pilots would get badly disoriented, then recover in a matter of seconds . . . Mike was messed up for minutes. He was very susceptible to that problem." Paul Bikle shook his head in sadness at the recollection. "Of course, the doctor probably never gave it a thought afterwards as to what happened to Mike. . . . When they adopted that test as the standard for screening pilots, you could say somebody should have gone back and looked to see whether any of the guys that had any trouble had gotten through, but they didn't." Mike Adams continued as a MOL astronaut, possibly not even aware himself his experience was out of the norm.

MOL had been an attempt to reinstate a stand-alone military presence in space after the Dyna-Soar space plane was canceled. Adams became one of eight astronauts announced on 12 November 1965 to fly the program. Right from the start MOL came under political fire for militarizing space. There was little opposition to unmanned spy satellites, but adding astronauts was a different matter. Political niceties aside, the Cold War was raging in full force. In that climate the air force continued to garner lukewarm support for the continuation of MOL.

Political haggling was left to those in Washington DC. At the Adams home, Freida immediately saw the difference in her husband after his astronaut selection. She said, "Everything was really exciting when he was chosen. . . . They were the top people at the time."

There was no upheaval for the family, as Mike's assignment kept him officially at Edwards throughout his time on MOL. He did often make the hop to the headquarters of the military space program, the Space Systems Division at Los Angeles Air Force Station. Unfortunately, it apparently didn't take too long for Adams's exuberance to fade as he figured out that his ride into space with the air force might be a long time coming. Mike's brother, George, said, "He definitely wanted to do more flying than what he was doing . . . but he had at least gotten onto MOL."

Apollo was moving quickly, and Mike expected that pace in the military too. Instead, the only fast-paced thing in this program was delays. From the beginning, launch dates got further away, not closer. Mike's boredom was palpable, so he felt he had to move on.

Being in an apparently dead-end job gave Adams all the impetus he needed to seek a new venue for his talents. George said, "[When] the spot opened up on the X-15, he jumped at that." Barely eight months after joining the astronaut team on the Manned Orbiting Laboratory, Adams departed for the X-15, officially joining the rocket plane team on 14 July 1966.

In an interview a month later for the *Sacramento Bee*, the newspaper Mike had delivered to neighbors in his youth, Adams was quoted as saying, "I had my choice of either [the X-15 or MOL]. . . . I felt I could be used more effectively now in the x-15 program." Clearly he was elated to get flying to-ward space in the x-15 as soon as possible. "I'm not particularly anxious to set any records. But if there are any, that's just a cheerful fact that goes with the mission. . . . The speeds and altitudes exceed anything I've seen before."

Mike's private predictions concerning the long-delayed flights of MOL were well founded, with the military making only one test flight of an un-manned Gemini-B before shutting down the program completely in 1969.

Once he shifted to the x-15, Mike perked up again. Freida explained, "It was like living in a unique world because that was a special thing. They treated you differently." The excitement from his wife and brother was much more visible than from Mike himself. "Mike wasn't subject to open-ing up too much about things," recalled George. "A lot of times you had to drag things out of him. . . . He didn't espouse a lot of opinions about things. He just wasn't that kind of an individual." George summed up his older brother as "quietly competent. If you didn't ask him a lot of ques-tions, you weren't going to find out a lot." Freida explained Mike's single-minded purpose: "When he set his mind to something, he did not waver. It would get done."

Adams went to work preparing for his first flight, which had originally been assigned to Jack McKay. When McKay left the program in September 1966, the flight was transferred to Adams. The first attempt was proceeding well until it was canceled by bad weather. A week later a second abort again forestalled his familiarization flight. This time it was the cockpit pressuriza-tion system, a nagging problem on all three x-15s as far back as June 1964. A new design was finally completed and installed, and only one x-15 flight was aborted during the remainder of the program due to this problem.

Two days later, on 6 October 1966, Mike Adams was ready to launch on his first flight, mission 1-69-116, the first step on what he hoped would be his road into space. Heading toward Nevada, Adams and the x-15 were firmly attached to the mothership's right-wing pylon. The launch lake was just inside the California border at Hidden Hills. After the turn to get the flight onto the correct heading back to Edwards, Adams dropped away from B-52 no. 003.

It was less than a second until Adams got the engine out of idle and start-ed to push forward on the throttle, getting the aircraft set up on profile as he pulled away. The joy of that flight was tempered just under ninety seconds later when the LR-99 quit more than thirty seconds early.

With the rocket thrust set at 50 percent to keep the speed down on this initial flight, he had climbed to 75,000 feet, then pushed over to level out for the remainder of the engine burn. Unbeknownst to the pilot, as he opened the fuel jettison valve during the checks prior to drop, a pressure differential caused a rupture between the center and forward compartment bulkheads in the ammonia tank. Due to this failure, the fuel could not flow out of the center tank, reducing what was available by one-third.

Both the bulkhead failure and the zero-g condition at pushover caused a momentary starvation in the manifold, forcing the engine to shut down. Adams's report stated, "That was when there was a loud bang behind me, and the engine quit." He attempted to restart the rocket, but the starvation caused a turbopump overspeed, preventing it from reigniting.

With ninety seconds on the clock, mission rules dictated Adams had to land about thirty miles short of Edwards by going into Cuddeback Dry Lake. NASA 1 radioed, "Okay, Mike, you look in real good shape," to which Adams replied, "Looks like I could make Edwards." The controller made sure the rules were understood: "Let's make it Cuddeback, Mike." Even with the emergency landing he was to face, Mike's demeanor was unfazed as he said, "This thing is fun to fly even if I have to go to Cuddeback." Getting him to focus on the landing, NASA 1 said, "Well, let's talk about it after you get it on the ground."

Because of the damaged fuel tank, the ammonia that was unable to feed the engine was also unable to be jettisoned. This created a dangerous situation, not recognized until post-landing analysis. At touchdown, the aircraft was more than a metric ton overweight, even for the standards of an emergency landing. This accounted for Adams's report that when the skids hit the lakebed, the shock was higher than normal. "I thought it was kind of hard, but I guess you can look at the traces and see how it looks." Those traces were not recorded by the flight instrumentation. Because of the two aborted flights prior to flight 1-69, the data film on board the X-15 was depleted thirty seconds before the landing at Cuddeback.

One of the chase plane pilots followed him in and said Adams had made a perfect centerline touchdown, about a half mile into the three-mile-long runway. NASA 1 passed along, "Excellent." Adams was exuberant, saying,

"I'll still say it's fun!" Now that he was safely on the ground, the controller agreed, "Very good, it's fun now, Mike. We can all laugh."

Most pilots might be disappointed when their first flight ran into difficulties and forced an emergency landing at a remote lakebed. For the normally stoic Mike, however, it was obvious he got a kick out of flying this rocket plane and was very happy being in the cockpit. Fellow x-15 pilot, Bill Dana, praised Mike's performance during the emergency: "He did an excellent job with that. . . . You could tell by the way he was talking that he wasn't uptight about having to go into a strange field on his first flight."

Afterward, in his post-flight report, Mike admitted, "In the launch itself, I was really a little bit behind, because I was getting the vertigo, which I expected from Pete [Knight] telling me, and from previous flight reports." Using the word "vertigo" here is telling for Mike, especially when added to additional reports over the next year. Other pilots had spoken of the shock of being dropped off the B-52 pylon on their first flight, but it was believed this was the first time anyone had used that specific term to describe a personal reaction to launch.

The second flight for Adams was for further familiarization with the dynamics and handling characteristics of the x-15. The biggest change was to open up the speed envelope to approximately Mach 4.5 by increasing the engine thrust from 50 to 75 percent. The increased power of the LR-99 was evident in that this higher velocity was achieved even though burn time was reduced by thirty seconds.

Flight 3-57 went flawlessly in the late morning of 29 November, with the small exception that Mike's radio cut out right after launch. Voice contact was not regained until he was passing just east of Cuddeback, already slowing through Mach 3 and descending toward the landing pattern at Edwards. After the flight Adams compared his first and second touchdowns: "I think the first one you make . . . you are just going to have to work extra hard. . . . The second one you try a little harder and you control the trajectory a little bit better, but gee, the thing will almost land itself."

Winter weather and wet lakebeds put off any further flights for anyone on the x-15 for nearly four months. Aircraft no. 1 was brought back to Edwards on 3 February 1967, and drier weather permitted flights to resume in March. On flight 1-70, Adams opened up the throttle to 100 percent, re-

40. B-52 mothership no. 008 carries X-15 no. 3 aloft. The T-38 chase plane flies close by awaiting the order for the X-15 to launch. Courtesy of NASA Headquarters.

cording his highest speed flight in the X-15 at Mach 5.59. Launch this time was from the Mud Dry Lake area, nearly twice as far from Edwards as his first two missions.

The next flight for Adams expanded his altitude experience to 167,200 feet, slightly lower than the planned 180,000 feet. Coasting after engine shutdown, he kept hearing a noise that he initially thought was coming from the ballistic control system. By that time he was low enough to no longer need the tiny steam jets to control his attitude, so he switched off the system. The noise persisted and may have been associated with the troublesome Western Test Range experiment, making its debut checkout on this flight.

By 15 June, and after three successive aborted attempts, Adams was ready for a much larger leap in altitude, topping out at 229,300 feet in his final flight in X-15 no. 1. In this instance he lofted an extra 10,000 feet. As he ascended through the 200,000-foot mark, Pete Knight radioed it was time to operate the experiment package. "I extended it, heard the clank, [then] saw the lights come on," Mike said. With the extra altitude also came extra time at zero g as the aircraft drifted through its ballistic arc at nearly Mach

5. Adams got somewhat annoyed by one aspect of having no gravity to hold things in place. "My checklist would keep flopping up and over, and it was waving around in front of me. . . . I knocked it down, [but] it immediately stood straight up . . . so I quit fighting that." Mike allowed the pages to do as they wanted until gravity returned as he began reentry.

During descent, Mike's speed topped out at Mach 5.14. Soon after this, air friction began to slow the descent of the x-15. After a successful reentry from 43 miles, Mike set up on final approach to the Rogers lakebed. He extended his flaps, brought the nose up to flare out, then released the rear skids and nose wheel. During his debriefing Adams was critical of his flight performance, stating, "Compared to what I could do in the simulator, and what I was doing in flight, I wasn't doing quite as well. Maybe we should fly more often."

Mike was correct that the flight rate of the x-15 had dropped to half the peak rate of thirty-two flights in 1965. During that year, there were six pilots who had flown for the research program. By this point in mid-1967 there were only three: Knight, Dana, and Adams. Overall, the flight rate for each pilot was relatively in line with previous years, but this would not have changed the fact Mike still craved a more aggressive flying schedule. After all, this was why he left MOL.

It took more than two months for Adams to get back into the x-15 cockpit and ready for flight 3-62. On the previous mission of no. 3, Bill Dana had experienced a guidance computer malfunction. After much discussion and several changes of plans, the decision was made to simply redo the same flight plan for Adams's mission. During final checkout of the computer, further problems were found, so it was pulled from the aircraft. The flight plan changed yet again.

Even though it had taken too long by Mike's reckoning, he was certainly ready to go. It was the first time since entering the program he accomplished two back-to-back flights with no aborts in between. Returning to Hidden Hills for the drop on 25 August, Mike got ahead on his checklist and was obviously anxious to get under way. A radio call from one of the chase planes questioned the cloud cover and asked Mike for his input: "Can you see Hidden Hills up there at one o'clock?" Adams let his irritation show in his reply: "Can't see out of this thing at any time." Four minutes later NASA 1 confirmed the weather was okay for launch. "Looks good, Mike." Adams went for the launch, saying, "I hit the switch and threw the throttle on as fast as I could—and got a vibration malfunction shutdown."

The x-15 continued its unpowered drop toward the ground as Mike ran through his restart checklist. "I got a malfunction shutdown. So I went throttle off, reset, prime, igniter ready, throttle on. . . . I waited and I waited." Sixteen very long seconds later and more than 4,000 feet lower in altitude: "About that time I was getting very discouraged—and it started." The familiar hard acceleration of the LR-99 finally kicked in.

Adams compensated for most of the altitude loss at launch, leveling out just 600 feet below the planned 85,000-foot target. Coming down quickly he prepared to land, later telling his debriefers, "I don't know where I touched. It goes and goes. . . . The thing really floats. You can hold it off a long time. . . . It's a good airplane."

As the dust on the lakebed began to settle and the friction of the skids brought the aircraft to a full stop, the clock read 1:35 p.m. on the afternoon of 25 August 1967. After six flights, Mike Adams felt he was just getting the feel of this rocket plane. He looked forward to new and more meaningful missions, and he still had his sights on space. He had no way of knowing he had just completed his final successful mission in the x-15.

Moving to Edwards from Louisiana had originally been a shock for Freida. She had been raised amid the lush greenery and the muggy air of the bayous. There was life everywhere. The high, dry desert of California seemed like being dropped onto Mars. "I got there and I thought, 'This is the end of the world.' Then I learned to love it." She found the climate and stark beauty of the area had a lot to offer. The three children, all born in different states as their father moved from assignment to assignment, seemed to thrive in the adventure of it all. For the senior Adams, it was a new place to explore, as he found not only the desert but the nearby mountains and rivers. Freida said, "It was just a real fun time. We had good friends. We were always getting together. He would love to ride through the countryside. We saw more strange areas that no one else could ever see or find. We'd go up into Tehachapi and see these strange-looking quail, or up toward Kern County, where there was a beautiful stream. Wherever Mike was, he was outdoors."

Adams also loved his music. Freida remembers one of his favorites as being trumpeter Al Hirt, who was at the top of his fame in 1966 and 1967 for recording the theme from *The Green Hornet* television series. "Then, the kids came along, and we grew up on the Beatles and all that. . . . Mike liked lots of different things. He played the guitar, played the piano, and the accordion. He was talented. [The] kids used to play in bands, and we used to say they got it from their dad. . . . They don't get far away from their dad and his achievements—or his life."

With Mike's long hours and piloting duties in the air force, Freida did the bulk of child rearing. When he was home, he tried to make good use of the time for family outings and truly doted on his daughter, Liese. It was his influence, and his memory, that later led her into a technical career when she finished school. "She's got that mathematical mind he had," Freida said.

Forty years later, on a late spring evening, sitting at a dining-room table in her home in Monroe, Louisiana, it was easy to see how vivid these memories still were for Freida. One memory would bring forth another of her husband, whom she called "Big Mike," differentiating from their son "Little Mike." Freida smiled and said, "One interesting thing about Big Mike . . . he'd go TDY and would sit in the room and make model airplanes while other guys, like Dave Scott and them, would go to the Bunny Club. Big Mike had all his money when he came home because he hadn't bought anything. He would sit there and make those model planes."

Early in the x-15 program, the pilots often got together for huge parties. Any excuse and the beer would flow and the barbecues were lit. Late in the program, this didn't tend to be the case any longer. They were still a unique fraternity that shared a dangerous job, but the closeness had somewhat disappeared. According to Freida, one of her husband's best friends was the base dentist, Dr. Paul Swenson. The doctor stood out in her memory as a man with a large mustache—one deserving of a finger twirl, as if from an evil villain in a silent movie. Mike and Paul hunted together, as would another of Mike's close friends, his former commander from the Test Pilot School, Chuck Yeager.

The trips were his best way to unwind, although Mike's idea of a good time did not always work out best for Freida. "I went with him once up into the Sierras. . . . That light air just knocked me out. . . . He'd arrange these vacations [where] all of us had to go whether we wanted to or not," she joked. One place that stood out in her memories was Lost Lake in southwest Colorado. "It was very remote. That's what he liked."

Big Mike instilled the love and adventure of wilderness in his son Brent, along with the responsibility. Brent recalled, "Every time he'd come home, he'd be cleaning a gun, always polishing those guns—which I have now. He probably had thirty guns. . . . They all looked brand new. He was real meticulous."

Mike rarely spoke of his flights or his work on the x-15. For the most part he was regarded as a very serious man. In photos he is never seen with a smile. All the way back to his high school yearbook, or photos with his buddy Charles while pheasant hunting, show the same dour look. A rare exception is a single photo from 1955. Freida is in her white lace wedding gown, a small string of pearls around her neck. With her short, dark hair she looks very much like Noel Neill of 1950s *Superman* fame. Mike is in a suit and tie, a white carnation in his lapel. With one arm around Freida, knife in their clasped hands, cutting their wedding cake, Mike has a huge grin. But at work in the military, or sitting in their yard with the kids, he was always back to that stoic expression. Mike's brother, George, said, "He had a good sense of humor. He just didn't expose it unless you brought it out in him." This was exemplified in another photo from his days at Edwards. During the rainy season, when no x-15 flights were scheduled because of the wet lakebeds, Mike had an image snapped of him standing on the rain-soaked dirt, holding his pressure suit helmet—and an umbrella.

Sheri McKay-Lowe, daughter of x-15 pilot Jack McKay, recalled, "I would say Mike Adams became my father's closest friend." Early in November 1967 Jack and Mike went hunting together over a weekend. A week or two previous to that, Adams had stopped in to see his family in Sacramento and decided to take a quick outing with his childhood buddy Charles, who remembered, "We took my dad's boat out and went fishing. . . . We were sitting out there, and he said, 'You know, I'm flying faster than those bullets coming out of a gun.' It was just fascinating to hear him say that. That was when I last saw Mike."

On the same day as Pete Knight's Mach 6.7 record speed flight in the x-15A-2, Bill Dana and a second team had attempted a flight in no. 3. That mission never got underway due to technical problems. The following day, 4 October 1967, ten years after the Soviet Union launched the first Sputnik satellite, escalating the Cold War into the space arena, Dana was able to make his research flight attempting to capture meteorite dust. This was the tenth time two flights occurred just one day apart, but the last time this was accomplished in the x-15 program.

Two weeks later Pete Knight was back in the air flying no. 3. This flight became the highest altitude achieved by Knight. At 9:40 a.m. on the morning of 17 October, Knight dropped away from the b-52 and ignited the lr-99 rocket engine. Ten minutes and six seconds later, he brought the aircraft safely to a halt on Rogers Dry Lake. It was the final time x-15 no. 3 successfully completed a mission. Less than a month later this aircraft, and one of Knight's and Dana's friends and fellow research pilots, Michael J. Adams, would no longer exist.

Preparations for flight 3-65 began with changing the lr-99 rocket engine. Serial no. 103 had been installed on 4 April and stayed with the aircraft for seven missions. No major malfunctions appeared until a leak was found in the throat area of the rocket after Pete Knight returned from flight 3-64 on 17 October. Engine 104 was installed. After the change-out, the rocket shop technicians had to run the engine on the test stand. A limit switch refused to open, and the engine would not ignite. Engine 104 was out and 110 was in. On Halloween morning Adams made the first attempt at mission 3-65. In the final seconds prior to launch the lr-99 failed at igniter idle, creating aborted flight 3-A-96. Upon return engine 110 was pulled and 108 was installed in its place. A fuel valve leak was uncovered, necessitating another change, this time for serial number 111. The auxiliary power units, ballistic control system jets, and engine no. 111 passed their ground runs on 9 November.

Aircraft no. 3 went into final preparations, culminating with close-out of all systems on 13 November. The weather was not favorable, causing one more day of delay before it was mated to b-52 no. 008. x-15 tail no. 66672 was finally ready to launch on flight 3-65-97.

Col. Joseph P. Cotton spooled up the b-52's eight jet engines and began his taxi onto the Edwards runway. In the right seat was Sq. Ldr. John Miller, on temporary assignment from the Royal Air Force. This was Cotton's twelfth time piloting the mothership for the x-15, and Miller's second as copilot. Jack Russell was again in his position as launch panel operator, seated near the bubble window, about the midpoint on the right side of the

B-52's fuselage. Mike Adams was sealed into the X-15 cockpit, mounted under the right wing. At 9:13 a.m. on 15 November the B-52 and its X-15 cargo lifted into the air and headed outbound. Adams was just along for the ride at this point, telling Chase 1, "I am just a tourist."

The other three assigned chase pilots all flew F-104 Starfighters. Chase 2, with Hugh M. Jackson, formed up, calling, "Chase 2 has mothership." Adams jokingly asked, "How come we don't call it the father ship?" Pete Knight as NASA 1 interjected, "Keep it clean up there now." Adams kept the joke going, "I'd rather have Colonel Cotton as my father than my mother." To which Cotton deadpanned, "Because of your umbilical source?"

The group climbed to 24,000 feet and went into a holding pattern over the Edwards north base area, awaiting the C-130 support aircraft, which was delayed thirty minutes due to hydraulic problems in the number three and four engines. This aircraft, under the command of Capt. Floyd B. Stroup, had already flown a mission at 6:30 a.m. to deliver a fire truck and the NASA 8 jeep to the launch lake at Delamar. They would be there in case Adams had to make an immediate emergency landing after drop. The C-130 Hercules then returned to Edwards, picking up a second jeep (NASA 7) and the rest of the personnel who would orbit at the halfway point for the mission, flying in the vicinity of Hidden Hills, on hand to support an emergency.

Bill Dana was originally assigned as the roving chase and took off twenty-four minutes following the B-52. Two minutes after Dana became airborne, Pete Knight informed the group Dana would take the duties of Chase 3. Dana headed his F-104 toward Hidden Hills, to be near the orbiting C-130. Chase 4, with Maj. Ted Twinting, was to meet up with Adams near the final emergency landing zone at Cuddeback to accompany the X-15 back into Edwards.

Twinting ran into a low fuel situation due to the extended wait for the C-130, so he was directed to instead head for the "high key" point on the flight path, which meant staying in the local area. Also during this wait time, his radio started to act up. Capt. Thomas J. Davey Jr. was on an F-104 mission to provide chase duties for an SR-71 Blackbird test. After takeoff the SR-71 flight was delayed, so Davey checked in with NASA and found he could be useful as the backup to Chase 4, in case Twinting's radio failed. The two aircraft met up over the Fort Irwin area.

After the C-130 finally took off, the B-52 and two chase aircraft proceeded northwest toward the Delamar Dry Lake launch zone. Cotton noted that they reached their cruising altitude of 45,500 feet faster due to an air temperature of minus forty-two degrees Fahrenheit, considered lower than normal. Adams felt what he thought was rough air while riding on the B-52's wing. Cotton believed it was vibration due to Mach buffet, so he lowered the altitude by nearly 500 feet to slightly increase air density.

Fifteen minutes prior to the scheduled release, Adams was busy prepping the systems. When he test fired the ballistic jets in the nose, a right yaw jet had a small amount of residual vapor from a propellant valve leak. Lt. Col. Fred J. Cuthill told Mike, who pulsed the jets again, clearing the problem. Jackson moved Chase 2 into launch position about one hundred feet to the left and behind the X-15, while Cuthill moved directly off the B-52's right wing to obtain photo coverage of the drop and climb-out. The launch master switch inside the B-52 was activated, allowing Adams to now control his own drop. Knight called from NASA I, "Looks good here, Mike." Adams responded, "Rog . . . two . . . one . . . launch." At 10:30:07.4 a.m. he commanded the launch release. Falling away from the pylon, he immediately moved the LR-99 throttle out of the idle detent position, smoothly and quickly pushing the X-15 to 100 percent thrust. Four seconds later NASA I radioed, "Rog, we got a good light here, Mike." Adams did not respond to further radio calls for nearly two minutes.

The X-15 quickly outpaced Cuthill and Jackson in the chase planes, as Adams angled upward through the stratosphere, heading to the top of his ballistic arc in the mesosphere. When Mike passed through 140,000 feet, with eighty-two seconds on the mission clock, he delayed the LR-99 shutdown four seconds longer than planned. The extra seconds imparted an additional velocity of 136 feet per second, causing an overshoot in altitude of 16,000 feet.

This configuration of X-15 no. 3 had two six-foot-long pods mounted to the tip of each wing. On the left were two experiments that opened at high altitude. One was the infamous micrometeorite collection box, while the other was the JPL solar spectrum measurement device. The right tip pod held an extendable probe known as the bow shock standoff measurement experiment. It was later found that the drive motor on this probe had not been properly checked in an altitude chamber, and electrical arcing occurred above 80,000 feet. By the time the aircraft was climbing through 100,000

feet, the arcing extended to one inch across the exposed experiment termi-
nals, producing a hot blue coronal discharge. The disturbance caused noise
in the wiring of the X-15, affecting operation of several critical systems, including the MH-96 adaptive flight control.

Another casualty of the interference was the computer, which started to
continually dump and reset—a total of sixty-one times before the end of
the mission. The computer light came on, informing Adams of a problem.
He pressed the light six seconds after engine shutdown in an attempt to reset the system. It did not go out, as the light was refreshed each time a new
dump occurred.

Mike activated the two experiments in the left pod. The nose cap popped
down and the micrometeorite collection box extended, while a hatch on
the upper side of the pod opened to reveal the solar spectrum instrument
in a white rectangular box with numerous lenses and mirrors. Seconds later NASA I called up to remind Adams to "rock your wings, and extend your
experiment." With the radio apparently acting up, the B-52 relayed the same
message, but Mike had already taken the action. At this point, he was staying on the checklist timeline.

Telemetry received on the ground showed Adams was on the planned
heading and flight profile, although they could tell his trajectory was slightly
higher than planned due to the extra burn time of the LR-99. As these calls
went out through Pete Knight, and were repeated by the B-52 crew, Adams was realizing the computer glitch was not going away. He called back
to Pete for the first time since launch to explain his situation. "Okay, I am
reading you. I got a computer and instrument light now." From this point
through the end of the mission, the communication will speak for itself,
although information will be interjected concerning what was transpiring
with the aircraft. The time of transmissions are given to the nearest second,
Pacific Standard Time.

10:32:42	NASA I: *"Have you coming over the top. You're looking real good. Right on the heading, Mike."*
10:32:51	NASA I: *"Over the top at about 261 [thousand feet], Mike."*
10:32:54	NASA I: *"Check your attitudes."*
10:32:58	*[Mike takes the ballistic control system off automatic.]*
10:33:00	*[Maximum altitude of 266,000 feet.]*

10:33:02	NASA 1: *"You're a little bit hot [with a higher than expected velocity], but your heading is going in the right direction, Mike."*
10:33:09	NASA 1: *"Real good."*
10:33:10	*[The nose of the x-15 is now yawing eighteen degrees off course to the right. There is no telemetry channel to pass heading data to the control room. Pete Knight and the rest of the team have no indication the x-15 is deviating from its intended direction.]*
10:33:11	NASA 1: *"Check your attitudes. How do you read, Mike?"*
10:33:14	NASA 1: *"Okay, let's check your dampers, Mike." [Mike fires the ballistic control jets in the nose and increases his yaw rate to the right.]*
10:33:17	Adams: *"They're still on."*
10:33:18	NASA 1: *"Okay."*
10:33:20	*[x-15 yaw is now twenty-eight degrees to the right of the flight path.]*
10:33:24	NASA 1: *"A little bit high, Mike, but real good shape."*
10:33:30	*[Yaw rate is now at 5.6 degrees per second, and the x-15 heading is fifty-three degrees to the right.]*
10:33:33	NASA 1: *"And, we got you coming downhill now. Are your dampers still on?"*
10:33:37	Chase 1: *"Dampers still on, Mike?" [Squelch break as Adams possibly hits his microphone key.]*
10:33:38	*[Yaw is now at ninety degrees.]*
10:33:39	Adams: *"Yeah, and it seems squirrely."*
10:33:44	NASA 1: *"Okay, have you coming back through 230 [thousand feet]. Ball nose, Mike." [The Q-ball would be needed once into the denser atmosphere.]*
10:33:49	*[Yaw is at 180 degrees. Adams is flying backward along his flight path.]*
10:33:50	NASA 1: *"Let's watch your alpha, Mike."*
10:33:53	*[x-15 enters a high-altitude, hypersonic spin, which lasts for the next forty-three seconds, and three full rotations of the aircraft. Adams tries to correct the spin using the ballistic control system. During this time the aircraft drops 100,000 feet.]*

10:33:58	NASA 1: *"Let's not keep it as high as normal with this damper problem. Have you at 210 [thousand feet]. . . . Check your alpha, Mike."*
10:34:02	Adams: *"I'm in a spin, Pete."*
10:34:05	NASA 1: *"Let's get your experiment in and the camera on."*
10:34:10	*[Mike switches the ballistic control system back to automatic. Pitch, roll, and yaw jets fire in an attempt to stabilize the X-15.]*
10:34:13	NASA 1: *"Let's watch your theta, Mike."*
10:34:16	Adams: *"I'm in a spin."*
10:34:18	NASA 1: *"Say again."* *[The X-15 is now spinning with the nose pointed almost straight toward the ground.]*
10:34:19	Adams: *"I'm in a spin."* *[Last transmission from Mike.]*
10:34:21	NASA 1: *"Say again."*
10:34:27	NASA 1: *"Okay, Mike, you're coming through about 135 [thousand feet] now."*
10:34:34	NASA 1: *"Let's get it straightened out."*
10:34:36	*[X-15 hypersonic spin ends. Aircraft yaw is zero degrees, heading directly into the flight path. Altitude is 120,000 feet. Speed is Mach 4.7.]*
10:34:37	*[Two more radio squelch breaks, possibly from Adams's microphone.]*
10:34:42	NASA 1: *"Okay, you got theta equal zero now."*
10:34:44	NASA 1: *"Get some angle of attack up."*
10:34:47	*[The ballistic control system is turned on and off several times during the emergency. At this moment, it is turned off for the final time.]*
10:34:50	NASA 1: *"Coming up to 80,000 [feet], Mike."*
10:34:52	*[X-15 starts to break up as the airframe is overstressed, pitching up and down at fifteen gs. Speed is Mach 3.9, approximately 2,600 mph.]*
10:34:53	NASA 1: *"Let's get some alpha on it."*
10:34:55	*[Final ballistic control system pulse. Possibly the last act accomplished by Adams as the aircraft is disintegrating around him.]*
10:34:57	NASA 1: *"Get some g on it, Mike."* *[Altitude approximately 70,000 feet.]*
10:34:58.8	*[All telemetry lost as X-15 no. 3 breaks apart at 62,000 feet.]*

10:34:59	NASA 1: *"Let's get some g on it."*
10:35:02	NASA 1: *"We got it now, let's keep it there. Coming around."*
10:35:09	NASA 1: *"Okay, let's keep it up, Mike."*
10:35:14	NASA 1: *"Keep pulling it up. Do you read, Mike?"*
10:35:20	NASA 1: *"Let's keep pulling it up, Mike." [Approximate time when the X-15 impacts the desert in a hilly area four miles north of Johannesburg and eight miles south of Ridgecrest and the China Lake Naval Weapons Center. Wreckage is scattered over several miles of terrain.]*

As is obvious from the flight 3-65 radio transcripts, those in the control room and the aircraft aloft supporting the mission had no idea of the severity of the events unfolding. Even though Adams radioed three times in just seventeen seconds that he was in a spin, it appears to have never fully registered.

The X-15 had flown successfully since Scott Crossfield's first glide test on 8 June 1959. It had never killed anyone, although Jack McKay had been seriously injured in November 1962 when no. 2 rolled over on landing. No one expected a program this mature could suddenly turn deadly.

The initial reactions of disbelief were similar to what occurred as the Space Shuttle *Challenger* broke apart high in the Florida sky on 28 January 1986. The controller's voice from Houston was even and measured, stating, almost flatly, "Obviously, a major malfunction." Everyone was waiting for some sort of miracle, hoping to see an intact Space Shuttle come winging out of the fireball, gracefully swooping back for a landing at the Kennedy Space Center, all hands unhurt. Like with those people at Edwards on the fateful day with Mike Adams, the reality of the situation took time to be accepted.

As the seconds unfolded from Adams's final transmission, the hope was this might be a similar situation to Pete Knight's electrical failure, less than five months previously. Maybe Mike had a radio malfunction and couldn't tell them his situation. Maybe, even then, he was lining up for an emergency landing at a remote lakebed. If the X-15 had somehow been lost, it was assumed Mike made a successful ejection, floating under his parachute after his harrowing descent from space.

In support of the first idea, someone in the air near Cuddeback, possibly Major Twinting in Chase 4, radioed he had seen dust on the lakebed. Everyone's hopes immediately soared that both plane and pilot had safely landed.

41. At the crash site of x-15 no. 3 on 15 November 1967, investigators start to piece together the reasons for the accident that took the life of pilot Michael Adams. Milt Thompson is in the center of the photo in the white coat. Courtesy of the Armstrong Flight Research Center.

Captain Davey, who had been vectored to help with x-15 chase duties, was over Cuddeback at the time. He informed NASA 1 there was no aircraft on the lakebed, and the hopes were dashed. Davey took up the search and sometime in the next ten minutes was the first to see signs of the downed aircraft. He radioed the position, and Captain Stroup's c-130 was soon over the zone to direct a helicopter in to land. Davey and the other chase plane pilots continued to search the area, looking for signs of a parachute in the desert.

An army UH-1 helicopter, commanded by U.S. Air Force major Joseph

G. Basquez III, was the first to land at the scene, about twenty yards from the x-15's forward fuselage. The time was 10:57 a.m. Aboard the helicopter with him were the flight surgeon, Lt. Col. Robert E. Matejka; NASA technician Frank Fedor; and a fourth unidentified person. All four headed to the still smoldering wreck. Basquez stated, "The aircraft was on its left side and burning slowly at the mid-section." The flight surgeon proceeded to the cockpit area, while the others used small fire extinguishers. Moments later the surgeon verified Adams's body was still in the wreckage.

Everyone involved with the mission that day was putting their hopes into Mike still somehow being alive. At 11:01 a.m. on Wednesday, 15 November 1967, the flight surgeon made the radio call dispelling any doubt of a miracle. Mike Adams had been killed in the crash of x-15 no. 3.

This day started no different for Freida Adams from all other x-15 flights when her husband was the pilot. She was at the NASA Flight Research Center watching the proceedings. Also joining her was Mike's mother, Georgia (his father had died thirteen years previously). She was in town to spend the upcoming Thanksgiving holiday with Mike and Freida and her three grandchildren. This was the first time Georgia had an opportunity to see her son doing the job that made him famous back in Sacramento.

Freida sat quietly with Georgia in the NASA control room, not wanting to get in anyone's way as the flight slowly proceeded through all the delays prior to launch. With six previous times Freida was expert enough to whisper explanations of what was transpiring to her mother-in-law. Once Mike dropped from the b-52, there was an excitement to hear all the radio chatter. She thought it unusual there wasn't much being heard from Mike, but everything seemed fine. Then she heard her husband come back on the line, and he was talking about problems. Freida perked up, and possibly grabbed Georgia's hand. Within seconds she heard the fateful call from Mike, "I'm in a spin, Pete." The two women were sitting not far from Knight at that moment, and saw his reaction. He appeared calm, as he was trained to be under any circumstances. Like everyone else in the control room at that point, no one knew exactly what was going on, but Freida could tell the attitude had shifted. Everyone was sitting or standing a little stiffer, more attentive than normal. Something was wrong. "I knew," she confided to me years later.

Almost immediately, even as events continued to unfold, someone had the presence of mind to realize Freida and Georgia should no longer be in the room. She recalled that, without anyone saying a word, the message was: "Get them out of here." Freida said that moments later, "They whisked us away. We were standing, listening, and, of course, everything was going awry. They quickly got us out of there. I can remember I wanted to help [Georgia]. I wanted to be sure she was protected from . . . whatever. So I was trying to be the mother hen to her when I was probably falling apart." Freida doesn't fully recall where the two of them were taken or if anyone was with them. "Probably a briefing room," she said. They had no way of knowing anything of what was transpiring in the desert north of Edwards.

As soon as it was made known Adams was in the wreck of the x-15, all chase planes were released from further searching. Soon after, the c-130 was also sent back to Edwards. The fire truck stationed on Cuddeback was told to head to the crash scene to render what aid it could. The closest military installation was the naval station at China Lake. Some air force security personnel happened to be attending a meeting there. Immediately, they and some additional naval security drove to the site to keep it closed while the investigation got underway. By 11:20 a.m. the news release went out to both the military and civilian media wire services informing the world of the accident. The b-52 launch aircraft landed at 11:25 a.m., two hours and twelve minutes after they had lifted off with Mike Adams and the x-15.

Brent Adams and his big brother, Mike, were in the same school. After their father's crash a school official's wife, Mrs. Ardent, came to each of their classes and had them follow her out. Brent recalled, "She said something to the fact they thought somebody was sick. . . . She was pretty smooth." Obviously reluctant, Brent admitted, "I hate rehashing that whole day," then he went silent for a time. Finally, Brent went on, "We went straight to the house and saw all those cars. I knew something was wrong." The subterfuge to the boys disappeared in a flash. Their mother and grandmother were already home, and once inside Freida sat the children down and explained their father would not be returning.

Less than ninety minutes after the wire-service notification went out, a civilian Cessna was witnessed making a very low, unauthorized pass over the crash site, snapping photographs. Several more civilian and military aircraft were also observed flying overhead in the next minutes, but they maintained a more respectful altitude. Around 1:00 p.m. NASA security arrived and officially took charge of the area. Soon representatives from the highway patrol, county sheriff, and military were present. At 1:30 p.m. the San Bernardino county coroner arrived and began the grim task of removing Adams's body from the cockpit. Members of the press started arriving at the scene soon afterward but were denied access until the coroner's job was completed at approximately 3:00 p.m.

Mike Adams's remains were loaded aboard a helicopter and returned to Edwards for an autopsy, which began at 5:15 p.m. Based on that examination and the full investigation into the causes of the accident, the flight surgeon, Lieutenant Colonel Matejka, wrote in his report: "Deceleration forces upon ground impact was an absolute cause of death from multiple extreme injuries."

The wreckage was photographed, plotted on maps, then gathered up and returned to Edwards, where it was reassembled as well as could be expected to try to determine the root cause of the accident. The pieces had been spread over a tract of desert twelve miles long and two miles wide and were now laid out on the floor of what was then the Edwards heat facility at building 4820.

One item not immediately found that could shed a great amount of light on events was the camera attached to the cockpit canopy mounted just over Adams's right shoulder. The canopy had separated cleanly from the rest of the x-15 as the fuselage bent severely downward during the stresses of coming apart, and it was found relatively quickly. The only damage apparent was a crushed area at the right front due to ground impact, with three of the four inner and outer window panes smashed. Sheer forces from the fall separated the camera and film cassette, and they were not found with the canopy. This camera recorded all instrument readings and, if found, would enable investigators to reconstruct events.

Flight planner Johnny Armstrong recalled, "Many of us, all the crew, went through a search pattern trying to find the cockpit camera." Frustrating to all, the camera initially eluded searches but was then located a few days later. However, the critical film cassette was not in the camera. They asked for more volunteers to comb the desert.

Armstrong had been the engineer who correctly calculated the location of the scramjet that had burned off the x-15a-2 just weeks previous to this accident. He and NASA engineer Victor Horton theorized that since the camera and film weighed so little in comparison to the rest of the aircraft, it may have been caught by high altitude winds, traveling back north along the flight path before hitting the ground. Two weeks after the accident, on 29 November, a new search was organized. About an hour after they began, and more than a mile northeast from where the camera itself had been found, team member Willard E. Davies located the cassette.

Appearing intact, the cassette was rushed back to Edwards. Armstrong explained, "The chief of the photo lab took it back to New York for hand processing." Then the real work began. Armstrong continued, "As soon as [the film] got back into Lancaster, Jack [Kolf] and I took it to his living room, along with a data analyzer, [and followed] the time history of the flight." The film made clear the sequence of events that started to unfold as Adams was accelerating to the edge of space. "We pretty well decided right there, sitting on the floor, that he had driven the [x-15] out in sideslip." The reason Adams actively rotated the aircraft out of alignment with his flight path was a combination that added up to disaster. "But that's just like any accident," Armstrong concluded. "There's no one thing that causes an accident. It's not until three or four crop up on you that are additive, that get you in trouble."

Starting with the electrical disturbance caused by the faulty traversing probe experiment in the right wingtip pod, the computer continually dumped and recycled. Traveling at more than Mach 5, Mike attempted to troubleshoot a major problem. In addition, he had a very precise mission to follow with all the various other experiments. The computer glitch caused the x-15 dampers to go offline, increasing his workload even more. One last piece later fell into place, discovered in the aftermath of the investigation, and all but assuring catastrophe: vertigo.

Going back to Mike's very first x-15 flight, he used the word "vertigo" in his report to describe his feelings just after b-52 release. Some considered the word an unusual choice, but nothing truly out of the ordinary from what other pilots had reportedly experienced at that same moment.

Several people who had worked with Adams talked about what they witnessed. First was Stan Butchart, the NASA director of Flight Operations. "Milt Thompson and I had lunch one day after Mike's first or second flight.

. . . Adams came over and joined us, and he mentioned he had vertigo at burnout. It didn't ring a bell because all the guys reported a little bit of vertigo right at burnout for some reason." But it didn't end there, Stan remembered. "He kind of kept coming back to that and trying to make a point of it, and that didn't ring a bell with Milt or myself. . . . I think he realized himself that it was pretty bad. I wish he had talked to somebody else or made it more clear to us, because something might have been able to be done. . . . My feeling is that's what got him."

Paul Bikle, the NASA Flight Research Center director, shared his thoughts, saying, "[Mike] seemed to be confident, and we had no reason—until after the flight—to suspect there was any problem. . . . All of the guys had degrees of vertigo after initial shutdown." He went on to relate how it came to his attention after the accident that others had also heard Mike speak of vertigo. "The [flight] planner [would] get to know the pilots real well. . . . Mike told them on several occasions that he had gotten vertigo, but no one said anything about it. Obviously, they couldn't have thought it was too important . . . or else they would have said something. I think when guys work together closely like that they don't go passing tales out at school."

Johnny Armstrong confirmed what Bikle had to say: "Jack [McKay] and I talked to Mike a little bit, and he had just casually mentioned he had some strange feelings during ballistic flight. . . . That could have been trying to tell us something, but we weren't clever enough to interpret it." As Armstrong pointed out, "You have three gs pushing you, and all of a sudden it shuts down, and you get the feeling that you're tumbling."

The final smoking gun was when Mike's medical records were revealed from the time he had undergone the voluntary vertigo tests at the School of Aerospace Medicine. They showed conclusively that "Mike's response [was] completely abnormal. Eye motion was severe for twenty seconds, [and he] became nauseated."

Increased workload stemming from the electrical problem, combined with vertigo, narrowed Mike's focus so much he apparently forgot a very critical item. Vince Capasso explained, "There was an instrument that had two modes. One was precision attitude, which had two bars for pitch attitude and yaw. But we needed more sensitivity for the roll angle of this [solar spectrum] experiment, so we could switch the yaw needle to roll during the time at altitude when you were doing the experiment." Adams had switched the instrument mode from yaw to roll, but later he forgot the switch and was still reading the needle as a yaw indicator as he set up for reentry. When Mike saw the needle off-center, he tried to correct the situation, not realizing he was continuing to drive the x-15's nose sideways. From the radio transmissions and cockpit camera footage, it was apparent he never understood his predicament, at least not until it was too late and the aircraft had entered the hypersonic spin. This most likely further exacerbated his vertigo, so once the aircraft recovered out of the spin, Adams was still not able to regain control.

The x-15 was designed as an extremely stable vehicle. The evidence points to the fact Adams may have actually been fighting this inherent stability after the spin, which then led to a pilot induced oscillation. This was similar to what Scott Crossfield encountered as he came in to land on his first glide flight eight years previously and what Milt Thompson experienced while attempting to maintain a precision attitude on another flight. Thompson further expanded on this thought with regard to Adams: "On that flight . . . the oscillations got big enough, and it broke the plane apart." Milt felt if Mike had simply taken his hands off the aerodynamic controls, it would have settled down on its own. Recovery and safe landing may have been possible.

John McTigue, like Vince Capasso, was an operations engineer on the program. He was fairly harsh in his assessment: "If he'd just looked out the window, he could have figured it out because he could see the sun going round in the cockpit. If it had been any other pilot there besides him, I believe the airplane would have been recovered. McKay definitely would have recovered it, because he'd be looking out the window!" Of course, this presupposed that Adams was operating at full capacity, which is likely not the case due to his possible disorientation.

As an x-15 pilot, Bill Dana actually indicted himself to what transpired. "Both Pete [Knight] and I had flown the same mission the two previous flights and hadn't had any troubles with getting disoriented. Mike had some electrical troubles during the boost that I think disoriented him and got him confused. Jack Kolf and I designed that whole system of time sharing on that needle. After the accident we felt terrible about what a stupid thing we'd done." The thoughts had crept in that if it worked once it was fine to keep doing the same thing. "I don't know really what the hell we were doing. . . . It worked fine for Pete and me, but [we] didn't have any trouble on our flights. . . . Mike got in and had all these troubles and got a little confused. It eventually cost him his life. . . . Mike Adams damn near survived the spin, and the airplane damn near recovered itself—well, it did recover from the spin, it just killed Mike in the process. Had he had one switch in a different position he might have gotten the airplane home."

The final report on the accident took two months to prepare. It was a hefty volume of 316 pages. x-15 pilots Pete Knight and Milt Thompson signed their names to the document, along with six others from engineering and medical professions. The ultimate conclusion of the investigation was that the aircraft broke apart due to excessive g loads "induced by severe pitch oscillations." It went on to say, "there has been no evidence found that would indicate any inadequacy in the structural design of the airplane." Within the document, under the title "Flight Surgeon's Rationale," Lieutenant Colonel Matejka stated, "It is my opinion that the precision instrument tracking task, roll maneuvers, and two degree right bank were not excessive. However, the loss of damper control and numerous computer failures added to the above tasks, did result in a task over-saturation." He briefly discussed the findings that had been uncovered from the vertigo tests at the School of Aerospace Medicine and sounded incredulous when he wrote, "Yet Major Adams was found to be qualified for any special assignment."

Two other people expressed grave misgivings about the x-15 program following the tragedy. This stemmed from how they felt the program had somehow lost sight of its objectives. Harry Shapiro, an engineer on the external tank system, had this to say, "The electrical system we had on the x-15 was 1950 vintage, yet we were putting in this computer that was 1960 vintage. It needed precision power. The two things were not compatible. The pulses that were in that brute force system put spikes back into the computer and made the computer go off line." Every flight was a potentially hazardous situation, but after a while it became routine. Harry said, "North American was very concerned there was a lackadaisical measure about the way the people were handling the aircraft. . . . I think the company felt we were going to lose more if we had this lackadaisical manner. We really had to treat each flight as a critical item."

x-15 pilot Robert White shared some of this same sentiment, saying, "I often felt that sometime we went too far with these things. . . . When I was participating in the program, your eye was always on the ball. It was very intense. Now you're starting to take this as a routine thing. Boy, that's not routine stuff! Then there was the dual instrument presentation. You'd flip a switch and it would show you something's happening, and you'd flip it another way and that same needle would show you something else. I thought, 'Give me a break.'" Bob felt that if you were sitting in a 747, cruising along, that was okay, but in the x-15, "you're not in a stable condition. You've got to be careful. *That* NASA is a different kind of NASA than the one I operated with."

In the context of the opinions of both Shapiro and White, the historical roots to the later *Challenger* and *Columbia* Space Shuttle disasters can be seen as early as 1967 with the Adams accident. In the investigations for all three, people can say they saw the problems beforehand. Yet no one was able or willing to step forward to get the attention of those who could have broken the chain of events before they spiraled out of control. With regard to Mike, it will never be known if those doctors and researchers at Brooks AFB told Adams his response to the vertigo test was abnormal. They never gave him the opportunity to know the danger in which he might be placing himself by pursuing a career as a high-speed, high-altitude test pilot and astronaut. The fact he had no trouble to that point was accepted as the normal situation.

The traversing probe experiment produced electrical problems yet was found to have never gone through altitude testing before being mounted aboard a rocket plane that took it into space. Instead of questioning this lack of testing, someone assumed that since it had flown before it was good enough. As was stated in both Space Shuttle loss investigations, the deviation from normal was accepted as being the new normal. Any variance from expected parameters should never be considered routine and certainly not "normal." Good people died in all three cases because of complacency, not wanting to be the one to stand up and say anything was wrong.

The x-15 is often considered the most successful research aircraft ever constructed. This success may also have had a hand in what eventually killed Mike Adams. The planned goals of the program had been to reach Mach 6 and 250,000 feet in altitude. These had both been exceeded by July 1962, just three years after the first flight. No one wanted to stop flying the x-15, so a new job had to be found to make use of these very expensive machines.

One was found in flying experiments. The x-15 did this job well, providing a unique environment no other aircraft or rocket could attain. Even though the advanced x-15A-2 was still expanding aerodynamic research, both no. 1 and no. 3 were relegated as workhorse experiment carriers.

Even though the x-15 continued to fly for nearly a year, the heart of the program had been shattered, along with Mike Adams and x-15 no. 3. Funding existed through 1968, but no one wanted to push their luck anymore before it was finished.

The A-2 was at the North American plant in Los Angeles being rebuilt following the near disaster of Knight's Mach 6.7 flight on 3 October. It was not long after the events of flight 3-65 when it was decided the A-2 aircraft would permanently stand down. Out of three rocket planes, only the no. 1 ship was left to finish out the program. Once the investigation was completed, the broken remains of x-15 no. 3 were sanctioned to be buried in an unmarked grave in the desert.

Mike's legacy is that the flight test community learned a powerful lesson in safety. A decade before his last mission, pilots seemed to be throw-away items at the base. Almost every week someone was being replaced after one crash or another. A decade later losing a pilot to an in-flight accident was a rarity. Unfortunately, these lessons were apparently not transferred to the Space Shuttle program, later costing two very expensive vehicles—and two irreplaceable crews.

Directly as a result of Adams's accident, and the subsequent investigation and report, two specific changes were made in the way the x-15 program was run. First was that future flights would ensure the pilot had all of his "pitch, roll, heading, angle-of-attack, and sideslip information" and would include a telemetry channel to the NASA control room providing these same data in real time. An 8-ball type of indicator was shown on a television screen to give this information. The second was that doctors would "medically screen x-15 pilot candidates for labyrinth (vertigo) sensitivity."

When tragedy strikes, the exact moment becomes a fixed point in time, one that never diminishes in memory. Charles Gerdel, Mike's best friend, took a moment to gather his thoughts before saying, "I was coming back into Sacramento. I was out in this area where there weren't any houses, [and] I was very close to crossing the 8th Street bridge when I heard it on the news."

Charles had to stop as the memories came flooding back, his voice catching in his throat. He apologized before continuing, "I had to pull over to the side. That's when I heard he was killed in the x-15."

Mike's brother, George, was an accountant. "I was in an attorney's office, overseeing the matters of a client of mine. I got a phone call there from a military spokesman, and he informed me of the fatal accident. It was a severe shock." George's first reaction was: "I guess disbelief. . . . Suddenly you're having to cope with something that heretofore wasn't possible. Finally, day by day, you just improve on it. It's like anything else, you don't think it's going to happen to him or to you."

Family and friends headed south from Sacramento to Edwards for the memorial service, which was held at the base chapel on Saturday, 18 November, at 3:00 p.m. Chaplain Roy gave the eulogy, intoning, "The flight testing profession has lost a dedicated pilot, and the United States Air Force has sustained a great loss. He gave his life doing the thing he most wanted to do." It was a sentiment echoed by Charles, when he said, "Mike loved that plane. We hated to see him go, but he died doing something he loved. . . . Can't think of Mike being in an old man's home someday, living like that."

Also read at the memorial was a piece written two years previously by the Rev. Anthony M. Ferrari. He titled it "The Astronaut's Psalm." One line stood out, which read, "Make me proud of my endeavors, humble in my achievements." The memorial concluded, and the people filed outside for a last tribute, as a group of Edwards Flight Test Center pilots flew a formation of jet fighters over the chapel, one of the planes pulling up sharply as they passed overhead, in tribute to the Missing Man.

Freida did not want to discuss the memorial service but recalled that afterward, "They were all so gracious." She was there when the investigation began, and she could tell right off where they were headed. "It was later on when they would have the briefings, and I was asked to come. That's when I argued with them, because I didn't like the comments they made. They said, 'Could be vertigo,' but they knew he was getting erroneous readings." The evidence points to vertigo as a contributing factor, but Freida felt they wanted to pin the whole accident on Mike's shoulders, instead of admitting problems with the equipment that began the sequence of events. The direction was eerily similar to the collision investigation after Joe Walker died in June 1966. Freida continued, "It just infuriated me when they made these subtle comments about Mike being disoriented. Who wouldn't be disoriented in a spin? I don't suppose we'll ever know. . . . I wanted them to stand up for Mike and not let them say things like that about the vertigo. [There were] many more factors than that."

Freida was angry. Her husband was gone, and there was nothing she could do about it. As the board investigated, Freida focused on the fact they were looking at her husband, although in the final report the overall conclusion was balanced between the electronics problems and the vertigo. It was also apparent they didn't blame Mike for the vertigo but instead the medical system that allowed someone so susceptible to do a job where it put his life in jeopardy. Brent Adams thought the idea was overblown of vertigo being responsible, even in part, for his father's accident. "My mother [told] me that none of those test pilots, or anyone in training, would ever mention, nor even hint, they had experienced vertigo, for fear of being rejected. Vertigo seems to be an easy explanation for a lot of air mishaps, but I am not sure that's totally true here."

Within just a few weeks, Freida and her children had to vacate their home at Edwards. She chose to move the family to Louisiana rather than Sacramento. Her roots were in the South, and she had a sister in Columbia. "You go right back to your own people," she stated firmly. Mike also went with her to be interred close by. Freida thought it was the right thing to do. "The family in Sacramento didn't argue." She always remained close to Mike's family and often crossed the country to visit. For her children, Louisiana was not where the rest of her family wanted to go. "My children weren't too happy. The boys did not like the South. Didn't like the schools. They got in trouble because they didn't say, 'Yes, ma'am,' and 'No, ma'am.' They had a time adjusting, they really did. Now, Liese, she was younger, and being a girl, she managed to do better."

Freida settled in Monroe, Louisiana, and returned to college to finish her master's degree, meeting her future second husband, George, a college professor who specialized in theater arts. She quipped, "Quite a difference going from a test pilot's wife to a professor's wife. The engineering side of things versus the theatrical side of things. . . . I got plays and ballets." After leaving college Freida became a teacher, but in the end felt she didn't fit that mold. "It was too late for me to jump in and corral kids, so I worked for the state in placement and counseling." There was also something else she could do in Monroe. "I loved working in my flower garden. You didn't do that at Edwards!"

The oldest son, Mike, had a very difficult time adjusting, wandering through life, eventually working at the Monroe civic center. Years later, sadly, he committed suicide. His sister, Liese, became successful in petroleum engineering, working at Schlumberger, now living in Houston with her husband and children. Freida beamed when she spoke of her daughter. "She's amazing." As the middle child Brent seemed to be exactly that when it came to adjusting in life. He stayed in Monroe and opened an automobile maintenance shop called "Rocket Lube." The business is dedicated to his father and the x-15, as is evidenced by the large photographs and clippings on the walls of the waiting room. Behind a snack counter, Mike's orange flightsuit and white pilot's helmet are mounted, along with an American flag. Freida admits of Brent, "He's not much of a talker. He's like his dad."

Flight 3-65 had been scheduled for an altitude of 250,000 feet. When Mike Adams shut down the LR-99 rocket approximately four seconds later than planned, he added 16,000 feet to his peak altitude. These extra 3 miles made a lot of difference in one very specific respect. In the 1960s the U.S. Air Force held the official position that any pilot who exceeded 50 miles was to be considered an astronaut. On 15 November 1967 Mike Adams accomplished this goal in the x-15 with less than a half mile to spare. A month earlier, on 17 October, Pete Knight had achieved the same feat in the same no. 3 aircraft.

Freida explained, "Pete Knight and I were to go to Washington DC. He was going to get his [astronaut wings], and I would get Mike's, posthumously. Before you knew it I was not to go to Washington [but instead] would go over to Shreveport, [Louisiana]. Some general there would give us ours." She felt the military wanted to keep her out of the limelight, even appearing resistant to presenting the wings at all. Freida said, "They finally thought, 'Well, we'll have to do something for that lady,' [but only] as an afterthought."

On Tuesday, 16 January 1968, Freida made the eighty-mile trek west to Barksdale AFB. There, with little ceremony, she received her husband's silver air force astronaut wings, which incorporates the same circle pierced by three spires topped with a star, used by NASA for their recognition of astronaut status. It was a status for her husband that Freida would have preferred to forego, if Mike could have instead returned alive from his flight. "You can't help but be bitter when you have to go through something like that. Just think of what could have been. . . . They whisked me out of there in no time."

As she looked back at her years with Mike, she had some regrets but also wonderful memories. "I would have done it a little bit better. I would have kept more in touch with him and what he was doing. . . . We had a good life and all, but we had our ups and downs. . . . When he would go out early, he'd fly over our house and wave those wings. I was out waiting to see him go by."

Back in the high desert of southern California, Highway 14 cuts a nearly straight line north and south through the Antelope Valley and the towns of Lancaster and Palmdale. A few miles north of the outskirts of this dual city is the main turnoff to Edwards at Rosamond Boulevard. Heading east along this route into the base, a road branches off to the south to go to the Rod and Gun Club. As can be imagined from his love of hunting and other outdoor activities, this was a favorite hangout for Mike while stationed at Edwards. Sometime after the memorial service, George recalled he and his mother drove back down to the base for a special ceremony at the club. At this event the access road was renamed "Adams Way," an excellent tribute for Mike. George said, "They were very good about everything for him. They gave him ample recognition."

One evening, not long before his final flight, Mike Adams decided to take his two boys, Mike Jr. and Brent, out to the NASA hangar at Edwards for a private tour of the area where their father worked. Brent recalls his dad saying it was the "secret hangar." This was not technically true, although, for his kids the idea just made the visit all the more exciting. Brent recalled, "I was probably nine—this was pretty close to the end."

At the time of this nocturnal visit the hangar was filled with a wide variety of experimental and operational aircraft in all sorts of strange shapes and colors. There were the deep black X-15s, silver F-104 Starfighters, the white and silver bathtubs of the various lifting bodies, maybe a gray F-4 Phantom, and even an orange F-5D Skylancer, which had been used for pi-

42. Michael Adams jokes about the wet conditions on Rogers Dry Lake after a rainstorm halts flight operations. Courtesy of the Armstrong Flight Research Center.

lot training for the x-20 Dyna-Soar.

Adams let his boys explore and opened up some of the aircraft so they could sit in the cockpits to pretend they were flying at high Mach numbers themselves. Remembering his long-ago childhood, a giant grin appeared as Brent recalled that magical night with his father. "I believe all three [x-15s] were in there that particular night. . . . I'll never forget that. . . . I knew what the x-15 was, but the other planes looked real strange. I remember sitting in there. [My dad] had to lift us up into the cockpit."

I asked Brent which image is the first to come to mind when he speaks of his father. "It would be with a flightsuit on, the orange one, out there standing on the lakebed. If there was a second image it would be fishing or something at Lake Tahoe."

This image of Mike out in the forest or on a lake is the way so many of his family and friends remember him, although the public image was always of a serious man, standing stiff and unsmiling next to the x-15. Mike was much more than that. Bill Dana related, "He was just a great big bear of a guy. Mike was probably as likable as anybody in the program. . . . He was really laid back compared to the average test pilot, who was running at max gain all the time. . . . He was just different than the other guys on the program."

With all Mike had gone through, from losing the NASA astronaut selection due to the F-104 ejection injury, becoming frustrated with the lack of progress on MOL to the point of leaving the program, and finally being selected to fly as one of the elite test pilots on the x-15, maybe in the end he found his heart being drawn in a different direction. Perhaps inner peace for Mike Adams would not have been in outer space, but in the backcountry of America. Dana shared a thought that opened a tiny window into the soul of the man. "He talked like he was ready to get out of the air force and go be a forest ranger. I don't think he actually would have done it, [but] that's what he claimed he wanted to do, go be a forest ranger." Maybe Mike would have chosen this new path, and been exceptionally good at it, if fate had not intervened so tragically.

13. Snow at Edwards

I think people should move forward, not backward,
with their ideas. And don't take the new technology just
because it's new. You take what you need for what you
want to accomplish.

Harrison A. Storms

Thirteen days prior to x-15 flight 199, America's Apollo program finally
kicked into high gear. *Apollo* 7 launched on 11 October 1968, with the crew
of Walter M. Schirra, Donn F. Eisele, and R. Walter Cunningham. After the
triumphs of the Mercury and Gemini programs, Apollo had started horribly with the loss of three astronauts in a launch pad fire on the evening of
27 January 1967. Virgil I. "Gus" Grissom, Edward H. White II, and Roger
B. Chaffee lost their lives, and the space program nearly lost its direction.
It took twenty-one months of investigations and recriminations, resigna-
tions and redesign, before Schirra's spacecraft was on the pad and ready to
launch from Complex 34, at what was then called the Cape Kennedy Air
Force Station.

What followed was a nearly flawless eleven-day check-out mission of the
Block II Apollo capsule. Splashdown in the Atlantic Ocean was southwest
of Bermuda, early in the morning of Tuesday, 22 October. A murdered president's audacious goal of taking the United States to the moon before the
end of the 1960s suddenly looked possible again.

Two days later Bill Dana flew what was to become the last x-15 mission.
Hardly anyone seemed to notice the final act of what had started as the
American answer to the Soviet challenge. Wings flying into space and returning for a controlled runway landing had been replaced by thundering
rockets with manned conical capsules at their apex.

But the x-15 was not quite finished. At least that had been the intent after Dana's return to the Rogers lakebed on 24 October. Funding for the program remained in place through the end of 1968, leaving more than two months to succeed with flight 200. It seemed like a nice, round number and a great way to end the nine-year history of the world's most productive rocket plane. Preparations started immediately, although mission 1-82 was not ready until 26 November. Attempts were made eleven times to get it off the ground.

On 21 November the first flight plan was released. Pete Knight was shooting for Mach 4.9 and 250,000 feet in a last try to capture a missile launch from Vandenberg. Test runs of the LR-99, auxiliary power units, and ballistic control system were accomplished on 18 November. An APU was in need of replacement, as was the LR-99. The day after the flight plan was nailed down everything was running properly, and aircraft no. 1 was cleared to be mated to B-52 no. 008. Once connected to the pylon on 25 November, various leaks were detected and fixed. The original choice of Smith Ranch as the launch lake was found to be incapable of supporting the mission, so a switch was made to Railroad Valley, about fifty miles east of Mud. This, and other factors, caused the first slip in the official schedule on 26 November.

The second came the next day. That morning, Knight entered the x-15 cockpit and started his pre-takeoff checks. A malfunction in a right-hand yaw ballistic control rocket contributed to cancellation before the B-52 started its taxi. The rocket plane was demated and sidelined because mothership no. 008 had another commitment, and no. 003 was already loaded with an HL-10. Jerry Gentry finally cleared the B-52's wing on 9 December with the lifting body's fourteenth mission. x-15 no. 1 was mated later that day for a flight attempt the next morning. Then weather stopped tries on 10 and 11 December.

Knight climbed the servicing stairs and entered the x-15 once again on 12 December, and the mothership actually got airborne. Maybe this was to finally be the day. Railroad Valley was socked in when the group of aircraft arrived in the area. Then, just to make matters worse, the inertial guidance system went down. All aircraft turned for home. Three more weather aborts followed, plus one for lack of a c-130 support aircraft.

Things were getting down to the wire with less than two weeks left in the year. By this point it appeared as if desperation was setting in. A new flight plan was drawn up for 20 December, with the justification of checking out a new experiment from the Autonetics Division of North American Aviation. The speed and altitude parameters for the mission were reduced to Mach 3.9 and 162,000 feet. The flight path distance was also cut in half, with a launch lake change to Hidden Hills.

Robert Rushworth explained that "the people from the air force and NASA all agreed they were just trying to stretch it out and get 200 flights—pushing to get some objectives that were more personal than business."

It turned out the earlier 12 December attempt with mission 1-A-142 was to go into the books as the last time an x-15 was operational and in the air ready to launch. The very last try occurred on Friday, 20 December, but it never left the tarmac. Pete Knight remembered, "Everybody was disappointed. . . . There were a number of things aboard the airplane to commemorate the 200th flight. There was a block of brass, as I remember, some kind of metal that they were going to stamp out medallions afterward, indicating the number of flights . . . the records, and so forth. Each one of the pilots was going to get one."

The winter weather was turning quickly as Pete made ready to go. "On this particular attempt," he said, "I was sitting in the servicing area and Bill Dana was the [NASA 1] controller. Everything was going fine, but I was looking outside, and Bill was in the control room. He couldn't see what I saw. I finally asked him, 'You're not serious about this flight, are you, Bill?' And he said, 'Yeah, everything's looking good in here.' I said, 'It's looking good here too—except I can't see the hangars across the ramp because it's snowing so bad!' Bill just couldn't believe it." There was little anyone could do at this point, so Paul Bikle got on the network and told everyone the situation. According to Pete, Paul had this to say, "We're aborting this flight. Somebody's trying to tell us something. We've tried all these times to get this flight off, and we've aborted it every time. I think 199 is enough."

Project engineers Perry Row and Vince Capasso completed their final report, writing, "The operations and instrumentation crews, as well as supporting shop and inspection personnel, are to be commended for maintaining their good spirits and high quality work through this flying season."

Huddled together against the cold, everyone involved on the tarmac that morning gathered around the base of the x-15 entry stairs for a final group photo. Jackets were pulled tight and hoods were drawn up. Snow pounded down. Streaks of white blurred in the photograph. An obvious wind made the chill even worse. But even through all this, the faces were smiling—this was family.

The x-15 was demated for the final time later that day and placed into what was termed "indefinite" storage for final disposition as a historical artifact instead of a hypersonic research air and space craft.

The primary response from everyone I spoke with, especially those present that day, was regret. Reasons ranged from simply wanting the flight to succeed to achieve number 200, while others were extremely disappointed the program itself was coming to an end. Roger Barniki said, "I was out there on the ramp. One of my people was suiting up Pete Knight. I remember driving down there and seeing how things were going, [how] bad the weather was getting. When they canceled it, I went, 'Oh, why couldn't we have just gotten off the ground one more time?'"

Crew chief Larry Barnett always stayed close to the x-15 when it was being prepared. "That's where I hung out most of the time," he laughed. "If you're going to be in charge of something like that, you'd better be there, not sitting in the hangar looking through a telescope." Watching the snow flurries was a different experience for people used to a dry, and often very hot, desert. Larry continued, "I think everybody kind of kicked back and enjoyed it. It's something you don't see very often. You don't fly airplanes in a snowstorm anyway. It was a real fluke. I don't think they had much of a choice, [and] those guys out there had shut it down way before [Bikle] thought about it."

Even a snow storm was not enough to deter some people from proposing the idea of making another attempt before the end-of-year holidays. Dean Bryan from the rocket shop recalled, "They wanted that 200th flight so bad. There was even a suggestion from Bill Dana to launch the plane from Cuddeback—which is the [dry lake] closest to Edwards—just to make a short flight." Bikle quickly put an end to further wishful thinking.

Dana was definitely upset by the turn of events, saying, "I was brokenhearted, of course. The end of the program was what had me so distressed. We thought we were going to have a delta wing version of the airplane, and I think we would have if we had a little more ambitious management. I think our director at that time had kind of tired of the program. Paul Bikle [seemed] willing to go along with all these ideas to keep the program going until Mike Adams got killed, then that soured him. Once [that happened] the x-15 was on borrowed time." Bill asked himself if he could have done more. "I've always regretted I didn't get in and stomp a little harder to continue the program. I think we could have persuaded Bikle." Considering the funding situation, there is little anyone could have done to keep the x-15 from cancellation.

The distinction of being the last pilot to fly the x-15 fell to Bill Dana. He admitted, "I'm glad to have the last flight, but I wasn't at the time. I wasn't thinking, 'Gee, I hope Pete aborts.' But, yeah, I'm glad to have the last one." He felt there was so much left to do and many more years to come of useful life in the program.

Wade Martin, in quality control at the propulsion system test stand, recalled, "I thought they were going to do more, I really did. . . . I'd have liked to have seen them get that last flight in. . . . I guess they shouldn't have gone that day, but I sure would have liked to see it go, to see them just take the darn thing up and fly it around the block."

It took flight planner Bob Hoey a while to understand the rationale for the conclusion of flight testing. Originally, he thought the decision was premature, then said, "Looking back, it was probably a wise decision to shut it down. We were never going to get that scramjet. . . . It was just as well we didn't want to play with it anymore [because] we'd run out of experiments."

I never previously heard anyone use the term "play" with regard to the x-15; however, in some respect, the word was appropriate. It was a unique aircraft, where every day was a new challenge, often with literally new frontiers to achieve. For those who craved this type of career, what they did could hardly be called "work." This thought was reiterated on many occasions during my conversations. The x-15 was a special vehicle, and people wanted to be on the job every day they could, often putting in large amounts of unpaid overtime. And the x-15 usually needed it. Hoey explained, "If you go back and look at the x-15 program compared to other research airplanes . . . every single thing in it was brand new. . . . Every actuator, every component. . . . In actuality, it was not a very reliable airplane—terribly unreliable, as a matter of fact. We always had something come back broken, something that didn't work the way it was supposed to. It was very complex."

Ralph Richardson said it was unfortunate the program was over, but "it was getting difficult to get support aircraft; it was getting more difficult to do anything. All the money was being siphoned off to the space program, so I guess it was reasonable to assume that it was beginning to outlive its usefulness. They had a warehouse full of data, so much data they didn't know what to do with it." x-15 research pilot Bob White agreed, saying, "The x-15 was not really canceled; the program accomplished all of its original objectives—and far more. After 199 successful flights, it was time to bring it to an orderly conclusion."

The morning after the decision was made to stand down the x-15 program, on 21 December 1968, a Saturn V rocket left Earth, boosting the *Apollo* 8 astronauts on humankind's first journey toward another world. By 24 December Frank F. Borman II, James A. Lovell Jr., and William A. Anders were in lunar orbit, beaming back holiday greetings. Two more test missions in the Apollo series took place in March and May 1969, paving the way for former x-15 pilot Neil Armstrong, in command of *Apollo* 11, to land the lunar module *Eagle* on the moon on 20 July.

While this was transpiring, permanent homes had to be found for the two x-15s that had survived through the entirety of the rocket plane research program. The obvious choices were the Smithsonian Institute in Washington DC and the U.S. Air Force Museum at Wright-Patterson AFB, Ohio (now known as the Smithsonian National Air and Space Museum and the National Museum of the U.S. Air Force). This decision was made even prior to the final flight attempt, but neither aircraft was moved until later in 1969.

Pete Knight personally drove the tug taking x-15 no. 1 to the Douglas c-133 Cargomaster, which had been assigned to carry the aircraft eastward. Maj. Phil Brandt (U.S. Air Force, ret.) contacted me about the small role he played in the x-15 retirement in May and October 1969. He said, "[I was] the navigator of the Dover-based MAC [Military Airlift Command] c-133A that was diverted on the way back from [Southeast Asia] to Edwards. . . . The wingless fuselage (the wings were packed separately) was loaded, and the next morning [10 May] we hauled the assembly . . . from Edwards to Andrews AFB, [Maryland]. . . . We MAC crewdogs had hoped to take each other's picture sitting in the x-15 cockpit, but when we raised the canopy, it was crammed full with spare parts and removed plumbing."

43. X-15 no. 1 arrives at the Smithsonian Institute in June 1969. It is being towed underneath the 1903 *Wright Flyer*, and the *Spirit of St. Louis* is behind it on the right side just above the two people walking. Courtesy of the Armstrong Flight Research Center.

Amazingly, he and his crew also got the job for the X-15A-2. "We picked up X-15 no. 2 later that fall [on 16 October]. . . . On approach [to Wright-Patterson] we lost an engine—but no big deal. While the engine was being replaced, we were given a behind-the-scenes VIP tour of the research and restoration area, and we were allowed to climb up and enter the cockpit of the recently arrived XB-70." Having the honor of being on the crew of both X-15 museum delivery flights was "purely by chance . . . [and the] luck of the draw," according to Phil.

Ed Nice was with the X-15 from almost the beginning to the end. "I started at NASA in 1959, [and] I was one of the guys who put the aircraft in the Smithsonian. It was our crew that went out there." Today X-15 no. 1 bears his name, along with several others. "You can't see it because it's underneath a panel. . . . It was myself, Herm Dorn, Ed Szabo, and Don Hall; we were the ones that put it back together in the Smithsonian."

For a time the aircraft was displayed on the floor of the museum, beneath

44. Many proposals were put forward to use the x-15 for additional research. The one that advanced the furthest, before being canceled after the crash of the no. 3 aircraft, was converting this airframe into a platform for hypersonic flight testing of a delta wing. © Thommy Eriksson.

the original 1903 Wright Flyer, starting with a special reception on 10 June. It was also transported back and forth to Edwards for various reasons over the years, before the brand new museum opened on the south side of the National Mall in time for the American Bicentennial in 1976. The x-15 was hung in the National Air and Space Museum's Milestones of Flight gallery, where it remains to this day.

On 7 May 1971 the x-15a-2 finally went on display at the air force museum in Ohio. The exterior burn-throughs were fully restored with all the proper markings for the aircraft in its original black livery. The A-2 sits at ground level, with support under the nose gear and the rocket plane's transport dolly beneath the rear. The museum has moved it from one building to another on occasion, but the aircraft finally found a permanent home in the Research and Development gallery in the new Hangar 4 at the museum.

There were many versions of the x-15 that were proposed but never materialized. The one closest to fruition was the delta wing. The wings and hori-

zontal tails were to be replaced with a long and narrow triangular wing surface to test theories if this was a better hypersonic design than the short,
straight wings on the base model x-15. Other modifications were also to be
accomplished, including lengthening the fuselage and possibly upgrading
the LR-99 rocket engine to higher thrust.

John V. Becker, chief of the aero-physics division at NASA's facility at Langley, Virginia, and one of the prime movers who created the idea of the x-15
in the early 1950s, wrote, "The highly swept delta wing has emerged from
studies of the past decade as the form most likely to be utilized on future
hypersonic flight vehicles in which high lift/drag ratio is a prime requirement, [such as in] hypersonic transports and military hypersonic cruise vehicles, and certain recoverable boost vehicles as well." Bob Hoey agreed with
Becker's assessment when he said, "The delta wing program would have
been good. I think it may have been beneficial all the way around, [and]
we could have learned a lot from it."

There was a feeling the B-52 might be overtaxed as the mothership to
carry the new and longer version of the x-15. This opened the question of
what to use instead. Top of the list was the XB-70 Valkyrie, which brought
the added bonus of possibly launching at supersonic velocity. However, the
question came up of the advisability of that option when plans were unveiled
with the x-15 carried in the belly of the bomber. Bob Rushworth pointed
out, "The x-15 ejection seat goes up, and they didn't leave any space to go
up through the B-70! I said, kind of facetiously, 'Well, it'll go on top just as
well as on the bottom.' They went back to the drawing board and figured
out how they were going to put it on top." Milt Thompson said, "We did
some studies. In fact, I think we even did some simulations to see whether
or not it was practical. But the cost was horrendous, because modifying a
one-of-a-kind airplane like the B-70 would have been way too much."

With Rushworth's suggestions and Thompson's simulations, there were
still more barriers to overcome. According to Bob, "They had a couple of
problems the NASA people couldn't figure out, such as how were they going to land the airplane because of the higher angle of attack. I suggested
to them, why not switch the position of the nose gear with the instrumentation bay. Christ, they never thought of a simple thing like that!"

Not all people agreed with flying an x-15 delta wing configuration, saying the idea was driving the program rather than any specific requirements.

Harrison Storms made his point, saying, "I'm not sure what they thought they were proving. . . . You see, people often mistakenly think there's something in that configuration. [They say,] 'Because I like a delta wing, therefore it should be a delta wing.' Instead, you ask the whole design to work as a unit."

This didn't mean Storms was completely against the idea of a hypersonic delta wing, just not necessarily the use of the x-15 to accomplish the goal. He continued, "Now, yes, if you want to start over and make a delta wing, that's fine, but you've got to ask, 'Why do I want to do that?' You can't worry about how it's going to look. It'll come out looking like what it's supposed to. Just trying to make something that's a specific geometrical shape for the purpose of doing it, to me, has very little practical value. Yes, it's interesting, maybe, but do you really have a mission for it? That's absolutely the wrong way to go about anything, in my estimation."

Nevertheless, the program was moving forward, as engineer Harry Shapiro of NAA explained, "I believe they built a wing, because I remember we had to put in pressure pickups and strain gauges. . . . I have a picture of it in my mind; I can remember all the stringers [internal wing framework]. They were specially made out of machined pieces." According to a flow chart from NAA, x-15 no. 3 was to stand down from active service to begin installation of the new wings in April 1968. Five months prior to that, the idea died along with Mike Adams in the sky above Ridgecrest. There was no longer a no. 3 aircraft to modify.

Two other x-15s were seriously proposed: a two-seat spaceflight trainer, and an orbital version. North American had invested a lot in the rocket plane, and if new uses could be found, then that meant more business. Neither idea got very far.

Bob Hoey said of the x-15B, "It was a two-seater, and they were going to sell it to the test pilot school. It would have been even worse than the NF-104." First designs for this can be traced back to a 1960 North American report that stated the x-15B could provide "full space training and biomedical research capability." It went on to suggest changing the rear skids to the same ablative wire brush–type designed for the x-20 Dyna-Soar to allow landing on a concrete runway instead of dry lakebeds. There were also military tests with radar-absorbent materials and other defensive countermeasures added to the proposal.

Around this same time, it appeared everyone had some crazy idea to get America into orbit first. This was fueled by the media frenzy that kept repeating the belief of a 100-mile altitude target for the x-15. This crazy idea didn't just come from movie, magazine, and television outlets hoping to cash in on the sensationalism. Milt Thompson said, "North American proposed an orbital x-15. Scott Crossfield was the one behind that. But that didn't really have an awful lot of credibility, because it would have taken so much modification to protect that airplane coming back from orbit. It just wasn't designed for that kind of thing." Harrison Storms confirmed, "It was about the same time we had to cancel the Navaho, which used the G-38 boosters, and there were about five or six of those that were surplus. The idea was— and it calculates out right—to put the x-15 on top of the G-38. But you'd have to change the skin [of the x-15] from Inconel-X to Rene-41, a higher-temperature material."

Milt said NAA went way beyond the smaller G-38. "They actually proposed to set it on top of a Titan. It was going to go around [the world] once." The proposals started to sound like the ones that caused the x-20 Dyna-Soar to be canceled. As Milt stated previously, one of the insurmountable problems was protecting the x-15, which was a structure designed to absorb heat, not dissipate it. "You could get away with that on a short flight, a flight to Mach 6, but anything above that, all that heat just started building up. . . . It would just kind of melt. So you can imagine how much [insulation] had to be put on to come back from orbit. . . . We saw how badly the ablators worked [on the Mach 6.7 flight]. I don't know how serious they were about it. Those programs, I was glad to see not happen."

The most bizarre, not to mention dangerous, of all proposals from that period was to create a one-way mission into space for the x-15. There was power to get into space, but the reentry always stumped the engineers. Some brainstorm session actually presented the case for allowing the x-15 to break apart while coming back to Earth, then the pilot was somehow going to survive to parachute to a safe landing. When Paul Bikle was asked of this, he said, "I'd never heard that, but I don't doubt that somebody suggested it. . . . People sit around and dream up those things all the time. There's no doubt in my mind that if it had been decided to go ahead that way, there would have been people willing to do it."

Storms agreed many of the ideas proposed were ludicrous. But, as far as the rocket plane actually built and the research accomplished, he had this to say: "The x-15 did the research job we set out to do. It did not go to the moon; it did not orbit; there are a lot of things it did not do. But it was never intended to do [those things] in the first place. I think for the dollars, it was one of the most successful programs ever put together."

By the 1970s theoretical research into these areas was tapering off. Follow-on programs to the x-15 were long since dead and buried. The last serious attempt to build on the technology of a manned hypersonic craft came around this time with the x-24C. The proposal was for a sustained cruise capability above Mach 5, and the name chosen for funding purposes was the National Hypersonic Flight Research Facility. In the age of government acronyms, this equated to NHFRF, simply pronounced "Nerf."

Scott Crossfield loved to fly. He denied it even when I spoke with him, but with all he did his actions showed the truth of the situation. I first caught up with him at a time when he didn't have his own airplane and was working in Washington DC. Years earlier, during the period when the x-15 was first taking to the sky, he often flew his distinctive V-tailed Beechcraft Bonanza from the North American plant in Los Angeles to Edwards for x-15 testing—until the company lawyers told him it was not safe to do so. He always laughed at how bizarre that sounded for a pilot regularly flying rocket planes.

When we talked, Scott said, "Until I get an opportunity to fly regularly, I won't. But I've never had a great compulsion to fly. Many pilots say they do. I can't see any real reason to go flying after the first hour—which is pure pleasure. . . . Just to sit up in an airplane and look around is awfully damn boring, particularly if you travel in a slow airplane. Travel cross-country in a Piper sometime. That'll get you."

Unfortunately, in the end that is what got Scott. It was the morning of 19 April 2006, and he was returning home to Virginia after giving a talk at Maxwell AFB, just west of Montgomery, Alabama. Scott was flying his Cessna 210A out of Autauga County Airport in Prattville, taking off at approximately 9:20 a.m., Central Time. His daughter, Sally, provided details of what happened next: "It was a 'soupy' day, but even as he took off and called from the plane to file his flight plan . . . and get the latest weather update (in the air), all was still 'clear,' so to speak. He missed the warning coming out by about one minute. It was only moments after his last update that the weather warnings were issued."

After fifty-five minutes in the air, as he was passing 11,500 feet above nearby Ludville, Georgia, air traffic control lost radar contact with the Cessna. Scott had unknowingly flown directly into a series of large thunderstorm cells, categorized as a level five to six, rated "intense" to "extreme" by the weather service. The small plane went out of control and crashed, killing Scott. Sally said, "The worst was the not knowing. I worried about him maybe just being injured and out in the dark rain. Of course, in your heart of hearts you knew, [but] you just still had hope."

Scott was found, still in his Cessna, by a Civil Air Patrol (CAP) search group a day later, near the northern Georgia town of Ranger. He was about six miles from his last reported radar contact position.

The official accident report later tried to lay blame on Crossfield. Sally saw it much differently, and she remained understandably bitter of the way her father was treated in that report. "If you knew him at all, you knew one important fact about him, and that was he planned, and planned well. He did not ever take chances and was never reckless. He was one of the most methodical, detailed men I have ever, and will ever, meet."

The reasoning of the National Transportation Safety Board (NTSB) was that Scott supposedly never checked for weather updates. However, the record showed he did everything he was supposed to do. No one on the ground during the time he was flying directly into the storm's path ever gave him a single warning. Sally said, "None of them issued the warning to him. [Someone] is even heard to question why the current controller wasn't warning him, but that came too late. He rose 500 feet and then went off radar. He went down 11,000 feet in less than sixty seconds." So why the assignment of pilot error by the NTSB? Sally explained, "It appears, if you are flying at the time of your accident, then you are at fault in some respect, regardless."

Losing her father was a painful experience, and she also saw what it did to her mother, Alice. "My mother essentially died that day. She has never been the same; her grief [is] just so raw. It is quite painful to watch. April 21, 2006, would have been their sixty-third wedding anniversary. . . . We've all been affected differently, but he was the glue that held our family together. No matter what happened, if he said stop, we stopped. If he said go, we went. We've fallen apart, I am sad to say. I don't feel like I have a family anymore. I've got my mom, but she is only sort of here. It's heartbreaking, actually. Maybe in time?"

There was more to endure for Sally and the family from a very unexpected source: Chuck Yeager. During our interview Scott commented about the friendly rivalry the two of them shared at Edwards during the 1950s and 1960s:

He and I are pretty good friends, [but] we are a whole world of different purpose. I'm a designer and a builder, and he's a flier. Yeager's view was quite opposite of mine. I'm a test pilot to find out what we can do to improve an airplane, and he's a pilot who finds out how we can conquer it. He will not admit that an airplane can have a characteristic you can't handle. I care if I handle it all right, but I'm going to fix it. We are two completely different pilots.

So all of this crap that's gone on, making us competitors, just has no merit to it at all. We were doing entirely different things. He was evaluating for air force purposes, while I was doing a lot of detailed data gathering to see why things happened the way they did and what we should do about it.

Within hours of her father's disappearance, Sally started hearing the reports on the news. "Listening to people speculate about the accident was pure torture. Many people were quite kind, but there are always those that can't stop themselves from blowing off opinions without facts. As if it isn't bad enough for him to be dead, to have people question his ability or preparedness was just such a terrible experience—and insulting to him. Once these things are written, you can't unwrite them."

Yeager was an obvious choice for media outlets to seek out for comment. And he had plenty of them—all very derogatory. Scott had considered him a friend, but in the end Yeager was a product of his own publicity. Sally's words said it all, including that she will now allude to him using only one specific variation on his name, carefully avoiding even capitalization: "The worst was the press giving 'yeagershit' the opportunity to spout off again. He said something like, he wasn't surprised, [and] that Crossfield was always reckless. Nothing could be further from the truth, but because shithead said it, it was printed and repeated. . . . In life, while competitors, they had a friendship, and while they would spar back and forth, [they were] never nasty. Guess 'yeagershit' only had the balls to say something like that because my dad wasn't here to defend himself. The man owes my father—and his family and friends—an apology."

Contrary to Sally's feeling about things not being able to be unwritten, not long after Yeager's original negative rhetoric was posted in online news stories, the Internet was cleansed of all his comments.

Neither Sally nor her siblings followed their father into aviation. "None of us have a pilot's license, though my brother Tony and I have taken lessons. I'd say, for me at least, life got in the way. My dad would say that means I didn't have the passion. We would disagree on that a bit, or maybe I would disagree just [for] argument's sake. One of my dad's favorite sayings to me was 'You'd argue with a stop sign!'"

Sally shared two very special honors that will live on for many years from her father's life. The first is "The A. Scott Crossfield Aerospace Education Teacher of the Year Award, which is presented annually to honor and reward a teacher for outstanding performance in aerospace education. This award was under the wings of the CAP [Civil Air Patrol] until 2007 and is now presented annually at the National Aviation Hall of Fame Enshrinement Weekend each July." The second happened while Scott was still alive. He had the distinct pleasure of having an elementary school named in his honor in Herndon, Virginia. "The naming of this school . . . was one of his proudest moments," Sally said. "He was always on hand to help them in any way possible. Each year he would also attend the sixth grade graduation, give a short talk, then shake the graduates' hands."

For many in the public, the x-15 is something dimly recalled from their youth, while others think of it fondly as an integral part of the early manned space program of the United States. The x-15 was the direct ancestor of the Space Shuttle. Although vastly different in design and construction, the two vehicles shared a surprisingly similar heritage. This included the prime contractor, North American Aviation, which later morphed into North American Rockwell, then finally Rockwell International, before it also was assimilated into The Boeing Company during the aerospace conglomerate buyouts of the late 1990s.

"I'd be very surprised to find the space program people would ever admit the x-15 existed," flight planner Jack Kolf jokingly suggested. "Seriously, it was quite a competition, particularly in our early days in the x-15. We were doing things they were doing with Mercury, and they never recognized our existence. And that's still pretty much true today. Anything they do with us is remarkable. That was their attitude. They worked in a vacuum as far as we were concerned."

The Space Shuttle was often criticized throughout its thirty-year history as never living up to its promise, yet, like the x-15, it did exactly what it was supposed to do for the money allocated and the design that was approved. Could it have done more, and done it better and safer? Of course—if it had been allowed to be the truly reusable fleet of space planes first envisioned and funded less than four years after the x-15 completed flight testing. Bill Dana related the two programs, saying, "I think the x-15 got data we needed to build the [Space] Shuttle. And I'm not saying we couldn't have built the shuttle without the x-15, but it would have been more risk and a longer time getting it flight tested. So I think the x-15 was a stepping stone in lifting entry orbital vehicles."

Astronaut Joe Engle can talk about the comparisons better than anyone, as the only pilot to fly both the x-15 and the Space Shuttle. "From a pilot-task standpoint, the entry and landing are very similar, performance-wise. . . . You fly roughly the same glide speed and same glide slope angle. [You] start the flare a little higher on the Space Shuttle, partly because it's such a big airplane, and it doesn't respond quite as swiftly as the x-15. . . . But the float and touchdown were very similar. The x-15 was a lot closer to the lakebed, and you got a lot better perspective on your height and your sink rate through the air."

Harrison Storms was, in so many ways, the father of the x-15 and later the founder of the Space Division of North American Aviation, which built the Space Shuttle. He was regarded as one of the best engineers to ever lay a hand on an airplane or a spacecraft. He died 11 July 1992 of a heart attack, four days shy of his seventy-sixth birthday. When we talked about the x-15 at his home in Palos Verdes, California, he was emphatic about one fact: "This is a damn dangerous business, and these people are in it, and they know it's dangerous. They're willing to take that chance because they want the glory that comes with it. And for that, you have to pay something. You can't expect to get it for nothing. . . . All I can say is that the mission can be done. You have to forget all the old, built-in ideas and accept some compromises. You can't do it the way we've been doing it. End of lecture."

Afterword

Many of the people associated with the x-15 are gone now. At the time of this writing, nearly half of the sixty-seven people I interviewed have died. I had the great fortune to speak with nine of the twelve pilots, and it is sad that eight of them are now lost to us. I never had the opportunity to meet the three who died at or near the time of the x-15: Joe Walker in the mid-air collision between his F-104 and the xB-70, Mike Adams in the only fatality directly from the x-15, and Jack McKay from the long and tortured aftermath of his rollover accident. However, I was able to talk with family and friends of each of these men, so I have hopefully been able to capture something of who they were in this narrative.

Forrest Petersen was the first who died later. He was diagnosed with a brain tumor and passed away 8 December 1990. Bob Rushworth and Milt Thompson were lost within five months of each other in 1993. Bob died of a heart attack on 17 March, and Milt on 6 August, after simply losing his will to live following the loss of his beloved wife, Terri.

Milt's death was especially bittersweet with regard to timing, in that later in the evening a dinner in his honor was being held in Lancaster. The NASA administrator at the time was Daniel S. Goldin, and the plan had been to award Milt the Distinguished Service Medal, the highest accolade bestowed by the space agency. Goldin issued a statement, saying, "During his long and distinguished career, he literally helped lead the way from our first faltering steps in space, through the successful flights of the Space Shuttle. He significantly enhanced not only the nation's flight testing and research programs, but also our capability to conduct flight research."

Eleven years passed before the next loss. Pete Knight died at the City of Hope Hospital in Duarte, California, during the evening of 7 May 2004. He had been diagnosed exactly one month previously with acute myelogenous leukemia. Reports were that his son David was able to be by his bedside in the final days. At the time of his death, Pete still held office as a California state senator, the position that had led to their estrangement. Pete had two other sons, William and Stephen, and both have remained close to David throughout the ordeals.

Just two years after this, Scott Crossfield crashed his Cessna in northern Georgia, his light now extinguished.

When I first started research and interviews for this book, Bob White still lived in Germany, having divorced his first wife, Doris, then getting remarried. My initial contact with Bob was through a series of written questions and answers sent back and forth across the Atlantic. It wasn't until many years later that we finally met in person, as I traveled to Florida where he settled with his new wife, Chris. Both of them seemed in excellent health, but just a year after we met, Chris died, and Bob followed three years later, passing away in his sleep on 17 March 2010.

As the final review of this manuscript was being completed the sad news was released that Neil Armstrong, seventh man to pilot the x-15 and first human to set foot on the moon, died on 25 August 2012. He had undergone heart bypass surgery close to his eighty-second birthday on 5 August, and all reports were that he had been recovering well. The news of his sudden and unexpected death shocked the world of aerospace, as well as everyone who shared in that "one priceless moment" on 20 July 1969. No longer will we live in an age where Neil Armstrong is a part of us. Armstrong would simply dismiss any adulation as undeserved. When asked about his achievements he made sure to give credit to all those who made his flights possible, without whom nothing he did meant a thing. Nonetheless, he was the commander of *Apollo* 11, the one human being who brought the lunar module *Eagle* in for a safe landing at Tranquility Base with only seconds of fuel remaining. Five other moon flight commanders also accomplished their missions, but Neil will forever be the first—and, in the opinion of many, the best.

Bill Dana died after a long battle with Parkinson's on 6 May 2014. The x-15 era truly came to an end on 10 July 2024, when the twelfth and final pilot who flew on these hypersonic wings into space passed. Joe Engle was ninety-one years old, and I counted him as a close friend as we shared many x-15 adventures together. The expertise and memories of all these men will be impossible to replace. The same can also be said for the mechanics, engineers, designers, and all the other people fortunate enough to have worked with this magnificent vehicle.

John Bodylski was looking for his Eagle Scout Leadership Service Project. Maj. Greg Frazier of the Civil Air Patrol was a friend of John's, as well as an admirer of the x-15. He came up with the perfect suggestion: a memorial to fallen x-15 astronaut, Michael J. Adams. After a lot of hard work, that idea came to fruition on 8 May 2004, with the unveiling of a truncated concrete pedestal constructed at the site where the forward fuselage of aircraft no. 3 came to rest on 15 November 1967.

The memorial is in the California desert, between the tiny communities of Red Mountain and Johannesburg to the south and the much larger city of Ridgecrest to the north. The site is on public land, overseen by the Bureau of Land Management (BLM). The agency offered their full cooperation to Bodylski's project.

John directed the effort of designing the memorial, then trucking in the materials, which consisted of two tons of donated concrete with rebar reinforcement and the sonotube form for the shape. The pillar is cut at an angle facing southward, the direction in which Adams was heading when fate overtook him. The angle is topped with an inset of Inconel-X sheet, etched with Mike's photograph and an inscription explaining what occurred that day. Besides the additional information, the marker recognized Adams as "The first in-flight fatality of the American space program."

Considering the remote location, the area was crowded that day. There were several members of Mike's family present, along with an honor guard from CAP Squadron 68, x-15 pilot Bill Dana, and invited members of the general public. It was a touching experience to witness this small piece of aerospace history unfold.

Speakers included Major Frazier from the CAP and Bill Lovelace from the BLM. John Bodylski explained the project and made special mention of his sister, Becky Hughes: "The picture that was originally going to go in there, you couldn't really see Michael Adams's face. It was in there, but it was pretty small. And [my sister] said, 'No, you've got to change that.' She went in and did some intense graphics . . . so that it could be engraved."

Bill Dana came forward to say a few words himself. "I just wanted to tell this gathering how pleased I am that Mike is finally getting a little recognition. Mike was the fifth air force pilot to fly the x-15 . . . and since Mike lost his life so early in his career he has never gotten the recognition that the other air force x-15 pilots got. I'm glad to see that's being corrected, to a small extent, today."

Finally, Mike's brother, George, stood next to the memorial to speak. "This is a very special day for the family." George paused, taking a moment to wipe tears from his eyes. He then continued, "I appreciate the effort very much, and thanks again for everything." When we spoke later, George said, "I'm happy for the family, because the kids are obviously very proud of their father."

The spot where the memorial stands seems lonely and isolated, but when I visited again several years later, it was obvious that others come by regularly to pay their respects to Mike Adams. Thanks primarily to the work of Rob Enriquez, the site has been expanded. The dirt road has been improved, an identifying sign marks the entrance, a fence surrounds the location, and a giant concrete "x" has been laid. Eight large display panels, that I created, border the "x" and tell the story of the x-15 and Mike Adams.

As I drove home on the afternoon of the original Adams memorial dedication ceremony, there was a devastating announcement on the radio. Pete Knight had died.

Less than five months later I was again out in the desert, not far from Edwards AFB. This trip took me to the town of Mojave, headquarters of the company Scaled Composites. It was 4 October, and that morning marked the first time anyone surpassed Joe Walker's August 1963 x-15 record altitude of 354,200 feet. The accomplishment came at the hands of pilot Brian Binnie, flying Burt Rutan's creation, SpaceShipOne. At 7:49 a.m. the craft was launched from the belly of the droplet-shaped WhiteKnightOne mothership's center fuselage.

The impetus for SpaceShipOne was the Ansari X Prize, modeled on the Orteig Prize won by Charles Lindbergh in 1927 with his first crossing of the Atlantic. In this case the craft had to fly into space successfully, then repeat that mission within two weeks. The kicker with the X Prize was they had to be fully funded, from design through execution, without money from any government. On this day of the final X Prize qualification, Binnie powered upward until rocket burnout at eighty-three seconds. This time is identical to the average LR-99 flights of the x-15. Brian then coasted upward to arc over the top at 367,500 feet, more than 2.5 miles higher than Walker accomplished in x-15 no. 3 forty-one years previously. As Brian reentered, his speed pegged at Mach 3.25; Pete Knight's record of Mach 6.7 was still safe. Both Knight and Walker probably would have asked why it had taken anyone so long to break their long-standing marks.

SpaceShipOne had a direct connection to the x-15, in that engineer and flight planner Bob Hoey worked with Rutan as a consultant on his project to achieve the first civilian flights into space. Bob recalled, "When they started flying it [Rutan] asked me to come up and be a test consultant. I was in the control room on all the flights . . . [and] I did some aerodynamic analysis of the test data, so I wasn't just watching. That program was really high risk. . . . But it was high risk in kind of the same way the x-15 was. . . . The airplane is really squirrelly. It doesn't have any dampers and it's prone to [pilot induced oscillations]. Everything is happening very, very fast—just like the x-15."

When I spoke with Bob about SpaceShipOne, we sat in his living room, and on the table was a homemade model rocket. It was an x-15, and Bob had been out launching it into the air earlier that day. I had my own history with model rockets, so I asked him about what he'd built. Bob said, "It's getting a lot of scars now from all the grass and stuff. I don't know how many more times I'm going to fly it, but it seems to be holding up." To me, it sounded just like the real x-15 in its heyday.

Nearly one more year passed after SpaceShipOne. During that time, much had been made in the press about how Rutan's small, private company had upstaged NASA with its missions into space. Three flights from the airport at Mojave exceeded the 1960s recognition of 50 miles. This began an effort to have NASA finally recognize the three civilian rocket plane pilots who had all achieved astronaut status by the 50-mile definition on the x-15. For whatever reason, this had never officially been done. On 23 August 2005 the error was finally rectified.

The five air force x-15 pilots were recognized by their military service at the time of the missions from 1962 through 1967. They were issued specially crafted astronaut wings for their uniforms, although the one for Adams was posthumously presented to Freida. However, civilian NASA x-15 pilots

Joe Walker, Jack McKay, and Bill Dana had to wait four decades until the space agency realized their error.

A crowd of several hundred people gathered inside the auditorium at the Dryden Flight Research Center on that warm summer day. Family members from both Walker and McKay were present, including Joe's widow, Grace Walker-Wiesmann. Bill Dana was the only one of the three pilots still alive to receive his honor in person. Because of his presence, Bill took a lot of ribbing from those who spoke that day. Dr. Victor Lebacqz, NASA associate administrator for aeronautics, did his part when he joked, "I'm a bit overwhelmed by the whole thing, especially seeing Bill Dana get an award for doing anything right!" Dana laughed with the audience as Dr. Lebacqz shook his head and smiled. He went on, saying, "Those of us who are proud to work for NASA know that NASA is a place of dreamers. . . . I like to think that we stand on the shoulders of giants." Then referring to the twelve men who flew the x-15 from 1959 through 1968, he said, "I wouldn't call you 'The Dirty Dozen'—I'd call you 'The Distinguished Dozen.' [They] are the shoulders we stand on today."

The daughter of Jack McKay, Sheri McKay-Lowe, told impassioned stories of her father from the podium. Some criticized her later for taking as long as she did to talk, but she rightly felt this was the closest Jack would ever have to speak of his own life, and Sheri took the opportunity. She and her brother John recognized that it was not only their father who deserved the long-overdue recognition but everyone from the x-15 program. Sheri said, "My brother John, who worked here [at Edwards] for twenty-nine years, has a request . . . he'd like to see this award—these astronaut wings—dedicated to the engineers, to the mechanics, and to the avionic instrumentation people who are really responsible for making this program happen. . . . My father had great respect for [these] people. . . . He knew that without the mechanics doing their jobs, he wouldn't . . . have a safe flight, or without the engineers or avionics people, that he wouldn't even have [had] the opportunity to go up in these planes."

It was a heady experience for Sheri to connect with her father in this way, and with all these people. It's something she wished she'd been able to do better when Jack was going through the painful turmoil in the aftermath of his accident. But on that day, and at that time, just one thing came to the forefront of her thoughts. Sheri hesitated a moment as she grasped the as-

45. Bill Dana, in his NASA jacket, is surrounded by X-15 pilots Bob White on the left and Neil Armstrong and Joe Engle on the right. The occasion is the belated presentation of Dana's astronaut wings on 23 August 2005. Bill originally earned his wings on 1 November 1966. Photo by author.

tronaut wings, then shared, "It's hard to believe that he's an astronaut. My father, the astronaut. I mean, it's just amazing. . . . I just thought my dad had this cool job."

We may never again hear of new exploits of men and women out in the high desert breaking new Mach barriers, pushing the frontiers of knowledge farther upward. The powerful blast of a rocket engine no longer roars from a test stand or high in the atmosphere, as it once did for so many years at Edwards.

Less than sixty-four years after the Wright brothers slid into the air above the sands of Kitty Hawk, North Carolina, the X-15 was streaking hundreds of thousands of feet in the air at over 4,500 miles an hour. In the more than fifty years that have passed since the beginnings of the X-15, we rarely have a jet surpass 1,000 miles an hour. It is a sad testament to our lack of courage, our aversion to risk, our skittishness of pursuing the unknown in the way these people once did.

I miss the excitement I had as a child growing up, "knowing" these people would take us out into the solar system and make spaceflight commonplace long before I reached middle age. Now that dream is past. Yet I hope that, in some way, these words may rekindle the flames of exhilaration that those of us born into the x-15 and Apollo generation felt. It is patently unfair I was able to watch Neil Armstrong that day preparing for a flight in the x-15 simulator, or that I could sit in front of a television a few short years later to see him step down onto the dusty gray surface of the moon, while those born today only read of these things as ancient history, questioning if they even happened in the first place.

The pilots of these rocket planes were a special breed that is hard to envision today. The same can be said of all the people associated with the x-15. Their attitudes were to make it happen, not deflect to others to cover possible failures. Everyone understood the unusual nature of this hypersonic research vehicle. Without the tragic loss of Michael Adams, these people may have taken the x-15 to even greater heights of achievement.

Today, the x-15 program is greatly admired for its accomplishments; however, the pilots who routinely flew into space as the first winged astronauts are often neglected when it comes to recognition of their individual achievements by other chroniclers of spaceflight. By the time of the *Apollo 11* lunar landing, thirty-one American astronauts had flown into the deep black. Astronaut-pilots of the x-15 accounted for eight of these men, yet they were unrecognized by the general public and shunned for many years by NASA. Even when they finally received some of their much-deserved recognition, it seemed to emerge grudgingly.

The men of the x-15 flew high above the atmosphere, and some observed our planet from the equivalent altitude of low orbit. I owe a debt of great gratitude to their observations and commentaries, and to the recollections of the people who worked next to them each day to make their flights possible. What they accomplished may never be repeated. No more testing our limits, no new frontiers to achieve. I hope that maybe the story of the x-15, and the people who made it happen, may inspire a new generations to take flight on a quest for the unknown.

Glossary

AFB. Air Force Base.

alpha. *See* "angle of attack."

ALT. Approach and Landing Tests. First glide tests of the Space Shuttle.

angle of attack. Angle relative to the direction of flight.

A/P-22S. U.S. Air Force designation for the advanced X-15 pressure suit.

APU. Auxiliary power unit. Two APUs supplied power to the X-15.

ARPS. Aerospace Research Pilot School.

BCS. Ballistic control system. A set of small attitude rockets used on the X-15.

BLM. Bureau of Land Management.

CAP. Civil Air Patrol.

CSM. Command and Service Modules (Apollo spacecraft).

CSS. Control stick steering. A constant acceleration response system on the X-15.

EVA. Extra-vehicular activity (spacewalk).

FAI. Federation Aeronautique Internationale. Organization that certifies all aviation and space records.

FRC. Flight Research Center. The NASA center designation at Edwards from 27 September 1959 through 26 March 1976, when "Dryden" was added to honor Hugh L. Dryden. On 1 March 2014, the center was renamed the Armstrong Flight Research Center to signify the achievements of X-15 pilot Neil Armstrong.

g. Force of gravity at the surface of the earth (plural is "gs").

High Range. Radar tracking corridor for the X-15.

HSFRS. High Speed Flight Research Station. NACA center designation at Edwards from 14 November 1949 through 30 June 1954.

HSFS. High Speed Flight Station. NACA, then NASA center designation at Edwards from 1 July 1954 through 26 September 1959.

hypersonic. Above Mach 5.

ICBM. Intercontinental ballistic missile.

Inconel-X. Nickel alloy used in construction of the outer skin of the X-15.

JPL. Jet Propulsion Laboratory. NASA center in Pasadena, California.

lift-to-drag. Ratio of downward movement to forward movement.

LOX. Liquid oxygen.

LPO. Launch panel operator. B-52 crew member who monitored X-15 status prior to launch.

LR-11. Short designation of the XLR11-RM-13 rocket engine. Two of these units mounted atop each other composed the interim powerplant for the X-15.

LR-99. Short designation of XLR99-RM-1 rocket engine. This was the primary powerplant for the X-15.

Mach number. Velocity relating to the speed of sound. This number varies based on altitude and temperature. At sea level it is approximately 760 mph, while it drops to around 660 mph at 50,000 feet.

MC-2. NAA designation for the original X-15 pressure suit.

MH-96. Flight control system blending aerodynamic and ballistic controls.

MOL. Manned Orbiting Laboratory. A proposed military space reconnaissance observatory crewed by astronauts.

mothership. B-52 bombers converted for use by the X-15.

NAA. North American Aviation. The X-15 prime contractor.

NACA. National Advisory Committee for Aeronautics.

NASA. National Aeronautics and Space Administration. The NACA became NASA on 1 October 1958 and is the civilian agency in charge of space exploration programs in the United States.

NASA 1. Flight controller during X-15 missions.

NTSB. National Transportation Safety Board.

PIO. Pilot induced oscillation.

PSTA. Propulsion System Test Article. This was a set of tanks and plumbing used to test the LR-99 rocket engine without an X-15.

PSTS. Propulsion System Test Stand. This was the facility where the X-15 rocket engine was tested prior to flight.

Q. Dynamic atmospheric pressure.

Q-ball. Mechanism to sense the X-15's direction into the airflow using dynamic pressure sensor.

RMI. Reaction Motors, Incorporated (later Thiokol). The prime contraor for the LR-11 and LR-99 rocket engines.

SAS. Stability augmentation system.

scramjet. Supersonic combustion ramjet. Creates thrust by compressing the airflow through the engine at hypersonic velocities.

SST. Supersonic transport. A passenger airliner that would have an average speed above Mach 1.

TDY. Military temporary duty assignment.

theta. Aircraft pitch.

USAF. U.S. Air Force.

USN. U.S. Navy.

WTR. Western Test Range (Vandenberg AFB).

XLR11-RM-13. *See* LR-11.

XLR99-RM-1. *See* LR-99.

Sources

Books

Bergman, Jules. *Ninety Seconds to Space: The Story of the x-15*. Garden City NY: Hanover House, 1960.

Caidin, Martin. *x-15: Man's First Flight into Space*. New York: Scholastic Book Services, 1959.

Crossfield, A. Scott. *Always Another Dawn: The Story of a Rocket Test Pilot*. With Clay Blair Jr. Cleveland OH: World Publishing Company, 1960.

Godwin, Robert, ed. *Dyna-Soar: Hypersonic Strategic Weapons System*. Burlington, Ontario: Apogee Books, 2003.

———, ed. *x-15: The NASA Mission Reports*. Burlington, Ontario: Apogee Books, 2000.

Gubitz, Myron B. *Rocketship x-15: A Bold New Step in Aviation*. New York: Julian Messner, Inc., 1960.

Hallion, Richard P. *Supersonic Flight: Breaking the Sound Barrier and Beyond*. New York: The MacMillan Company, 1972.

———. *Test Pilots: The Frontiersman of Flight*. Garden City NY: Doubleday and Company, Inc., 1981.

Hansen, James R. *First Man: The Life of Neil A. Armstrong*. New York: Simon & Schuster, 2005.

LaPierre, Carolena. *Joseph's Road*. Rogersville MO: Titus Home Publishing, 2006.

Miller, Jay. *The X-Planes: x-1 to x-29*. Marine on St. Croix MN: Specialty Press, 1983.

Powers, Sheryll Goecke. *Women in Flight Research at NASA Dryden Flight Research Center from 1946 to 1995*. Washington DC: NASA, 1997.

Stambler, Irwin. *Space Ship: The Story of the x-15*. New York: G. P. Putnam's Sons, 1961.

Stillwell, Wendell H. *x-15 Research Results*. Washington DC: NASA, 1965.

Thompson, Milton O. *At the Edge of Space: The x-15 Flight Program*. Washington DC: Smithsonian Institution Press, 1992.

———. *Flight Research: Problems Encountered and What They Should Teach Us*. Washington DC: NASA, 2000.

Tregaskis, Richard. *x-15 Diary: The Story of America's First Space Ship*. New York: E. P. Dutton & Co., 1961.

Wollheim, Donald A. *Mike Mars Flies the x-15*. New York: Doubleday and Company, Inc., 1961.

———. *Mike Mars South Pole Spaceman*. New York: Doubleday and Company, Inc., 1962.

Interviews and Personal Communications

Adams, Brent. 15 May 2006. Monroe LA.

Adams, Freida. 15 May 2006. Monroe LA.

Adams, George. 27 October and 17 December 2004. Sacramento CA.

Albrecht, William. 24 June 1985 and 25 October 2004. Lancaster CA. 3 November 2004. Palmdale CA.

Armstrong, Johnny. 26 March 1984. Edwards CA.

Armstrong, Neil A. 3 May 1984. Lebanon OH.

Barnes, Doris. 3 February 2005. Henderson NV.

Barnes, Thornton D. 3 February 2005. Henderson NV.

Barnett, Florence. 9 March 2007. Lancaster CA.

Barnett, Lorenzo. 9 March 2007. Lancaster CA.

Barniki, Roger. 9 March 2007. Lancaster CA.

Bikle, Paul F. 15 June 1985. Lancaster CA.

Brandt, Phil. 7 January 2008. E-mail correspondence.

Bryan, Dean. 3 February 2006. Tehachapi CA.

Butchart, Stan. 15 December 2004. Lancaster CA.

Capasso, Vince. 25 October 2004. Lancaster CA. 3 November 2004. Palmdale CA.

Crossfield, A. Scott. 30 April 1984. Washington DC.

Crossfield Farley, Sally. 8 August 2008, 11 September 2009, and 11 January 2010. E-mail correspondence.

Dana, William H. 16 March 1984. Edwards CA. 1 November 2005. Tehachapi CA.

DeGeer, Meryl. 1 November 2005. Tehachapi CA.

Engle, Joe H. 26 January 1986. Long Beach CA.

Fedor, Frank. 3 November 2004. Palmdale CA.

Fulton, Fitzhugh L. 8 March 1985. Edwards CA.

Furr, Billy. 16 December 2004. Squaw Valley CA.

Gerdel, Charles. 17 December 2004. Sacramento CA.

Gibbs, Byron. 3 November 2004. Palmdale CA.

Hallberg, Don. 3 November 2004. Palmdale CA.

Hoey, Bob. 3 February 2006. Palmdale CA.

Knight, David. 24 November 2009. E-mail correspondence.

Knight, Gail. 25 April 1991. Palmdale CA.

Knight, William J. 26 March 1984 and 24 June 1985. Palmdale CA.

Kolf, Jack. 22 September 1983 and 24 June 1985. Edwards CA.

Kordes, Eldon. 1 November 2005. Tehachapi CA.

Larson, Terry. 1 November 2005. Tehachapi CA.

Martin, Wade. 2 February 2006. Quartz Hill CA.

McKay, Charlie. 4 November 2005. Lancaster CA.

McKay, John. 4 November 2005. Lancaster CA.

McKay, Mark. 4 November 2005. Lancaster CA.

McKay, Milt. 4 November 2005. Lancaster CA.

McKay-Lowe, Sheri. 1 February 2006. Santa Barbara CA.

McTigue, John. 2 February 2006. Lancaster CA.

Moore, Phil. 18 April 2005. E-mail correspondence.

Nice, Edward. 3 November 2004. Palmdale CA.

Painter, John. 12 April 2011. E-mail correspondence.

Petersen, Forrest S. 28 April 1984. McLean VA.

Revert, Bob. 20 December 2004. Beatty NV.

Richardson, Ralph. 23 February 1985. Camarillo CA.

Riegert, Daniel. 1 November 2005. Tehachapi CA.

Robertson, Jim. 15 June 1984. Culver City CA.

Rushworth, Robert A. 5 August 1984. Camarillo CA.

Shapiro, Harry. 25 May 1985. Yorba Linda CA.

Smith, Glynn. 25 October 2004. Lancaster CA.

Stoddard, Dave. 3 November 2004. Palmdale CA. 4 November 2005. Lancaster CA.
    2 February 2006. Edwards CA.

Storms, Harrison A. 5 December 1987. Palos Verdes CA.

Szuwalski, Bill. 3 November 2004. Palmdale CA.

Thompson, Milton O. 22 September 1983 and 24 June 1985. Edwards CA.

Townsend, Daryl. 3 February 2006. Tehachapi CA.

Townsend, Jim. 3 February 2006. Tehachapi CA.

Veatch, Donald. 1 November 2005. Tehachapi CA.

Walker-Weissmann, Grace. 2 November 2005. Reedley CA.

Waltman, Gene. 3 November 2004. Palmdale CA.

Webb, Lonnie Dean. 3 November 2004. Palmdale CA.

White, Ray. 3 November 2004. Palmdale CA.

White, Robert M. 3 September and 3 October 1984. Correspondence. 22 May
    2006. Sun City Center FL.

Williams, Walt. 4 August 1985. Tarzana CA.

Wilson, Jim. 3 November 2004. Palmdale CA.

Other Sources

An American Adventure: The Rocket Pilots. Directed by Darold Murray. NBC News Special Report. 28 September 1981.

Barnes, T. D. "NASA X-15 Program." http://area51specialprojects.com/x15.html.

Bond, Johnny. "X-15." New York: Republic Label Records, 1960.

Crossfield Farley, Sally, ed. Scott Crossfield Foundation. http://www.scottcross-fieldfoundation.org.

Donoghue, Eddie. "Lindberg Fever Hits St. Thomas." *Virgin Islands Daily News* (St. Thomas). 15 July 2010.

Evans, Michelle. "The X-15 Rocket Plane: Index of Supplemental Material: Downloads and Photographs." http://www.mach25media.com/x15index.html.

Frontiers of Flight: The Threshold of Space. Directed by John Honey. Washington DC: Smithsonian National Air and Space Museum, 1992.

Gathering of Eagles: X-15 50th Anniversary Event. Lancaster CA. 16 October 2009.

Gibbs, Yvonne, ed. "Dryden Photo Gallery." http://www.nasa.gov/centers/dryden/multimedia/imagegallery/X-15/index.html.

"Joe Walker." *This Is Your Life.* Hollywood CA: Ralph Edwards Productions, NBC. 4 June 1961.

Jones, Eric M., and Ken Glover, eds. *Apollo Lunar Surface Journal.* http://www.hq.nasa.gov/alsj.

My Favorite Martian. Directed by Sheldon Leonard. Hollywood CA: Jack Chertok Television, Inc., CBS. 29 September 1963.

"Neil Armstrong." NASA Johnson Space Center Oral History Project. 19 September 2001.

"The Premonition." *The Outer Limits.* Directed by Gerd Oswald. Hollywood CA: Daystar Productions, ABC. 9 January 1965.

Raveling, Paul. *Sierra Foot.* http://www.sierrafoot.org/x-15/x-15.html.

2001: A Space Odyssey. Directed by Stanley Kubrick. Borehamwood, England: Metro-Goldwyn-Mayer, 1968.

Uchu Daisenso. Directed by Ishiro Honda. Tokyo, Japan: Toho Studios, 1959. Released in the United States as *Battle in Outer Space.* Culver City CA: Columbia Pictures, 1960.

X-15. Directed by Richard D. Donner. Hollywood CA: Essex Productions, 1961.

X-15 First Flight 30th Anniversary Symposium. Dryden Flight Research Facility, Edwards AFB CA. 8 June 1989.

X-15: The Edge of Space. Columbus OH: Spacecraft Films, 2006.

Index

Page numbers in italics refer to illustrations.

In the Outward Odyssey: A People's History of Spaceflight series

Into That Silent Sea: Trailblazers of the Space Era, 1961–1965
Francis French and Colin Burgess
Foreword by Paul Haney

*In the Shadow of the Moon: A Challenging Journey
to Tranquility, 1965–1969*
Francis French and Colin Burgess
Foreword by Walter Cunningham

To a Distant Day: The Rocket Pioneers
Chris Gainor
Foreword by Alfred Worden

Homesteading Space: The Skylab Story
David Hitt, Owen Garriott, and Joe Kerwin
Foreword by Homer Hickam

*Ambassadors from Earth: Pioneering Explorations with
Unmanned Spacecraft*
Jay Gallentine

Footprints in the Dust: The Epic Voyages of Apollo, 1969–1975
Edited by Colin Burgess
Foreword by Richard F. Gordon

Realizing Tomorrow: The Path to Private Spaceflight
Chris Dubbs and Emeline Paat-Dahlstrom
Foreword by Charles D. Walker

The X-15 Rocket Plane: Flying the First Wings into Space
Michelle Evans
Foreword by Joe H. Engle

Wheels Stop: The Tragedies and Triumphs of the Space

Colin Burgess
Foreword by Don Thomas

The Ultimate Engineer: The Remarkable Life of NASA's Visionary Leader George M. Low
Richard Jurek
Foreword by Gerald D. Griffin

Beyond Blue Skies: The Rocket Plane Programs That Led to the Space Age
Chris Petty
Foreword by Dennis R. Jenkins

A Long Voyage to the Moon: The Life of Naval Aviator and Apollo 17 Astronaut Ron Evans
Geoffrey Bowman
Foreword by Jack Lousma

The Light of Earth: Reflections on a Life in Space
Al Worden with Francis French
Foreword by Dee O'Hara

To order or obtain more information on these or other University of Nebraska Press titles, visit www.nebraskapress.unl.edu.